Assessment and Nonlinear Modeling of Wave, Tidal and Wind Energy Converters and Turbines

Assessment and Nonlinear Modeling of Wave, Tidal and Wind Energy Converters and Turbines

Editors

Madjid Karimirad
Maurizio Collu

MDPI • Basel • Beijing • Wuhan • Barcelona • Belgrade • Manchester • Tokyo • Cluj • Tianjin

MDPI

Editors
Madjid Karimirad
Queen's University Belfast
UK

Maurizio Collu
University of Strathclyde
UK

Editorial Office
MDPI
St. Alban-Anlage 66
4052 Basel, Switzerland

This is a reprint of articles from the Special Issue published online in the open access journal *Energies* (ISSN 1996-1073) (available at: https://www.mdpi.com/journal/energies/special_issues/Wave_Tidal_Wind_Converters).

For citation purposes, cite each article independently as indicated on the article page online and as indicated below:

LastName, A.A.; LastName, B.B.; LastName, C.C. Article Title. *Journal Name* **Year**, *Article Number*, Page Range.

ISBN 978-3-03936-912-6 (Hbk)
ISBN 978-3-03936-913-3 (PDF)

Contents

About the Editors

Madjid Karimirad (BSc, MSc, PhD, Postdoc, CEng, MIMechE, M.ASME, PGCHET, FHEA, and FIMarEST) is a Senior Lecturer (Associate Professor) in Marine and Coastal Engineering in the School of Natural and Built Environment at Queen's University Belfast. Dr. Karimirad is the Associate Editor of the Journal of Ocean, Offshore, and Arctic Engineering, Editor of Ocean Engineering Journal (Elsevier), member of IMechE and ASME, and a fellow of IMarEST. He was Scientist in MARINTEK (Norwegian Marine Technology Research Institute) and SINTEF Ocean, Norway. Dr. Karimirad has been researching marine structures and offshore technology for more than 13 years. His knowledge covers salient aspects of offshore mechanics, hydrodynamics, and structural engineering, and he has more than 80 scientific publications (Citations 1522, h-index 22). He has been part of NOWITECH (Norwegian Research Center for Offshore Wind Technology) for eight years. While working at MARINTEK/SINTEF (Norway), he has been the key researcher and research manager in collaborative projects such as H2020 LIFES50+, EU FP7 DTOcean, and EU IPRWind. Recently, he was the investigator on Floating Solar Energy and a contributor to the decarbonization project (granted by Invest Northern Ireland).

Maurizio Collu (BSc, MSC, PhD, CEng, MRINA, FHEA) is a Reader (Associate Professor) in Offshore Renewable Energy Systems in Naval Architecture, Ocean, and Marine Engineering Department at the University of Strathclyde, Glasgow, UK. He has been a chartered engineer since 2011, member of the Royal Institution of Naval Architects, member of the ITTC Hydrodynamic Modelling of Marine Renewable Energy Devices Committee (2017-21), associate editor of the Journal OMAE, Special Issue guest editor for Energies (MDPI), and Ocean Engineering (Elsevier). His research has focused on coupled dynamics of offshore renewable energy devices since 2009, with more than 60 international peer-reviewed papers, and four book chapters on this topic. He led the development of a coupled model of dynamics for a wind-wave hybrid renewable energy device as work package leader in the EU project H2Ocean (EU FP7 288145). Currently, Dr. Collu is the co-investigator and work package leader on the 3.8M GBP grant EP/P009743/1 "HOME Offshore", leading the development of numerical model of dynamics for offshore wind farms, and principal investigator of the 0.8M GBP INNO-MPP project (EP/R007497/1, EP/R007497/2), a UK—China project focusing on the development of multipurpose offshore platforms, managing 19 investigators from 3 UK and 3 Chinese research institutions. He is also the work package leader in the EU H2020 project "The Blue Growth Farm (774426)", coordinating the development of a coupled model of dynamics for offshore multi-purpose platforms.

Preface to "Assessment and Nonlinear Modeling of Wave, Tidal and Wind Energy Converters and Turbines"

Offshore renewable energy (ORE) sources, such as offshore wind turbines, wave energy converters, and tidal and current turbines, have experienced rapid growth in the past decade. The combination of wave, wind, and current energy devices in hybrid marine platforms that use synergies through proper combinations has been a recent scientific focus. The new concepts and structures being investigated require developing new design and analysis approaches that implement novel numerical modeling tools and simulation methods, thus advancing science, technology, and engineering. ORE structures may be subject to complex loads and load effects, which demand comprehensive and accurate numerical modeling representations of the physics underpinning the problem. Important factors that affect design, functionality, structural integrity, and performance of offshore structures include (but are not limited to): fluid–structure interactions, controller actions, intense dynamic effects, nonlinear loadings, extreme and harsh weather conditions, and impact pressure loads. Furthermore, these factors cannot be considered in isolation, since each factor is potentially coupled with another, requiring fully coupled models. To enable further growth in reliable ORE technologies, more advanced numerical tools and nonlinear modeling are needed.

Madjid Karimirad, Maurizio Collu
Editors

energies

MDPI

Article

Study of Generalized Interaction Wake Models Systems with ELM Variation for Off-Shore Wind Farms

Mingcan Li [1,†], Hanbin Xiao [1,†], Lin Pan [1,2,3,*] and Chengjun Xu [1,†]

1 School of Logistics Engineering, Wuhan University of Technology, Wuhan 430063, China; mclichina@126.com (M.L.); xhbchina@126.com (H.X.); xcjwhut@163.com (C.X.)
2 National Engineering Research Center for Water Transport Safety (WTS Center), Wuhan 430063, China
3 State Key Laboratory of Ocean Engineering, Shanghai Jiaotong University, Shanghai 200240, China
* Correspondence: lin.pan@whut.edu.cn
† These authors contributed equally to this work.

Received: 20 January 2019; Accepted: 28 February 2019; Published: 5 March 2019

Abstract: This paper reports a novel frandsen generalized wake model and its variation model-frandsen generalized normal distribution wake model for off-shore wind farms. Two different new wake models in off-shore wind farms have been studied comparatively. Their characteristics have been analyzed through mathematical modeling and derivation. Meanwhile, simulation experiments show that the proposed two new wake models have different properties. Furthermore, the distributions of wind speed and wind direction are modeled by the statistical methods and Extreme Learning Machine through the off-shore wind farms of Yangshan Deepwater Harbor in the Port of Shanghai, China. In addition, the data of wind energy are provided to verify and test the correctness and effectiveness of the proposed two models. Wind power has been demonstrated by wind rose and wind resources with real-time data. These techniques contribute to enhance planning, utilization and exploitation for wind power of off-shore wind farms.

Keywords: off-shore wind farms (OSWFs); wake model; wind turbine (WT); Extreme Learning Machine (ELM); wind power (WP); large-eddy simulation (LES)

1. Introduction

The global warming and the climate changes lead to a gradual shift from conventional to renewable energy sources which are more reliable clean resources. The main reason is the growing trend of global energy demands accompanied with the detrimental effects of overuse conventional fossil fuel sources namely [1–3].

Wind is one of the fastest growing energy sources, and it is also pollution-free, renewable and abundant. At present, some researchers have done a lot of research work on wind energy [4,5]. For example, based on short term wind speed forecasting of variable weight, Li et al. provided the research and application of a combined model [6]. Romanic et al. studied wind and tornado climatologies and wind resource modelling for a modern development situated in 'Tornado Alley' [7]. Ahmed Shata Ahmed investigated wind energy characteristics and wind park installation in Shark El-Ouinat, Egypt [8]. Using standard exergy and Extended Exergy Accounting (EEA) approaches, Aghbashlo et al. studied performance assessment of a wind power plant [9].

Wind energy has a low carbon footprint, which is a type of renewable energy. Some researchers and experts believe that local power generation sources like micro-grids on wind farm or micro power sources are better, in view of transmission efficiency and grid laying cost [10]. Some of the methods by which sustainable energy can be harvested are solar, wind, vibration, tidal, etc. Ideally, a power source should be sustainable, reliable, economical and eco-friendly [11,12].

Among renewable energy sources, Wind Power (WP) is considered to be one of the most prospective renewable energy technologies, and its usage has been increased immensely over recent decades. Usually, Horizontal Axis Wind Turbines (HAWTs) as types of wind energy converters are a kind of medium to large scale rotating machinery and their technologies have been developed substantially during the last decades, resulting in significant performance improvement. Despite the considerable achievements in the aerodynamic improvement of the HAWTs, wind energy still has certain drawbacks that make the popularity of the wind turbine technology difficult.

Due to the location optimization caused by operation in the unsteady flow condition and dynamic loading exerted on the Wind Turbine (WT), one of its main problems is how to increase performance of the control system as decision-making, which makes the most effective use of the WT controller. The WT location optimization is an interesting topic, which provides a predominant approach to increase the total WP of the Wind Farm (WF) and decreases the wake effect of WTs [13–15]. In recent years, the size of WTs has been more extensive from a few kilowatts to several megawatts. Lots of experiences have shown that the larger the WTs, the lower the cost per kilowatt installed [16]. Furthermore, their total costs of production, installation and maintenance are less than the total of smaller WTs achieving the same WP [17]. In 2013, Chen et al. investigated tower height matching optimization for WT positioning in the wind farm [18]. One scheme for properly handling these aerodynamic interactions is to use and promote wake models in the optimization and distributed algorithms control. An alternative approach is to present an online control method where each WT adjusts its own sense model coefficients, which is in line with the communication of local Wind Farms (WFs) [19–21]. For example, V. Seshadri Sravan Kumar and D.Thukaram presented accurate modeling of doubly fed induction generator [22] based WFs in load flow analysis. Farajzadeh et al. proposed statistical modeling of the power grid from a wind farm standpoint [23]. Tian et al. developed and verified a new WT wake model with two dimensions in 2015 [24–27].

It is worth mentioning that Wind Energy (WE) has developed rapidly in China in recent years. For example, Figures 1 and 2 show off-shore wind farms (OSWFs) of Yangshan Deepwater Harbor in the Port of Shanghai, China, which is one of the largest wide range or scope WFs in China [28,29].

From the control point of view, the research of WFs has attracted a great deal of interest from researchers. Recently, Ebrahimi et al. proposed a new optimizing power control scheme based on a centralizing WF control system [30]. Song et al. presented the decision model of WF layout design with three dimensions [31]. Sadegh Ghani Varzaneh et al. have studied a novel simplified model for assessment of power alteration of Doubly-fed Induction Generator (DFIG) [32] based wind farm participating in frequency control system. Hossain has presented a nonlinear controller for transient stability enhancement of DFIG based new bridge type fault current limiter [33] for WFs. Yao et al. studied coordinated control of hybrid WFs [34] in a Permanent Magnet Synchronous Generator (PMSG) based and fixed-speed induction generator (FSIG) for WTs during an asymmetrical grid fault. Li et al. also proposed adaptive fault lenient control of WTs with safeguard transient performance considering active power control of WFs [35]. From met-mast and remote sensing techniques, Chaurasiya et al. proposed comparative analysis of Weibull parameters with wind data measured [36]. Hamid Atighechi et al. presented an effective load shedding remedial action strategy [37] for WF generation. Suganthi et al. proposed an Improved Differential Evolution algorithm [38] based on congestion management in the presence of wind turbine generators.

Figure 1. China's first off-shore wind farm in Yangshan Port.

Figure 2. Off-shore wind farm of Yangshan Deepwater Harbor in the Port of Shanghai, China.

Recently, there are some machine learning techniques used in such prediction of wind direction and wind speed for OSWFs. For example, Wan et al. proposed an Extreme Learning Machine (ELM) based method of probabilistic forecasting for wind power generation [39]. Vlastimir Nikolić et al. presented a novel wake model based upon ELM for sensor-less computation of wind speed based on WT parameters in WFs [40]. Lazarevska presented an alternative approach to forecasting the wind speed based on Extreme Learning Machine [41]. Wu et al. presented a real-time precise wind speed estimation approach and sensor-less control for variable pitch and variable speed Wind Turbine Power Generation System (WTPGS) [42].

Based on the above discussion, in general, the main contributions of this paper are as follows. Firstly, a Frandsen Generalized Wake Model (FGWM) and its variation model-Frandsen Generalized Normal Distribution Wake Model (FGNDWM) for WFs have been analyzed and presented with mathematical derivation forms. Then, comparisons of these two different wake models of OSWFs have been presented. Furthermore, comparative experiments of both Wake Models have been studied. Finally, focused on the OSWFs of Yangshan Deepwater Harbor in the Port of Shanghai, China, wind rose, wind Weibull probability density distribution and ELM prediction are elaborated and discussed through the OSWFs of Yangshan Harbor. Simulation figures are also provided to show the effectiveness of the proposed approach.

The structure of the paper is as follows: FGWM and FGNDWM are derived and studied in Sections 2 and 3. Meanwhile, in Sections 4 and 5, comparative analysis and experiments for the proposed two novel different wake models are studied. Furthermore, in Section 6, the wind rose, wind Weibull probability density distribution and ELM prediction are proposed for OSWFs of Yangshan Deepwater Harbor in the Port of Shanghai, China. Finally, in Section 7, the conclusions are summarized.

2. A Frandsen Generalized Wake Model (FGWM) for OSWFs

Frandsen Generalized Wake Model (FGWM) in the ideal state is shown in Figure 3, where the far wake area is described by the sideways trapezoidal region. The near field is denoted WT_{1ij} (radius is r_{1ij}) and can be seen as a turbulent wake. In the down-wind distance x_{ij}, the wind speed denoted P_{1ij} and P_{2ij} are assumed to be equal v_{0ij}, and the wind speed on S_{xij} is given by v_{xij}. The circular cross-section radius is r_{xij}, where i and j are row vector and column vector for Wind Turbine in large-scale Wind Farms, respectively.

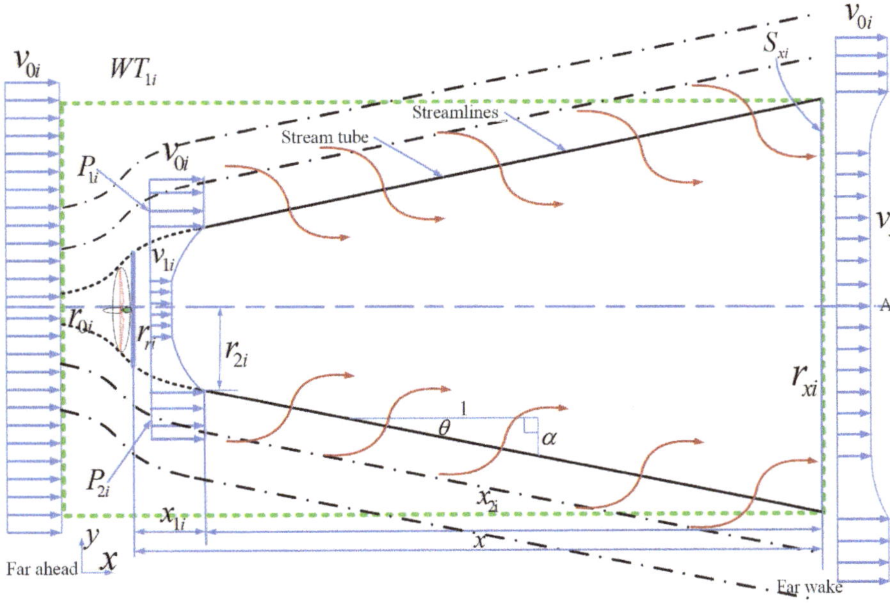

Figure 3. The portrait of Frandsen Generalized Wake Model (FGWM) with stream tube [43].

In an ideal state, the model assumes that the far wake region spreads with a linear approach, and the distribution of wind speed is homogeneous on every cross-section. In FGWM, the tube includes the near wake area, which is described in the green rectangular region in Figure 3. The fluid inlet mass flow rate in the tube is equal to $\sum_{i=1}^{n}\sum_{j=1}^{m}\rho\pi r_{xij}^{2}v_{0ij}$ and it goes through the WT. The fluid outlet mass flow rate is equal to $\sum_{i=1}^{n}\sum_{j=1}^{m}\rho\pi r_{xij}^{2}v_{xij}$. The FGWM assumes

$$\sum_{i=1}^{n}\sum_{j=1}^{m}\dot{M}_{ij} = \sum_{i=1}^{n}\sum_{j=1}^{m}\frac{\partial M_{ij}}{\partial t_{ij}} = \sum_{i=1}^{n}\sum_{j=1}^{m}\rho\pi r_{xij}^{2}v_{xij}, \tag{1}$$

for $n = 1, 2, \cdots, N$; $m = 1, 2, \cdots, N$.

By conservation of momentum, we obtain

$$\begin{aligned}\sum_{i=1}^{n}\sum_{j=1}^{m}T_{ij} &= \sum_{i=1}^{n}\sum_{j=1}^{m}\left(\dot{M}_{ij}v_{0i} - \dot{M}_{ij}v_{xij}\right)\\ &= \sum_{i=1}^{n}\sum_{j=1}^{m}\left(\frac{\partial M_{ij}}{\partial t_{ij}}v_{0ij} - \frac{\partial M_{ij}}{\partial t_{ij}}v_{xij}\right),\end{aligned} \tag{2}$$

for $n = 1, 2, \cdots, N$; $m = 1, 2, \cdots, N$.

We assume that the radius of the row vector ith and column vector jth actuator disk is r_{rij}, then the area of ith $\times$ jth actuator disk is given as $A_{rij} = \pi r_{rij}^2$. According to the definition of thrust coefficient C_{Tij}, one can have

$$\begin{aligned}\sum_{i=1}^{n}\sum_{j=1}^{m} C_{Tij} &= \sum_{i=1}^{n}\sum_{j=1}^{m} \frac{T_{ThrustForceij}}{T_{DynamicForceij}} \\ &= \sum_{i=1}^{n}\sum_{j=1}^{m} \frac{T_{ij}}{T_{ijmax}} \\ &= \sum_{i=1}^{n}\sum_{j=1}^{m} \left[\frac{\frac{1}{2}\rho(v_{0ij}^2 - v_{xij}^2)A_{rij}}{\frac{1}{2}\rho v_{0ij}^2 A_{rij}}\right] \\ &= \sum_{i=1}^{n}\sum_{j=1}^{m} \left[\frac{2\rho v_{0ij}^2 A_{rij} a_{ij}(1-a_{ij})}{\frac{1}{2}\rho v_{0ij}^2 A_{rij}}\right],\end{aligned} \tag{3}$$

which is equivalent to

$$\sum_{i=1}^{n}\sum_{j=1}^{m} T_{ij} = \sum_{i=1}^{n}\sum_{j=1}^{m} C_{Tij} T_{ijmax}. \tag{4}$$

Substituting Equations (1), (3) and (4) into Equation (2), we obtain the following equation:

$$\sum_{i=1}^{n}\sum_{j=1}^{m} \left[v_{xij}^2 - v_{0i}v_{xij} + \frac{1}{2}C_{Tij}(\frac{r_{rij}}{r_{xij}})^2 v_{0ij}\right] = 0. \tag{5}$$

Solving Equation (5), one can have the following equation:

$$\begin{aligned}\sum_{i=1}^{n}\sum_{j=1}^{m} v_{xij} &= \sum_{i=1}^{n}\sum_{j=1}^{m} \frac{v_{0ij} \pm \sqrt{v_{0ij}^2 - 2C_{Tij}(\frac{r_{rij}}{r_{xij}})^2 v_{0ij}^2}}{2} \\ &= \sum_{i=1}^{n}\sum_{j=1}^{m} \left[\frac{1}{2} \pm \frac{1}{2}\sqrt{1 - \frac{2C_{Tij}}{(r_{xij}/r_{ri})^2}}\right] v_{0ij} \\ &= \sum_{i=1}^{n}\sum_{j=1}^{m} \left\{1 - \frac{1}{2}\left[1 \pm \sqrt{1 - \frac{2C_{Tij}}{(r_{xij}/r_{rij})^2}}\right]\right\} v_{0ij}.\end{aligned} \tag{6}$$

Using the physical solution of Equation (6), then we obtain

$$\sum_{i=1}^{n}\sum_{j=1}^{m} v_{xij} = \sum_{i=1}^{n}\sum_{j=1}^{m} \left\{1 - \frac{1}{2}\left[1 - \sqrt{1 - \frac{2C_{Tij}}{(r_{xij}/r_{rij})^2}}\right]\right\} v_{0ij}. \tag{7}$$

The above equation gives the main variables and results of the FGWM.

3. A Frandsen Generalized Normal Distribution Wake Model (FGNDWM) for OSWFs

The Frandsen Generalized Normal Distribution Wake Model (FGNDWM) called Frandsen Generalized Gaussian Distribution Wake Model (FGGDWM) for OSWFs is illustrated in Figure 4 where the two dotted lines A_{ij} and C_{ij} are selected to be the boundaries of the FGNDWM tube. The far wake region is confined to the dotted line tube region while the farthest boundary is extended to infinity. The wind speed on P_{1ij} and P_{2ij} are recognized as v_{0ij} and the wind speed on S_{xij} is v_{xij}. Here, i and j are row vector and column vector for Wind Turbine in large-scale offshore Wind Farms, respectively.

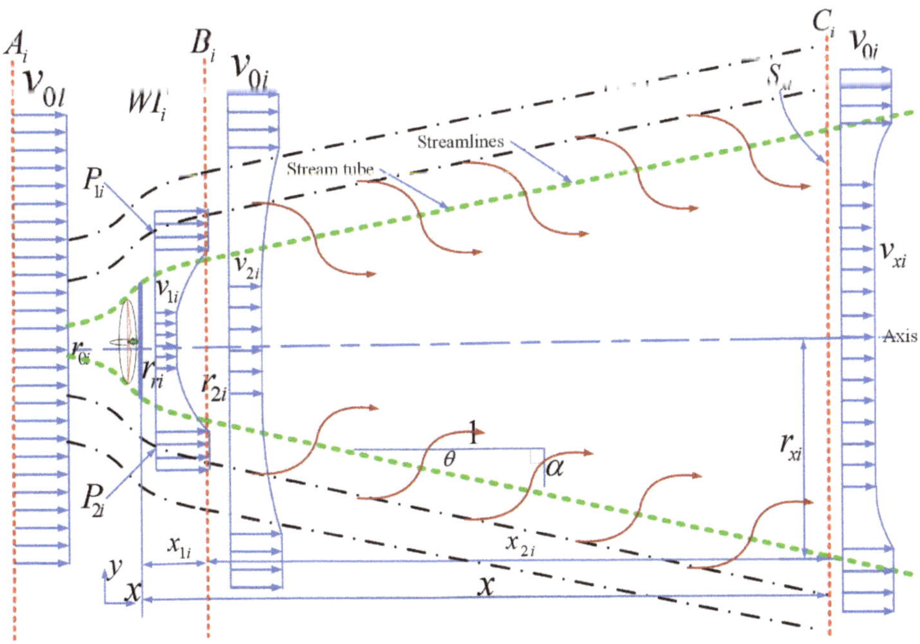

Figure 4. The portrait of Frandsen Generalized Normal Distribution Wake Model (FGNDWM) with stream.

The conservation of mass does not hold comparing to the FGWM. The following equation is considered as the outlet mass flow rate

$$\sum_{i=1}^{n}\sum_{j=1}^{m}\dot{M}_{ij} = \frac{\partial M_{ij}}{\partial t} = \sum_{i=1}^{n}\sum_{j=1}^{m}\int_{0}^{+\infty} 2\pi r_{xij}\rho v_{x_{ij},r_{xij}} dr_{xij},$$
$$n = 1,2,\cdots,N;\ m = 1,2,\cdots,N. \tag{8}$$

The FGNDWM satisfies the following equation:

$$\sum_{i=1}^{n}\sum_{j=1}^{m} v_{x_{ij},r_{xij}} = \sum_{i=1}^{n}\sum_{j=1}^{m}\left[1-\Psi(x_{ij})\exp\left(-\frac{r_{ij}^2}{2\sigma_{ij}^2}\right)\right]v_{0ij}, \tag{9}$$

where σ_{ij} is the standard deviation, which is also called the characteristic width of FGNDWM and $\Psi(x_{ij})$ is a coefficient related to x_{ij}.

According to the momentum conservation law, one can obtain the following result

$$\sum_{i=1}^{n}\sum_{j=1}^{m}\int_{0}^{+\infty} 2\pi r_{xij}\rho v_{x_{ij},r_{xij}}\left(v_{0i} - v_{xij,r_{xij}}\right) dr_{xij} = \sum_{i=1}^{n}\sum_{j=1}^{m} T_{ij}, \tag{10}$$

which is equivalent to

$$\sum_{i=1}^{n}\sum_{j=1}^{m}\lim_{b\to+\infty}\int_{0}^{b} 2\pi r_{xij}\rho v_{x_{ij},r_{xij}}\left(v_{0ij} - v_{xij,r_{xij}}\right) dr_{xij}$$
$$= \sum_{i=1}^{n}\sum_{j=1}^{m} T_{ij}. \tag{11}$$

From the definition of the thrust coefficient, one can have the following equation:

$$\begin{aligned}\sum_{i=1}^{n}\sum_{j=1}^{m} T_{ij} &= C_{Tij}T_{ij_{max}} = \frac{1}{2}\sum_{i=1}^{n}\sum_{j=1}^{m} \rho A_{rij} v_{0i}^2 C_{Tij} \\ &= \frac{1}{2}\sum_{i=1}^{n}\sum_{j=1}^{m} \rho \pi r_{rij}^2 v_{0ij}^2 C_{Tij}.\end{aligned} \tag{12}$$

Substituting Equations (9) and (12) into Equation (11), we obtain

$$\begin{aligned}&\sum_{i=1}^{n}\sum_{j=1}^{m} \lim_{b\to+\infty}\int_0^b 2\pi r_{xij}\rho v_{x_{ij},r_{xij}}(v_{0i} - v_{xij,r_{xij}})dr_{xij} \\ &= \sum_{i=1}^{n}\sum_{j=1}^{m} C_{Tij}T_{ij_{max}}.\end{aligned} \tag{13}$$

From this equation, we obtain:

$$\begin{aligned}&\sum_{i=1}^{n}\sum_{j=1}^{m} \lim_{b\to+\infty}\int_0^b 2\pi r_{xij}\rho v_{0ij}^2 \left[1 - \Psi(x_{ij})\exp\left(-\frac{r_{ij}^2}{2\sigma_{ij}^2}\right)\right] dr_{ij} \\ &- \sum_{i=1}^{n}\sum_{j=1}^{m} \lim_{b\to+\infty}\int_0^b 2\pi r_{xij}\rho v_{0ij}^2 \left[1 - \Psi(x_{ij})\exp\left(-\frac{r_{ij}^2}{2\sigma_{ij}^2}\right)\right]^2 dr_{ij} \\ &= \sum_{i=1}^{n}\sum_{j=1}^{m} \frac{1}{2}\rho\pi r_{rij}^2 v_{0ij}^2 C_{Tij}, \quad n = 1,2,\cdots,N;\ m = 1,2,\cdots,N,\end{aligned} \tag{14}$$

which is also equivalent to

$$\begin{aligned}&\sum_{i=1}^{n}\sum_{j=1}^{m} \lim_{b\to+\infty}\int_0^b 2r_{xij} \left[1 - \Psi(x_{ij})\exp\left(-\frac{r_{ij}^2}{2\sigma_{ij}^2}\right)\right] dr_{ij} \\ &- \sum_{i=1}^{n}\sum_{j=1}^{m} \lim_{b\to+\infty}\int_0^b 2r_{xij} \left[1 - \Psi(x_{ij})\exp\left(-\frac{r_{ij}^2}{2\sigma_{ij}^2}\right)\right]^2 dr_{ij} \\ &= \sum_{i=1}^{n}\sum_{j=1}^{m} \frac{1}{2} r_{rij}^2 C_{Tij}, \quad n = 1,2,\cdots,N;\ m = 1,2,\cdots,N.\end{aligned} \tag{15}$$

Computing this equation, then we have

$$
\begin{aligned}
&\sum_{i=1}^{n}\sum_{j=1}^{m}\lim_{b\to+\infty}\left.\frac{r_{rij}}{2}\right|_0^b \\
&-\sum_{i=1}^{n}\sum_{j=1}^{m}\lim_{b\to+\infty}\int_0^b \Psi(x_{ij})\exp\left(-\frac{r_{ij}^2}{2\sigma_{ij}^2}\right) r_{ij}dr_{ij} \\
&-\sum_{i=1}^{n}\sum_{j=1}^{m}\lim_{b\to+\infty}\left.\frac{r_{xij}}{2}\right|_0^b \\
&+2\sum_{i=1}^{n}\sum_{j=1}^{m}\lim_{b\to+\infty}\int_0^b \Psi(x_{ij})\exp\left(-\frac{r_{ij}^2}{2\sigma_{ij}^2}\right) r_{ij}dr_{ij} \\
&-\sum_{i=1}^{n}\sum_{j=1}^{m}\lim_{b\to+\infty}\int_0^b \Psi^2(x_{ij})\exp\left(-\frac{2r_{ij}^2}{2\sigma_{ij}^2}\right) r_{ij}dr_{ij} \\
&=\sum_{i=1}^{n}\sum_{j=1}^{m}\frac{1}{4}r_{rij}^2 C_{Tij}, \quad n=1,2,\cdots,N;\ m=1,2,\cdots,N
\end{aligned}
\tag{16}
$$

or also

$$
\begin{aligned}
&\sum_{i=1}^{n}\sum_{j=1}^{m}\lim_{b\to+\infty}\int_0^b \Psi(x_{ij})\exp\left(-\frac{r_{ij}^2}{2\sigma_i^2}\right) r_{ij}dr_{ij} \\
&\quad-\sum_{i=1}^{n}\sum_{j=1}^{m}\lim_{b\to+\infty}\int_0^b \Psi^2(x_{ij})\exp\left(-\frac{r_{ij}^2}{\sigma_i^2}\right) r_{ij}dr_{ij} \\
&=\sum_{i=1}^{n}\sum_{j=1}^{m}\frac{1}{4}r_{rij}^2 C_{Tij}, \quad n=1,2,\cdots,N;\ m=1,2,\cdots,N,
\end{aligned}
\tag{17}
$$

which can be also written as

$$
\begin{aligned}
&-\sum_{i=1}^{n}\sum_{j=1}^{m}\lim_{b\to+\infty}\Psi(x_{ij})\sigma_{ij}^2\exp\left.\left(-\frac{r_{xij}^2}{2\sigma_{ij}^2}\right)\right|_0^b \\
&\quad+\sum_{i=1}^{n}\sum_{j=1}^{m}\lim_{b\to+\infty}\Psi^2(x_{ij})\frac{\sigma_{ij}^2}{2}\exp\left.\left(-\frac{r_{ij}^2}{\sigma_{ij}^2}\right)\right|_0^b \\
&=\sum_{i=1}^{n}\sum_{j=1}^{m}\frac{1}{4}r_{rij}^2 C_{Tij}, \quad n=1,2,\cdots,N;\ m=1,2,\cdots,N.
\end{aligned}
\tag{18}
$$

Then, we obtain

$$
\begin{aligned}
&\sum_{i=1}^{n}\sum_{j=1}^{m}\left[\Psi(x_{ij})\sigma_{ij}^2-\Psi^2(x_{ij})\frac{\sigma_{ij}^2}{2}\right]=\sum_{i=1}^{n}\sum_{j=1}^{m}\left(\frac{1}{4}C_{Tij}r_{rij}^2\right), \\
&\quad n=1,2,\cdots,N;\ m=1,2,\cdots,N,
\end{aligned}
\tag{19}
$$

which is

$$
\begin{aligned}
&\sum_{i=1}^{n}\sum_{j=1}^{m}\left[\Psi^2(x_{ij})-2\Psi(x_{ij})+\frac{1}{2}C_{Tij}\left(\frac{r_{rij}}{\sigma_{ij}}\right)^2\right]=0, \\
&\quad n=1,2,\cdots,N;\ m=1,2,\cdots,N,
\end{aligned}
\tag{20}
$$

or equivalently

$$\sum_{i=1}^{n}\sum_{j=1}^{m}\left[\Psi^2(x_{ij})-2\Psi(x_{ij})+\frac{C_{Tij}}{2\left(\frac{\sigma_{ij}}{r_{rij}}\right)^2}\right]=0, \quad n=1,2,\cdots,N;\ m=1,2,\cdots,N. \tag{21}$$

By solving Equation (21), we obtain

$$\begin{aligned}\sum_{i=1}^{n}\sum_{j=1}^{m}\Psi(x_{ij}) &= \sum_{i=1}^{n}\sum_{j=1}^{m}\frac{2\pm\sqrt{4-2\frac{C_{Tij}}{(\sigma_{ij}/r_{rij})^2}}}{2}\\ &=\sum_{i=1}^{n}\sum_{j=1}^{m}\left[1\pm\sqrt{1-\frac{C_{Tij}}{2(\sigma_{ij}/r_{rij})^2}}\right]\\ &=\sum_{i=1}^{n}\sum_{j=1}^{m}\left[1\pm\sqrt{1-\frac{C_{Tij}}{2}\left(\frac{r_{rij}}{\sigma_{ij}}\right)^2}\right],\\ & n=1,2,\cdots,N;\ m=1,2,\cdots,N.\end{aligned} \tag{22}$$

From Equation (22), the physical solution is

$$\sum_{i=1}^{n}\sum_{j=1}^{m}\Psi(x_{ij})=\sum_{i=1}^{n}\sum_{j=1}^{m}\left[1-\sqrt{1-\frac{C_{Tij}}{2}\left(\frac{r_{rij}}{\sigma_{ij}}\right)^2}\right], \quad n=1,2,\cdots,N;\ m=1,2,\cdots,N. \tag{23}$$

By substituting this solution into Equation (9), we obtain

$$\begin{aligned}&\sum_{i=1}^{n}\sum_{j=1}^{m}v_{x_{ij},r_{xij}}\\ &=\sum_{i=1}^{n}\sum_{j=1}^{m}\{1-\\ &\left[1-\sqrt{1-\frac{C_{Tij}}{2}\left(\frac{r_{rij}}{\sigma_{ij}}\right)^2}\right]\exp\left(-\frac{r_{ij}^2}{2\sigma_{ij}^2}\right)\Bigg\}v_{0ij},\\ &n=1,2,\cdots,N;\ m=1,2,\cdots,N.\end{aligned} \tag{24}$$

Here, σ_{ij} is recognized as a linear function of x_{ij} in FGNDWM. In FGWM and FGNDWM, owing to every plane is perpendicular to the axis, the rates of mass flow are equal to each other in both FGWM and FGNDWM. According to the law of mass conservation, we can calculate and get:

$$\begin{aligned}&\sum_{i=1}^{n}\sum_{j=1}^{m}\int_0^{+\infty}2\pi r_{ij}\rho_{ij}v_{x_{ij},r_{xij}}dr_{ij}\\ &=\sum_{i=1}^{n}\sum_{j=1}^{m}\int_0^{r_{xij}}2\pi r_{ij}\rho_{ij}v_{xij}dr_{ij}+\sum_{i=1}^{n}\sum_{j=1}^{m}\int_{r_{xij}}^{+\infty}2\pi r_{ij}\rho_{i}v_{0ij}dr_{ij},\\ &n=1,2,\cdots,N;\ m=1,2,\cdots,N,\end{aligned} \tag{25}$$

which is

$$\begin{aligned}
&\sum_{i=1}^{n}\sum_{j=1}^{m}\lim_{b\to+\infty}\int_{0}^{b}2\pi r_{ij}\rho_{ij}v_{x_{ij},r_{xij}}dr_{ij} \\
&=\sum_{i=1}^{n}\sum_{j=1}^{m}\int_{0}^{r_{xij}}2\pi r_{ij}\rho_{ij}v_{xij}dr_{ij} \\
&+\sum_{i=1}^{n}\sum_{j=1}^{m}\lim_{b\to+\infty}\int_{r_{xij}}^{b}2\pi r_{ij}\rho_{ij}v_{0ij}dr_{ij}, \\
&n=1,2,\cdots,N;\ m=1,2,\cdots,N.
\end{aligned} \tag{26}$$

Substituting Equations (7) and (24) into Equation (26), we obtain

$$\begin{aligned}
&\sum_{i=1}^{n}\sum_{j=1}^{m}\lim_{b\to+\infty}\int_{0}^{b}2\pi r_{ij}\rho_{ij}\left\{1\right. \\
&\left.-\left[1-\sqrt{1-\frac{C_{Tij}}{2}\left(\frac{r_{rij}}{\sigma_{ij}}\right)^{2}}\right]\exp\left(-\frac{r_{ij}^{2}}{2\sigma_{ij}^{2}}\right)\right\}v_{0ij}dr_{ij} \\
&=\sum_{i=1}^{n}\sum_{j=1}^{m}\int_{0}^{r_{xij}}2\pi r_{ij}\rho_{ij}\left\{1\right. \\
&\left.-\frac{1}{2}\left[1-\sqrt{1-\frac{2C_{Tij}}{(r_{xij}/r_{rij})^{2}}}\right]\right\}v_{0ij}dr_{ij} \\
&+\sum_{i=1}^{n}\sum_{j=1}^{m}\lim_{b\to+\infty}\int_{r_{xij}}^{b}2\pi r_{ij}\rho_{ij}v_{0ij}dr_{ij}, \\
&n=1,2,\cdots,N;\ m=1,2,\cdots,N,
\end{aligned} \tag{27}$$

or equivalently

$$\begin{aligned}
&\sum_{i=1}^{n}\sum_{j=1}^{m}\lim_{b\to+\infty}\int_{0}^{b}r_{ij}\left\{1\right. \\
&\left.-\left[1-\sqrt{1-\frac{C_{Tij}}{2}\left(\frac{r_{rij}}{\sigma_{ij}}\right)^{2}}\right]\exp\left(-\frac{r_{ij}^{2}}{2\sigma_{ij}^{2}}\right)\right\}dr_{ij} \\
&=\sum_{i=1}^{n}\sum_{j=1}^{m}\int_{0}^{r_{xij}}r_{ij}\left\{1-\frac{1}{2}\left[1-\sqrt{1-\frac{2C_{Tij}}{(r_{xij}/r_{rij})^{2}}}\right]\right\}dr_{ij} \\
&+\sum_{i=1}^{n}\sum_{j=1}^{m}\lim_{b\to+\infty}\int_{r_{xij}}^{b}r_{ij}dr_{ij}, \\
&n=1,2,\cdots,N;\ m=1,2,\cdots,N,
\end{aligned} \tag{28}$$

which can be also written as

$$
\begin{aligned}
&\sum_{i=1}^{n}\sum_{j=1}^{m}\lim_{b\to+\infty}\int_{0}^{b} r_{ij}dr_{ij}-\sum_{i=1}^{n}\sum_{j=1}^{m}[1-\\
&\quad\sqrt{1-\frac{C_{Tij}}{2}\left(\frac{r_{rij}}{\sigma_{i}}\right)^{2}}\,\Bigg]\lim_{b\to+\infty}\int_{0}^{b} r_{ij}\exp\left(-\frac{r_{ij}^{2}}{2\sigma_{ij}^{2}}\right)dr_{ij}\\
&=\sum_{i=1}^{n}\sum_{j=1}^{m}\int_{0}^{r_{xij}} r_{ij}\left[\frac{1}{2}+\frac{1}{2}\sqrt{1-\frac{2C_{Tij}}{(r_{xij}/r_{rij})^{2}}}\right]dr_{ij}\\
&\quad+\sum_{i=1}^{n}\sum_{j=1}^{m}\lim_{b\to+\infty}\int_{r_{xij}}^{b} r_{ij}dr_{ij}\\
&\quad n=1,2,\cdots,N;\; m=1,2,\cdots,N,
\end{aligned}
\tag{29}
$$

or equivalently

$$
\begin{aligned}
&\sum_{i=1}^{n}\sum_{j=1}^{m}\lim_{b\to+\infty}\frac{r_{ij}^{2}}{2}\Big|_{0}^{b}-\sum_{i=1}^{n}[1-\\
&\quad\sqrt{1-\frac{C_{Tij}}{2}\left(\frac{r_{rij}}{\sigma_{i}}\right)^{2}}\,\Bigg]\lim_{b\to+\infty}\int_{0}^{b} r_{ij}\exp\left(-\frac{r_{ij}^{2}}{2\sigma_{ij}^{2}}\right)dr_{ij}\\
&=\sum_{i=1}^{n}\sum_{j=1}^{m}\frac{1}{2}\frac{r_{i}^{2}}{2}\Big|_{0}^{r_{xij}}+\sum_{i=1}^{n}\sum_{j=1}^{m}\frac{1}{2}\sqrt{1-\frac{2C_{Tij}}{(r_{xij}/r_{rij})^{2}}}\frac{r_{ij}^{2}}{2}\Big|_{0}^{r_{xij}}\\
&\quad+\sum_{i=1}^{n}\sum_{j=1}^{m}\lim_{b\to+\infty}\frac{r_{ij}^{2}}{2}\Big|_{r_{rij}}^{b}\\
&\quad n=1,2,\cdots,N;\; m=1,2,\cdots,N.
\end{aligned}
\tag{30}
$$

From Equation (30), we have

$$
\begin{aligned}
&-\sum_{i=1}^{n}\sum_{j=1}^{m}[1-\\
&\sqrt{1-\frac{C_{Tij}}{2}\left(\frac{r_{rij}}{\sigma_{ij}}\right)^{2}}\,\Bigg]\lim_{b\to+\infty}\int_{0}^{b} r_{ij}\exp\left(-\frac{r_{ij}^{2}}{2\sigma_{ij}^{2}}\right)dr_{ij}\\
&=\sum_{i=1}^{n}\sum_{j=1}^{m}\frac{1}{2}\frac{r_{xij}^{2}}{2}+\sum_{i=1}^{n}\sum_{j=1}^{m}\frac{1}{2}\sqrt{1-\frac{2C_{Tij}}{(r_{xij}/r_{rij})^{2}}}\frac{r_{xij}^{2}}{2}-\sum_{i=1}^{n}\sum_{j=1}^{m}\frac{r_{xij}^{2}}{2}\\
&n=1,2,\cdots,N;\; m=1,2,\cdots,N,
\end{aligned}
\tag{31}
$$

which can be expressed by the following simple equation:

$$
\begin{aligned}
&-\sum_{i=1}^{n}\sum_{j=1}^{m}\left[1-\sqrt{1-\frac{C_{Tij}}{2}\left(\frac{r_{rij}}{\sigma_{ij}}\right)^{2}}\right]\sigma_{i}^{2}\\
&=-\sum_{i=1}^{n}\sum_{j=1}^{m}\frac{r_{xij}^{2}}{4}+\sum_{i=1}^{n}\sum_{j=1}^{m}\frac{r_{xij}^{2}}{4}\sqrt{1-\frac{2C_{Tij}}{(r_{xij}/r_{rij})^{2}}}\\
&n=1,2,\cdots,N;\; m=1,2,\cdots,N
\end{aligned}
\tag{32}
$$

or also

$$\sum_{i=1}^{n}\sum_{j=1}^{m}\left[1-\sqrt{1-\frac{C_{Tij}}{2}\left(\frac{r_{rij}}{\sigma_{ij}}\right)^{2}}\right]\sigma_{ij}^{2} = \sum_{i=1}^{n}\sum_{j=1}^{m}\frac{r_{xij}^{2}}{4}-\sum_{i=1}^{n}\sum_{j=1}^{m}\frac{r_{xij}^{2}}{4}\sqrt{1-\frac{2C_{Tij}}{(r_{xij}/r_{rij})^{2}}} \quad n=1,2,\cdots,N;\ m=1,2,\cdots,N. \tag{33}$$

Now, we can deduce the following results from Equation (33):

$$\sum_{i=1}^{n}\sum_{j=1}^{m}\left[\sigma_{ij}^{2}-\sigma_{ij}\sqrt{\sigma_{ij}^{2}-\frac{C_{Tij}r_{rij}^{2}}{2}}\right] = \sum_{i=1}^{n}\sum_{j=1}^{m}\frac{r_{xij}^{2}}{4}-\sum_{i=1}^{n}\sum_{j=1}^{m}\frac{r_{xij}^{2}}{4}\sqrt{1-\frac{2C_{Tij}}{(r_{xij}/r_{rij})^{2}}} \quad n=1,2,\cdots,N;\ m=1,2,\cdots,N, \tag{34}$$

which leads to

$$\sum_{i=1}^{n}\sum_{j=1}^{m}\left[\sigma_{ij}^{2}-\frac{r_{xij}^{2}}{4}\left(1-\sqrt{1-\frac{2C_{Tij}}{(r_{xij}/r_{rij})^{2}}}\right)\right] = \sum_{i=1}^{n}\sum_{j=1}^{m}\sigma_{ij}\sqrt{\sigma_{ij}^{2}-\frac{C_{Tij}r_{rij}^{2}}{2}} \quad n=1,2,\cdots,N;\ m=1,2,\cdots,N. \tag{35}$$

Now, by squaring the two sides of Equation (35), we obtain:

$$\sum_{i=1}^{n}\sum_{j=1}^{m}\left[\sigma_{ij}^{4}-\frac{r_{xij}^{2}}{2}\sigma_{ij}^{2}\left(1-\sqrt{1-\frac{2C_{Tij}}{(r_{xij}/r_{rij})^{2}}}\right) +\frac{r_{xij}^{4}}{16}\left(1-\sqrt{1-\frac{2C_{Tij}}{(r_{xij}/r_{rij})^{2}}}\right)^{2}\right] = \sum_{i=1}^{n}\sum_{j=1}^{m}\sigma_{ij}^{2}\left(\sigma_{ij}^{2}-\frac{C_{Tij}r_{rij}^{2}}{2}\right) \quad n=1,2,\cdots,N;\ m=1,2,\cdots,N. \tag{36}$$

By arranging the different terms, we obtain:

$$\sum_{i=1}^{n}\sum_{j=1}^{m}\left[-\frac{r_{xij}^{2}}{2}\left(1-\sqrt{1-\frac{2C_{Tij}}{(r_{xij}/r_{rij})^{2}}}\right)\sigma_{ij}^{2}+\frac{C_{Tij}r_{rij}^{2}}{2}\sigma_{ij}^{2}\right] = \sum_{i=1}^{n}\sum_{j=1}^{m}\left[-\frac{r_{xij}^{4}}{16}\left(1-\sqrt{1-\frac{2C_{Tij}}{(r_{xij}/r_{rij})^{2}}}\right)^{2}\right] \quad n=1,2,\cdots,N;\ m=1,2,\cdots,N, \tag{37}$$

which gives the following equality:

$$\sum_{i=1}^{n}\sum_{j=1}^{m}\sigma_{ij}^2=\sum_{i=1}^{n}\sum_{j=1}^{m}\left[\frac{-\frac{r_{xij}^4}{16}\left(1-\sqrt{1-\frac{2C_{Tij}}{(r_{xij}/r_{rij})^2}}\right)^2}{-\frac{r_{xij}^2}{2}\left(1-\sqrt{1-\frac{2C_{Tij}}{(r_{xij}/r_{rij})^2}}\right)+\frac{C_{Tij}r_{rij}^2}{2}}\right] \tag{38}$$
$$n=1,2,\cdots,N;\ m=1,2,\cdots,N,$$

which can be rewritten as:

$$\begin{aligned}&\sum_{i=1}^{n}\sum_{j=1}^{m}\sigma_{ij}^2\\&=\sum_{i=1}^{n}\sum_{j=1}^{m}\left[\frac{-\frac{r_{xij}^2}{4}\times\frac{r_{xij}^2}{4}\left(1-2\sqrt{1-\frac{2C_{Tij}}{(r_{xij}/r_{rij})^2}}+1-\frac{2C_{Tij}}{(r_{xij}/r_{rij})^2}\right)}{-\frac{r_{xij}^2}{2}+\frac{r_{xij}^2}{2}\sqrt{1-\frac{2C_{Tij}}{(r_{xij}/r_{rij})^2}}+\frac{C_{Tij}r_{rij}^2}{2}}\right]\end{aligned} \tag{39}$$
$$n=1,2,\cdots,N;\ m=1,2,\cdots,N,$$

which is equivalent to:

$$\begin{aligned}&\sum_{i=1}^{n}\sum_{j=1}^{m}\sigma_{ij}^2\\&=\sum_{i=1}^{n}\sum_{j=1}^{m}\left[\frac{\frac{r_{xij}^2}{4}\left(-\frac{r_{xij}^2}{2}+\frac{r_{xij}^2}{2}\sqrt{1-\frac{2C_{Tij}}{(r_{xij}/r_{rij})^2}}+\frac{C_{Tij}r_{rij}^2}{2}\right)}{-\frac{r_{xij}^2}{2}+\frac{r_{xij}^2}{2}\sqrt{1-\frac{2C_{Tij}}{(r_{xij}/r_{rij})^2}}+\frac{C_{Tij}r_{rij}^2}{2}}\right]\end{aligned} \tag{40}$$
$$n=1,2,\cdots,N;\ m=1,2,\cdots,N.$$

Then, we deduce

$$\sum_{i=1}^{n}\sum_{j=1}^{m}\sigma_{ij}^2=\sum_{i=1}^{n}\sum_{j=1}^{m}\frac{r_{xij}^2}{4},\quad n=1,2,\cdots,N;\ m=1,2,\cdots,N, \tag{41}$$

which gives, for the real σ_i, the following value

$$\begin{aligned}\sum_{i=1}^{n}\sum_{j=1}^{m}\sigma_{ij}&=\sum_{i=1}^{n}\sum_{j=1}^{m}\frac{r_{xij}}{2}\\&=\sum_{i=1}^{n}\sum_{j=1}^{m}\frac{r_{0ij}+\alpha_{ij}x_{ij}}{2}\\&=\sum_{i=1}^{n}\sum_{j=1}^{m}\frac{r_{0ij}}{2}+\frac{\alpha_{ij}}{2}x_{ij},\\&\quad n=1,2,\cdots,N;\ m=1,2,\cdots,N,\end{aligned} \tag{42}$$

where $r_{xij}=r_{0ij}+\alpha_{ij}x_{ij}$, α_{ij} and r_{0ij} can be given and estimated empirically from

$$\sum_{i=1}^{n}\sum_{j=1}^{m}\alpha_{ij}=\sum_{i=1}^{n}\sum_{j=1}^{m}\frac{0.5}{\ln(z_{hij}/z_{0ij})},\quad n=1,2,\cdots,N;\ m=1,2,\cdots,N \tag{43}$$

and

$$\sum_{i=1}^{n}\sum_{j=1}^{m} r_{0ij} = \sum_{i=1}^{n}\sum_{j=1}^{m} \frac{4}{5} r_{rij}, \quad n = 1, 2, \cdots, N; \; m = 1, 2, \cdots, N \tag{44}$$

in [44], respectively. Finally, Equations (9), (23), (24) and (42) constitute the FGNDWM.

4. Comparisons and Analysis of Two Different Wake Models for OSWFs

In this section, we will give the definition of Wind Speed Deficit (WSD) and further discuss the relationship between FGWM and FGNDWM. Usually, comparing WSD is a very important approach in different Wake Models. The WSD of OSWFs is expressed as by the following equation:

$$\frac{\Delta v_{ij}}{v_{0ij}} = \frac{v_{0ij} - v_{xij}}{v_{0ij}}, \quad i = 1, 2, \cdots, N; j = 1, 2, \cdots, N. \tag{45}$$

Firstly, the WSD of FGWM is derived by Equations (7) and (45):

$$\begin{aligned} \Delta v_{FGWM_{ij}} &= \frac{v_{0ij} - v_{x_{ij}}}{v_{0ij}} \\ &= \frac{v_{0ij} - \left\{1 - \frac{1}{2}\left[1 - \sqrt{1 - \frac{2C_{Tij}}{(r_{xij}/r_{rij})^2}}\right]\right\} v_{0ij}}{v_{0ij}} \\ &= \frac{1}{2}\left[1 - \sqrt{1 - \frac{2C_{Tij}}{(r_{xij}/r_{rij})^2}}\right], \\ & i = 1, 2, \cdots, N; j = 1, 2, \cdots, N. \end{aligned} \tag{46}$$

Meanwhile, we obtain the WSD of FGNDWM. With the condition: $r_{ij} = 0$, the WSD of FGNDWM on the axis is interpreted based on Equations (24) and (45):

$$\begin{aligned} \Delta v_{FGNDWM_{ij}} &= \frac{v_{0ij} - v_{x_{ij}, r_{ij}}}{v_{0ij}} \\ &= \frac{v_{0ij} - \left\{1 - \left[1 - \sqrt{1 - \frac{C_{Tij}}{2}\left(\frac{r_{rij}}{\sigma_{ij}}\right)^2}\right] \exp\left(-\frac{r_{ij}^2}{2\sigma_{ij}^2}\right)\right\} v_{0ij}}{v_{0ij}} \\ &= 1 - \sqrt{1 - \frac{C_{Tij}}{2}\left(\frac{r_{rij}}{\sigma_{ij}}\right)^2} \\ &= 1 - \sqrt{1 - \frac{C_{Tij}}{2\left(\frac{\sigma_{ij}}{r_{rij}}\right)^2}}, \quad i = 1, 2, \cdots, N; j = 1, 2, \cdots, N. \end{aligned} \tag{47}$$

Substituting $\sigma_{ij} = \frac{r_{xij}}{2}$ into Equation (47), we obtain

$$\begin{aligned}\Delta v_{FGNDWM_{ij}} &= \frac{v_{0ij} - v_{x_{ij},r_{ij}}}{v_{0ij}} \\ &= 1 - \sqrt{1 - \frac{C_{Tij}}{2}\left(\frac{r_{rij}}{\frac{r_{xij}}{2}}\right)^2} \\ &= 1 - \sqrt{1 - \frac{2C_{Tij}}{(r_{xij}/r_{rij})^2}}, \\ &i = 1,2,\cdots,N; j = 1,2,\cdots,N.\end{aligned} \tag{48}$$

The ratio of the WSD from FGWM to FGNDWM is calculated by Equations (46) and (48):

$$\begin{aligned}\frac{\Delta v_{FGWM_{ij}}}{\Delta v_{FGNDWM_{ij}}} &= \frac{\frac{1}{2}\left[1 - \sqrt{1 - \frac{2C_{Tij}}{(r_{xij}/r_{rij})^2}}\right]}{1 - \sqrt{1 - \frac{2C_{Tij}}{(r_{xij}/r_{rij})^2}}} \\ &= \frac{1}{2}, \quad i = 1,2,\cdots,N; j = 1,2,\cdots,N.\end{aligned} \tag{49}$$

We can find out the WSD of FGWM is half as small as that of FGNDWM on the axis.

5. Experimental Comparisons and Analysis of Two Different Wake Models for OSWFs

In this section, we collect and use data of five cases to confirm the different characteristics between FGWM and FGNDWM. Usually, using miniature WT with the Large-Eddy Simulation (LES) data was known as the standard case in the literature.

The main data and parameters of the five cases (OSWFs in Yangshan port, Shanghai) are shown in the following Table 1, in which z_{hij} is the height of the Hub, and z_{0ij} is the rate of surface sea or roughness. These roughness lengths shown for $Cases(b-e)$ in Table 1 are representative of different sea surface types, including very rough terrain, for instance, islands with different sizes ($z_{0ij} = 0.1$ m), sea surface with reefs, rocks and shoal rocks ($z_{0ij} = 0.01$ m), sea surface with medium waves and large waves ($z_{0ij} = 0.001$ m), and sea surface with small waves ($z_{0ij} = 0.00001$ m). d_{rij} is the diameter of the rotor, α_{ij} is the axial induction factor, r_{0ij} is the downstream rotor radius, C_{Tij} is the thrust coefficient, and v_{0ij} is cut-in wind speed. Here, i and j are row vector and column vector for Wind Turbine in large-scale offshore Wind Farms, respectively.

LES are applied in many fields of flow simulations. The initial conditions have a very significant influence on the LES results. $x_{ij}/d_{rij} = 3$ and $\Delta v_{ij}/v_{0ijmax} = 0.5$ were chosen as the initial conditions used for LES simulation conducted in this study.

Table 1. Different experiments and Large-Eddy Simulation (LES) case to validate the Generalized model of OSWFs in Yangshan port, Shanghai.

Cases	d_{rij} (m)	z_{hij} (m)	v_{0ij} (m/s)	C_{Tij}	z_{0ij} (m)	α_{ij}	r_{0ij} (m)
$Case_a$	0.15	0.125	2.2	0.4194	0.00003	0.119	0.066
$Case_b$	66	65	6	0.3916	0.1	0.11	38.08
$Case_c$	66	98	6	0.2944	0.01	0.08	40.48
$Case_d$	82	75	6	0.2256	0.001	0.06	43.52
$Case_e$	70	65	6	0.2256	0.00001	0.062	41.12

Note: $Case_a$ represents the Large-Eddy Simulation (LES) data; $Case_b$ represents OSWF-b in Yangshan port, Shanghai; $Case_c$ represents OSWF-c in Yangshan port, Shanghai; $Case_d$ represents OSWF-d in Yangshan port, Shanghai; $Case_e$ represents OSWF-e in Yangshan port, Shanghai.

Using some given data in Table 1, we conduct the simulation experiments with FGWM and FGNDWM. By analyzing Figures 5–9, from simulations and experiments, we can find the WSD of FGWM are half times smaller then that of FGNDWM on the axis. These simulations validate the results obtained in the last sections.

If the length of the near wake region x_{1i} is taken into consideration and the parameters are selected according to Table 1, the result is shown in Figure 5. The FGNDWM appeared superior to FGWM. The problem was that FGWM did not take x_{1i} into consideration, whereas the FGNDWM took it into consideration. In addition, the characteristic width of the FGNDWM was obtained by fitting the LES data in the experiment of this study.

It can be seen from Figure 8 that the accuracy of the FGNDWM and the FGWM is better than that in Figures 7 and 9. The maximum Wind Speed Deficit (WSD) of this FGNDWM can be proved to be twice as large as that of the FGWM; if α_{ij} and r_{0ij} are estimated by (43) and (44), respectively, the result is shown in Figure 9. As previously stated, (43) is not applicable to case a and case e. As a result, the accuracy of the FGNDWM and FGWM in Figure 6 is worse than that in Figures 5 and 7–9.

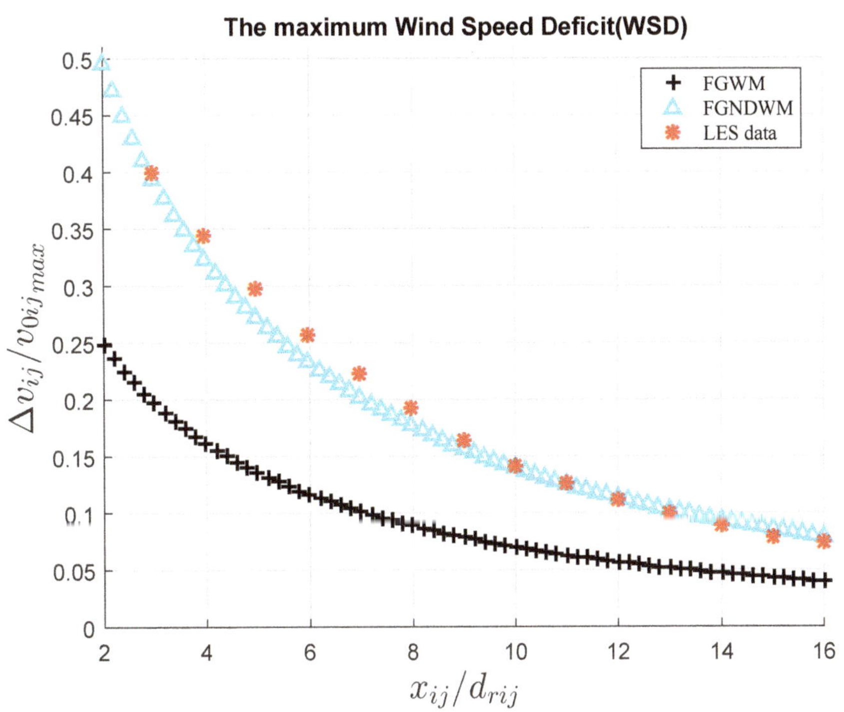

Figure 5. The portrait of the maximum Wind Speed Deficit (WSD) in $Case_a$ (fitted by LES data, $i = 1, 2, \cdots, N$; $j = 1, 2, \cdots, N$).

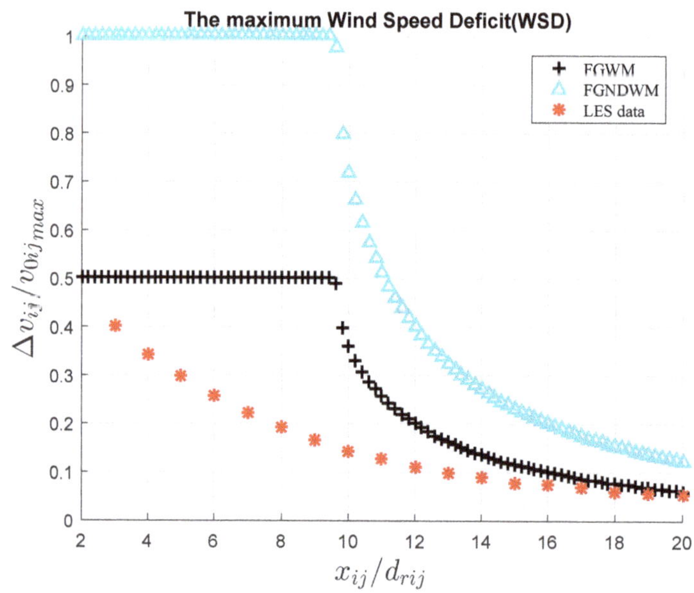

Figure 6. The portrait of the maximum Wind Speed Deficit (WSD) in $Case_b$ (fitted by LES data, $i = 1, 2, \cdots, N$; $j = 1, 2, \cdots, N$).

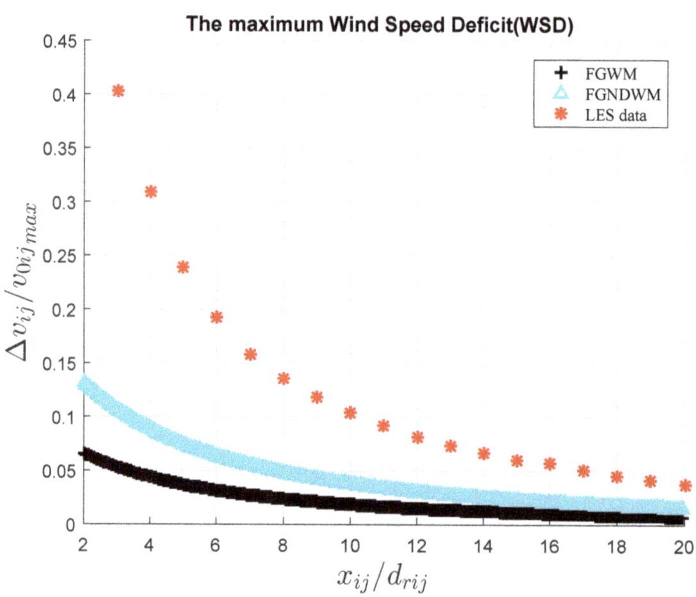

Figure 7. The portrait of the maximum Wind Speed Deficit (WSD) in $Case_c$ (fitted by LES data, $i = 1, 2, \cdots, N$; $j = 1, 2, \cdots, N$).

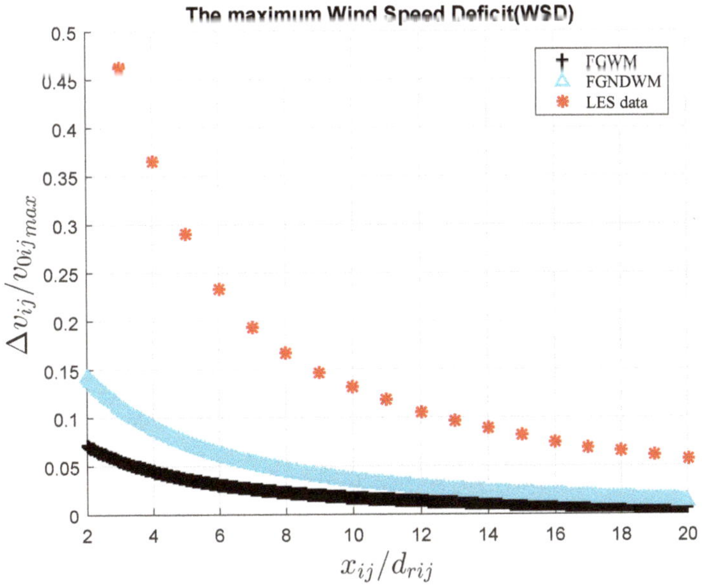

Figure 8. The portrait of the maximum Wind Speed Deficit (WSD) in $Case_d$ (fitted by LES data, $i = 1, 2, \cdots, N$; $j = 1, 2, \cdots, N$).

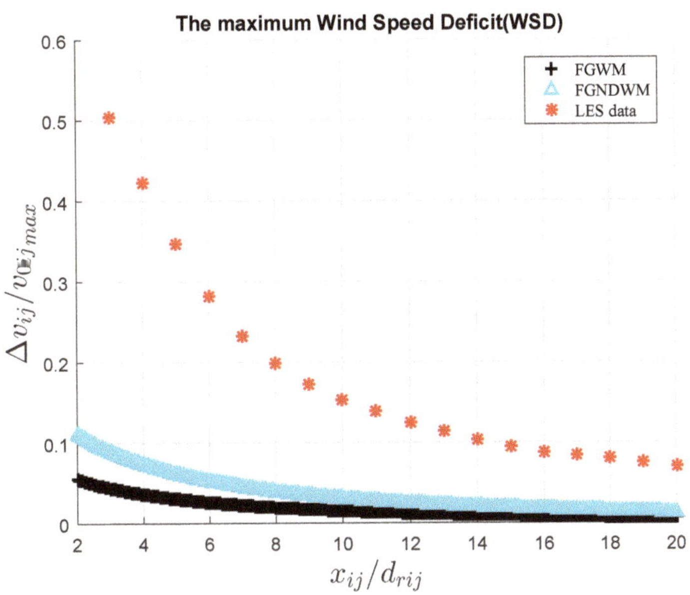

Figure 9. The portrait of the maximum Wind Speed Deficit (WSD) in $Case_e$ (fitted by LES data, $i = 1, 2, \cdots, N$; $j = 1, 2, \cdots, N$).

6. Analysis and Enlightenment of Wind Rose, Wind Weibull Probability Density Distribution and ELM Prediction

Based on the actual situation in OSWFs in Yangshan port, Shanghai, this section summarizes and describes the analysis of wind rose, wind Weibull probability density distribution and ELM prediction. We obtain the real-time and actual data from this website [45]. The WE cases are analyzed and studied based on mathematical models, and WRs are abstracted through the variable wind directions and wind speeds in OSWFs in Yangshan port, Shanghai. These data of WE resources are collected and shown in the following figures.

From 22 August 2017 to 22 August 2018, the portraits of wind direction and wind speed (m/s) in OSWFs of Yangshan port, Shanghai are shown with the wind rose in Figure 10. From the southwest direction, most of the wind mean speeds in a whole year are greater than 6 m/s, whereas, from the northwest direction, a small part of wind mean speeds are close to 6 m/s in a whole year.

From 22 August 2016 to 22 August 2017 with a whole year in OSWFs of Yangshan port, Shanghai, we keep obtaining the real-time and actual data from this website [45]. The wind speeds are collected and their Mean Wind Speed (MWS) is 3.4934 m/s. The details of them are shown in Figure 11.

From 22 August 2016 to 22 August 2017 with a whole year in OSWFs of Yangshan port, Shanghai, the rose portraits of Wind Direction (WD) and Average Wind Direction (AWD) are shown in Figure 12. Among them, the north direction is 0 degrees (North = 0 °C) . The probabilities of most of the parts of WDs from the southwest direction are greater than 5% and are close to 6%. Meanwhile, a small part of WD from the northeast direction is greater than 5% and close to 6%.

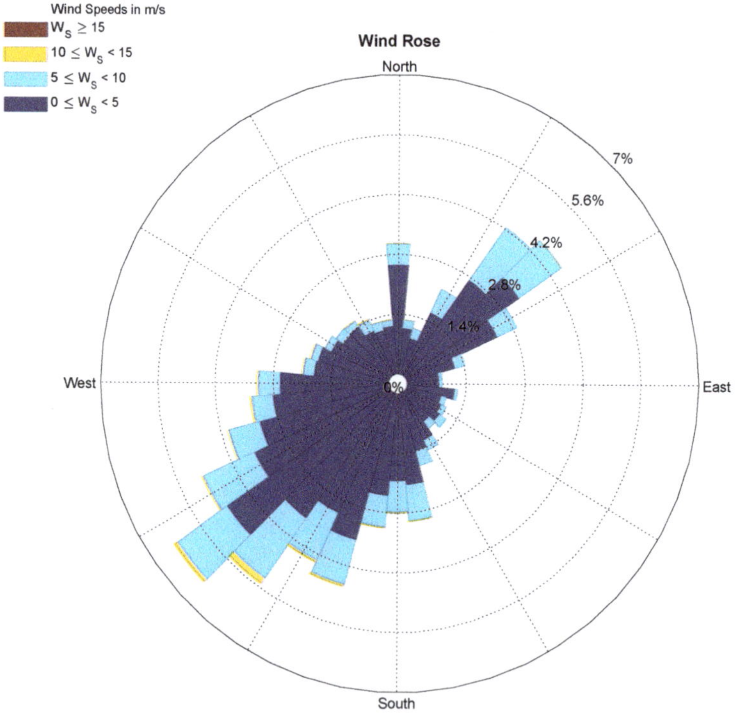

Figure 10. Rose portrait of WD (North = 0) and speed (m/s) in off-shore wind farms (OSWFs) in Yangshan port, Shanghai from 22 August 2017 to 22 August 2018.

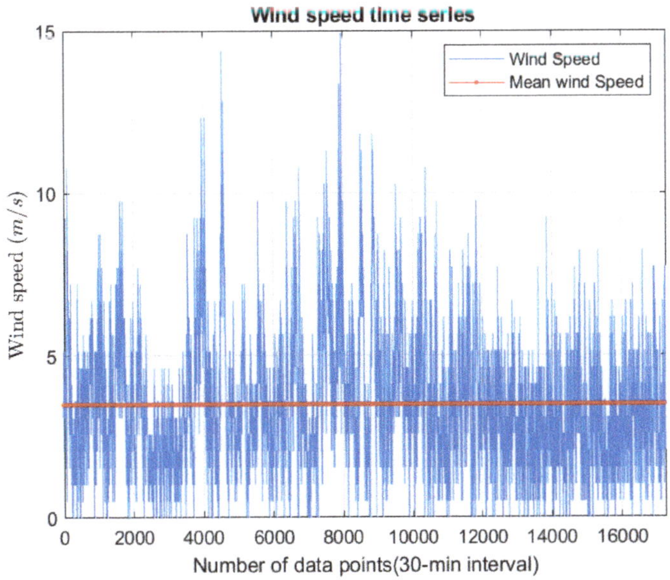

Figure 11. The portraits of wind speed time series and mean speed ($\bar{v}$ = 3.4934 m/s) in OSWFs in Yangshan port, Shanghai from 22 August 2017 to 22 August 2018.

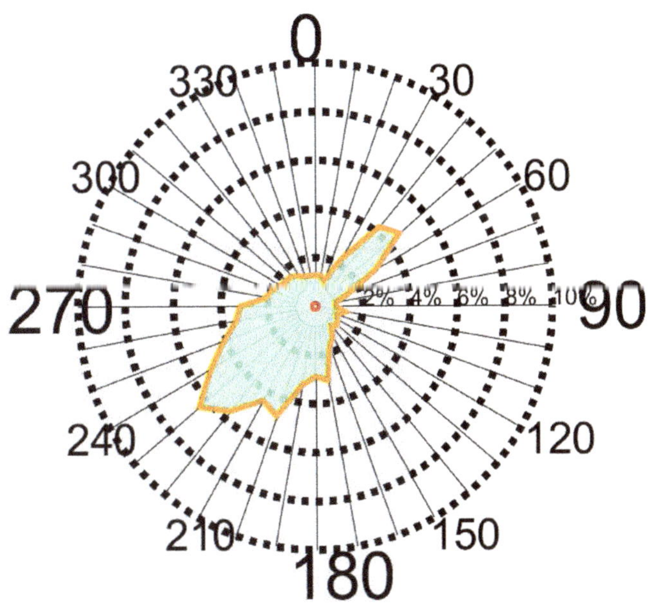

Figure 12. Rose portrait of WD (North direction = 0°) in OSWFs in Yangshan port, Shanghai from 22 August 2017 to 22 August 2018.

In Figure 13, the wind rose shows the portraits of wind mean speed (m/s) in OSWFs of Yangshan port, Shanghai from 22 August 2017 to 22 August 2018 with a whole year. The majority of the wind mean speeds from the southwest are greater than 5 m/s, whereas, a small amount of wind mean speed from the northwest is approximate to 5 m/s. Therefore, WTs should adjust the direction to the southwest in OSWFs of Yangshan port, Shanghai throughout the summer, for even more periods.

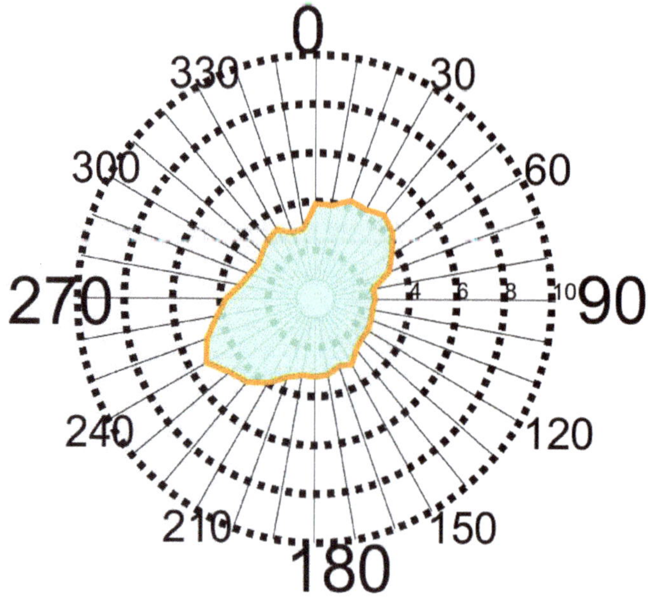

Figure 13. Rose portrait of Wind Mean Speed (m/s) in OSWFs of Yangshan port, Shanghai from 22 August 2017 to 22 August 2018.

Figures 14–16 show all the wind situations, including the time series of air relative humidity, wind direction and wind temperature in OSWFs in Yangshan port, Shanghai from 22 August 2017 to 22 August 2018, respectively. In this period, the mean air relative humidity is 82.0978%, the mean wind direction is 169.4492°(North direction = 0°), and the mean wind temperature is 10.0367 °C.

ELM is a learning algorithm, initially introduced to train a Single Layer Feedforward Neural network [46]. In ELM theory, the input weights are randomly generated according to any continuous distribution function, while the output weights are analytically computed by the minimum norm solution of a linear system.

Here, as shown in Figure 17, the proposed ELM can be seen as three hidden layer neural networks, trained using the ELM algorithm. ELM is applied to wind direction and wind speed prediction in OSWFs in Yangshan port, Shanghai from 22 August 2017 to 22 August 2018. The wind direction and wind speed of OSWFs are relevant to local air relative humidity and local wind temperature. Some simulation results show the corresponding rationality. Figures 18 and 19 show the portraits of comparison of wind direction and wind speed forecasting results with ELM in OSWFs in Yangshan port, Shanghai from 22 August 2017 to 22 August 2018, respectively. Here, the number of forecast data pieces is 2248. Figures 20 and 21 show the portraits of comparison of wind direction and wind speed forecasting results with ELM in OSWFs in Yangshan port, Shanghai from 22 August 2017 to 22 August 2018, respectively. At the same time, the number of forecast data pieces is 500.

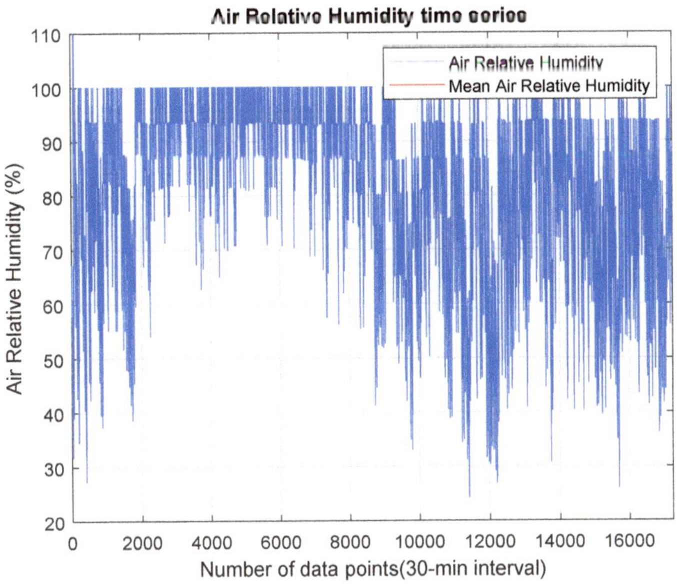

Figure 14. The portraits of air relative humidity in OSWFs in Yangshan port, Shanghai from 22 August 2017 to 22 August 2018.

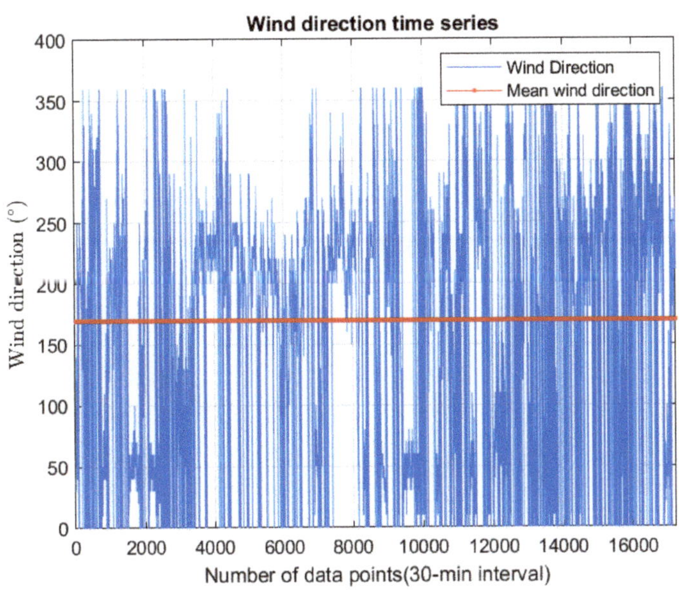

Figure 15. The portraits of wind direction (north direction = 0°) in OSWFs in Yangshan port, Shanghai from 22 August 2017 to 22 August 2018.

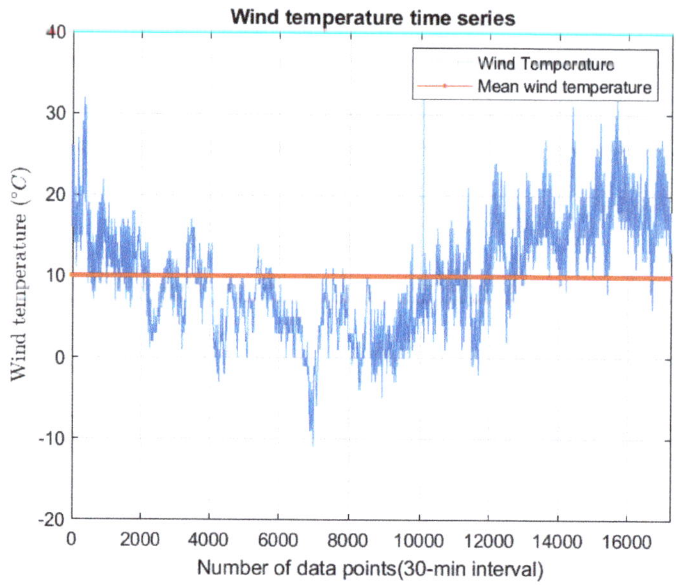

Figure 16. The portraits of wind temperature in OSWFs in Yangshan port, Shanghai from 22 August 2017 to 22 August 2018.

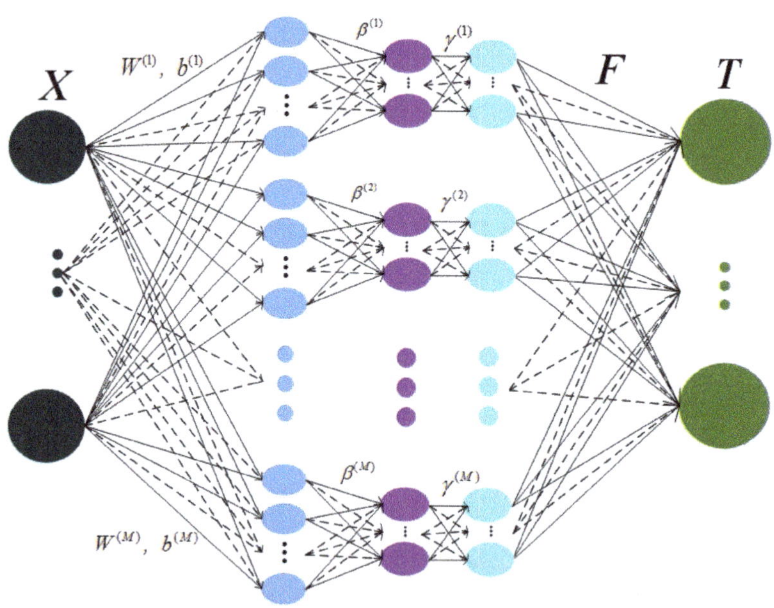

Figure 17. The proposed Extreme Learning Machine (ELM) can be seen as a three hidden layer neural network trained using the ELM algorithm.

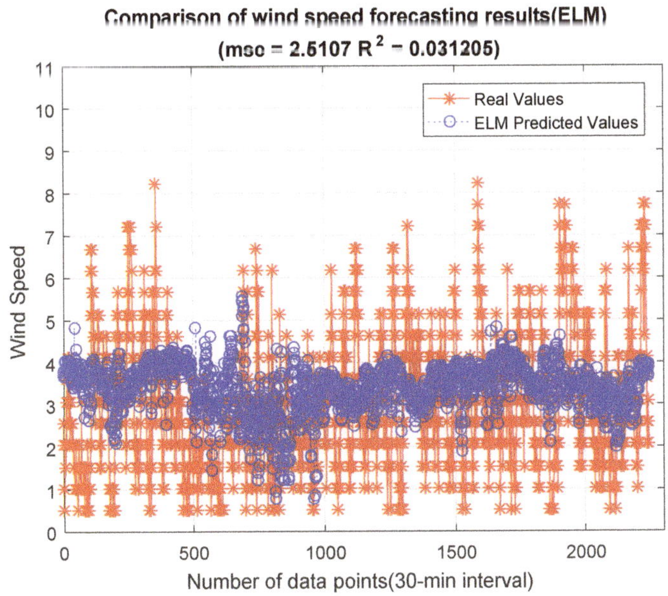

Figure 18. The portraits of comparison of wind speed forecasting results (ELM) in OSWFs in Yangshan port, Shanghai from 22 August 2017 to 22 August 2018 (number of predicted data pieces is 2248).

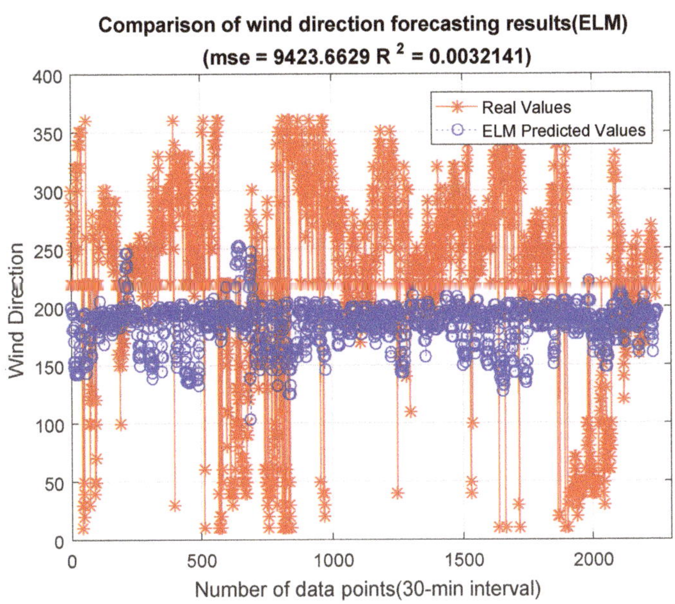

Figure 19. The portraits of comparison of wind direction forecasting results (ELM) in OSWFs in Yangshan port, Shanghai from 22 August 2017 to 22 August 2018 (number of predicted data pieces is 2248).

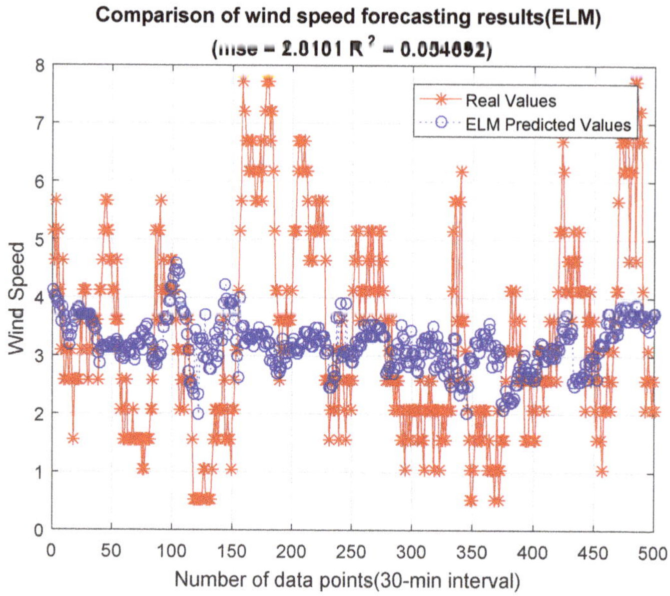

Figure 20. The portraits of comparison of wind speed forecasting results (ELM) in OSWFs in Yangshan port, Shanghai from 22 August 2017 to 22 August 2018 (number of predicted data pieces is 500).

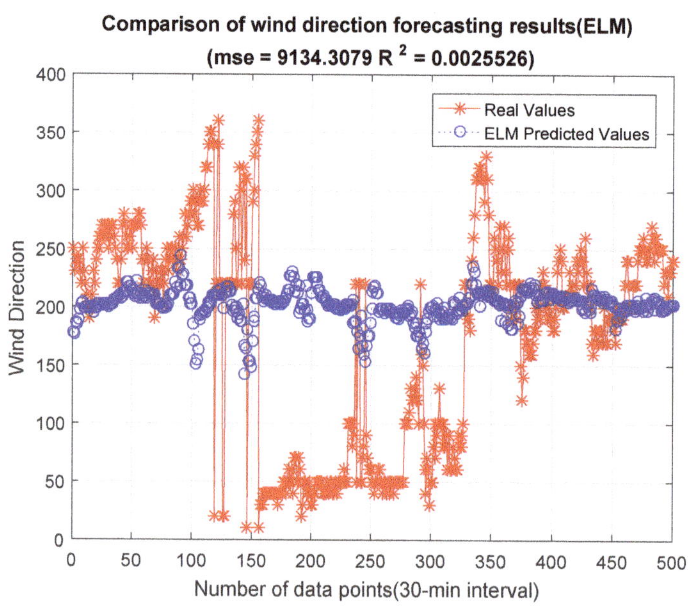

Figure 21. The portraits of comparison of wind direction forecasting results (ELM) in OSWFs in Yangshan port, Shanghai from 22 August 2017 to 22 August 2018 (number of predicted data pieces is 500).

Through the study of the Weibull Probability Distribution of Wind Velocity Data in OSWFs in Yangshan port, Shanghai, we obtain the following results. The cumulative distribution and linearized curve are plotted and shown in Figure 22, Linearized curve and fitted line comparison are shown in

Figure 23, Weibull probability density function and Cumulative Weibull probability density function are shown in Figure 24. In light of wind tower, measuring data and plotted wind speed histograms, we inferred and estimated two parameters of the Weibull distribution by maximum likelihood estimation method, i.e., $c = 3.7660$, and $k = 1.7153$. The histogram of wind speed at hub height with the fitted Weibull probability density distribution are plotted and shown in Figure 25. Wind speed at hub height conforms to the Weibull distribution.

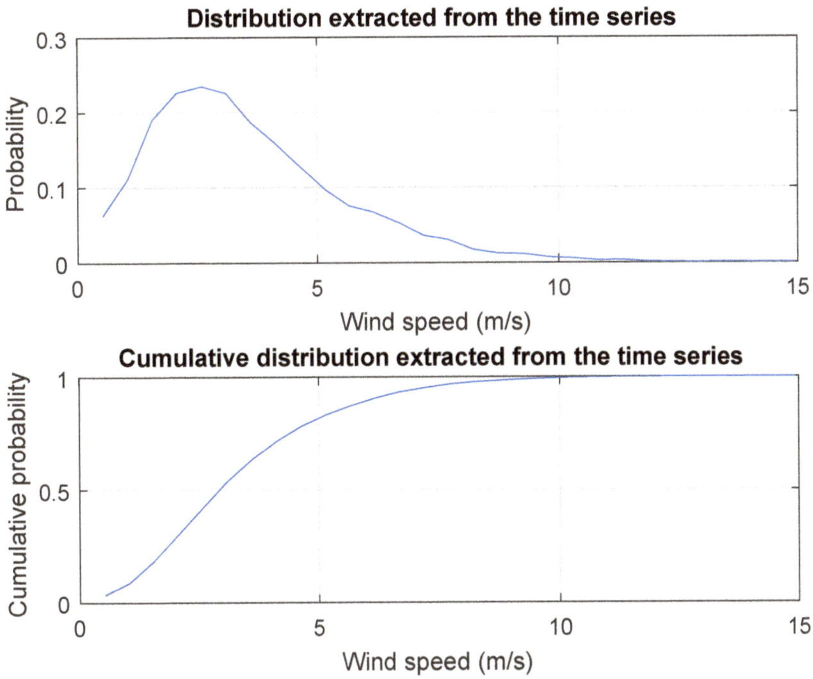

Figure 22. The portraits of distribution and cumulative distribution extracted from the time series in OSWFs in Yangshan port, Shanghai from 22 August 2017 to 22 August 2018.

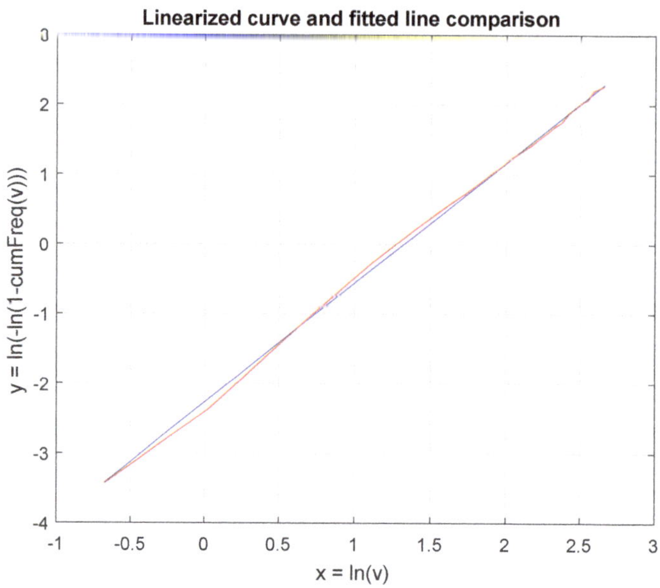

Figure 23. The portraits of linearized curve and fitted line comparison in OSWFs in Yangshan port, Shanghai from 22 August 2017 to 22 August 2018.

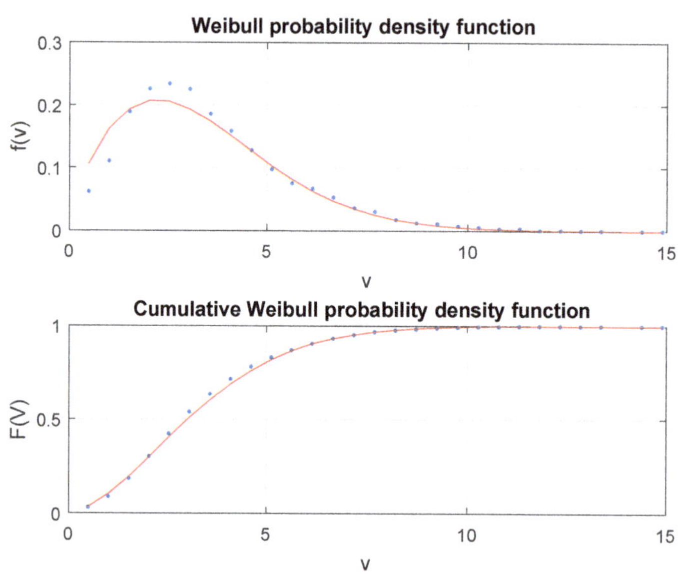

Figure 24. The portraits of Weibull and Cumulative Weibull probability density functions in OSWFs in Yangshan port, Shanghai from 22 August 2017 to 22 August 2018.

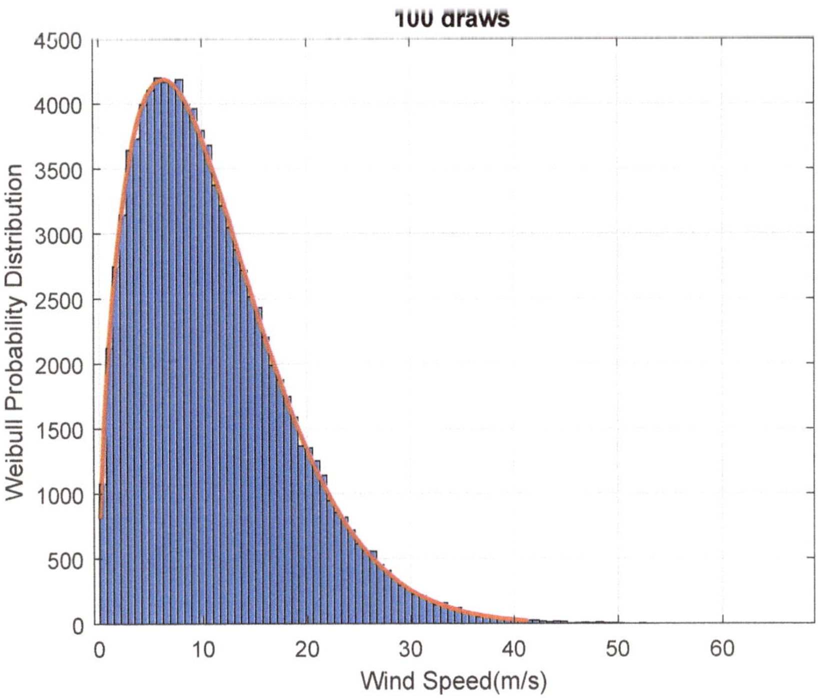

Figure 25. The portraits of wind speed histogram in hub height and the fitted Weibull probability density distribution in OSWFs in Yangshan port, Shanghai from 22 August 2017 to 22 August 2018.

7. Conclusions

From this study, we found that the feature of FGWM is simple and intuitive, while the feature of FGNDWM is complex and precise. The accuracy of the FGNDWM is inherently better than that of the FGWM. When describing the relationship between FGNDWM and FGWM, Equation (42) can reflect the essential characteristics of FGNDWM, whereas Equation (7) reflects the basic characteristics of FGWM. Equation (42) is more accurate than (7) in expressing the characteristics of the large off-shore wind farms. By taking x_{1i} into consideration, the accuracy of the FGNDWM and the FGWM can be improved. Their accuracy depends on the axial induction factor α_{ij}. The maximum Wind Speed Deficit (WSD) of this FGNDWM can be proved to be twice as large as that of the FGWM if α_{ij} and r_{0ij} are estimated by (43) and (44), respectively.

Currently, the experiments show that the accuracy of ELM predictions needs to be improved based on the actual situation in OSWFs in Yangshan port, Shanghai. In future research, we will work on hybrid wake models for near wakes and far wakes and improve the ELM predictions accuracy of large off-shore wind farms. The data assimilation and reduced order modelling will be provided in a future paper concerning induced large off-shore wind farms dynamics.

Author Contributions: Conceptualization, M.L. and H.X.; Methodology, L.P. and C.X.; Software, L.P. and C.X.; Validation, M.L., H.X. and L.P.; Formal Analysis, M.L. and H.X.; Investigation, M.L. L.P. and C.X.; Resources, M.L. and H.X.; Data Curation, M.L. and H.X.; Writing–Original Draft Preparation, L.P. and C.X.; Writing–Review and Editing, M.L. and H.X.; Visualization, L.P. and C.X.; Supervision, H.X. and L.P.; Project Administration, H.X.; Funding Acquisition, H.X..

Funding: This research was funded by the PhD research startup foundation of Wuhan University of Technology grant number 40120291, the Fundamental Research Funds for the Central Universities grant number 173118001, the Open Project Program of National Engineering Research Center for Water Transport Safety grant number A2019006, and the Open Project Program of State Key Laboratory of Ocean Engineering grant number 1812. The APC was funded by National Key Technology Research and Development Program of the Ministry of Science and Technology of China grant number 2015BAF06B01-2.

Conflicts of Interest: The authors declare no conflict of interest.

Abbreviations

The following abbreviations are used in this manuscript:

OSWFs	Off-Shore Wind Farms
WT	Wind Turbine
ELM	Extreme Learning Machine
WP	Wind Power
LES	Large-Eddy Simulation
EEA	Extended Exergy Accounting
WP	Wind Power
HAWTs	Horizontal Axis Wind Turbines
WF	Wind Farm
WE	Wind Energy
DFIG	Doubly-fed Induction Generator
PMSG	Permanent Magnet Synchronous Generator
FSIG	Fixed-Speed Induction Generator
WTPGS	Wind Turbine Power Generation System
FGWM	Frandsen Generalized Wake Model
FGNDWM	Frandsen Generalized Normal Distribution Wake Model
WSD	Wind Speed Deficit
WD	Wind Direction
AWD	Average Wind Direction

References

1. Marden, J.R.; Ruben, S.D.; Pao, L.Y. A Model-Free Approach to Wind Farm Control Using Game Theoretic Methods. *IEEE Trans. Control Syst. Technol.* **2013**, *21*, 1207–1214. [CrossRef]
2. Pao, L.Y.; Johnson, K. A tutorial on the dynamics and control of wind turbines and wind farms. In Proceedings of the American Control Conference, St. Louis, MO, USA, 10–12 June 2009; pp. 2076–2089.
3. Bitar, E.; Seiler, P. Coordinated control of a wind turbine array for power maximization. In Proceedings of the American Control Conference (ACC), Washington, DC, USA, 17–19 June 2013; pp. 2898–2904.
4. Remmers, T.; Cawkwell, F.; Desmond, C.; Murphy, J.; Politi, E. The Potential of Advanced Scatterometer (ASCAT) 12.5 km Coastal Observations for Offshore Wind Farm Site Selection in Irish Waters. *Energies* **2019**, *12*, 206. doi:10.3390/en12020206. [CrossRef]
5. Hübler, C.; Weijtjens, W.; Gebhardt, C.G.; Rolfes, R.; Devriendt, C. Validation of Improved Sampling Concepts for Offshore Wind Turbine Fatigue Design. *Energies* **2019**, *12*, 603. doi:10.3390/en12040603. [CrossRef]
6. Li, H.; Wang, J.; Lu, H.; Guo, Z. Research and application of a combined model based on variable weight for short term wind speed forecasting. *Renew. Energy* **2018**, *116*, 669–684. [CrossRef]
7. Romanic, D.; Parvu, D.; Refan, M.; Hangan, H. Wind and tornado climatologies and wind resource modelling for a modern development situated in Tornado Alley. *Renew. Energy* **2018**, *115*, 97–112. [CrossRef]
8. Ahmed, A.S. Wind energy characteristics and wind park installation in Shark El-Ouinat, Egypt. *Renew. Sustain. Energy Rev.* **2018**, *82*, 734–742. [CrossRef]
9. Aghbashlo, M.; Tabatabaei, M.; Hosseini, S.S.; Dashti, B.B.; Soufiyan, M.M. Performance assessment of a wind power plant using standard exergy and extended exergy accounting (EEA) approaches. *J. Clean. Prod.* **2018**, *171*, 127–136. [CrossRef]

10. Ahmad, T.; Basit, A.; Anwar, J.; Coupiac, O.; Kazemtabrizi, B.; Matthews, P.C. Fast Processing Intelligent Wind Farm Controller for Production Maximisation. *Energies* **2019**, *12*, 544. doi:10.3390/en12030544. [CrossRef]
11. Shakoor, R.; Hassan, M.Y.; Raheem, A.; Wu, Y.K. Wake effect modeling: A review of wind farm layout optimization using Jensen's model. *Renew. Sustain. Energy Rev.* **2016**, *58*, 1048–1059. [CrossRef]
12. Kuenzel, S.; Kunjumuhammed, L.; Pal, B.; Erlich, I. Impact of Wakes on Wind Farm Inertial Response. *IEEE Trans. Sustain. Energy* **2014**, *5*, 237–245. [CrossRef]
13. Pan, L.; Voos, H.; Li, Y.; Darouach, M.; Xu, Y.; Hu, S. A wake interaction model for the coordinated control of Wind Farms. In Proceedings of the 2015 IEEE 20th Conference on Emerging Technologies Factory Automation (ETFA), Luxembourg, 8–11 September 2015; pp. 1–7.
14. Pan, L.; Voos, H.; Pan, Y.; Darouach, M. A generalized interaction Wake Model with its variation for control in Wind Farms. In Proceedings of the 2016 35th Chinese Control Conference (CCC), Chengdu, China, 27–29 July 2016; pp. 8759–8764.
15. Pan, L.; Voos, H.; Li, Y.; Xu, Y.; Darouach, M.; Li, Z. A class of Improved Wake Interaction Model for the coordinated control of wind farms. In Proceedings of the Chinese Automation Congress (CAC), Wuhan, China, 27–29 November 2015; pp. 1322–1327.
16. He, P.; Arefifar, S.A.; Li, C.; Wen, F.; Ji, Y.; Tao, Y. Enhancing Oscillation Damping in an Interconnected Power System with Integrated Wind Farms Using Unified Power Flow Controller. *Energies* **2019**, *12*, 322. doi:10.3390/en12020322. [CrossRef]
17. March, V. Key issues to define a method of lightning risk assessment for wind farms. *Electr. Power Syst. Res.* **2017**, in press. [CrossRef]
18. Chen, K.; Song, M.; Zhang, X. The investigation of tower height matching optimization for wind turbine positioning in the wind farm. *J. Wind Eng. Ind. Aerodyn.* **2013**, *114*, 83–95. [CrossRef]
19. Sørensen, K.L.; Galeazzi, R.; Odgaard, P.F.; Niemann, H.; Poulsen, N.K. Adaptive Passivity Based Individual Pitch Control for Wind Turbines in the Full Load Region. In Proceedings of the 2014 American Control Conference, Portland, OR, USA, 4–6 June 2014; pp. 554–559.
20. Barreiro-Gomez, J.; Ocampo-Martinez, C.; Bianchi, F.; Quijano, N. Model-free control for wind farms using a gradient estimation-based algorithm. In Proceedings of the 2015 European Control Conference (ECC), Linz, Austria, 15–17 July 2015; pp. 1516–1521.
21. Long, M.; Becerra, M.; Thottappillil, R. On the attachment of dart lightning leaders to wind turbines. *Electr. Power Syst. Res.* **2017**, *151*, 432–439. [CrossRef]
22. Thukaram, D. Accurate modeling of doubly fed induction generator based wind farms in load flow analysis. *Electr. Power Syst. Res.* **2018**, *155*, 363–371.
23. Farajzadeh, S.; Ramezani, M.H.; Nielsen, P.; Nadimi, E.S. Statistical modeling of the power grid from a wind farm standpoint. *Electr. Power Syst. Res.* **2017**, *144*, 150–156. [CrossRef]
24. Tian, L.; Zhu, W.; Shen, W.; Zhao, N.; Shen, Z. Development and validation of a new two-dimensional wake model for wind turbine wakes. *J. Wind Eng. Ind. Aerodyn.* **2015**, *137*, 90–99. [CrossRef]
25. Park, J.; Law, K.H. A data-driven, cooperative wind farm control to maximize the total power production. *Appl. Energy* **2016**, *165*, 151–165. [CrossRef]
26. Marseglia, G.R.; Arbasini, A.; Grassi, S.; Raubal, M.; Raimondo, D.M. Optimal placement of wind turbines on a continuous domain: An MILP-based approach. In Proceedings of the American Control Conference (ACC), Chicago, IL, USA, 1–3 July 2015; pp. 5010–5015.
27. van Dam, F.; Gebraad, P.; van Wingerden, J.W. A maximum power point tracking approach for wind farm control. In Proceedings of the Science of Making Torque from Wind, Oldenburg, Germany, 9–11 October 2012.
28. Chen, J. Development of offshore wind power in China. *Renew. Sustain. Energy Rev.* **2011**, *15*, 5013–5020. doi:10.1016/j.rser.2011.07.053. [CrossRef]
29. Zhixin, W.; Chuanwen, J.; Qian, A.; Chengmin, W. The key technology of offshore wind farm and its new development in China. *Renew. Sustain. Energy Rev.* **2009**, *13*, 216–222. doi:10.1016/j.rser.2007.07.004. [CrossRef]
30. Ebrahimi, F.; Khayatiyan, A.; Farjah, E. A novel optimizing power control strategy for centralized wind farm control system. *Renew. Energy* **2016**, *86*, 399–408. [CrossRef]
31. Song, Z.; Zhang, Z.; Chen, X. The decision model of 3-dimensional wind farm layout design. *Renew. Energy* **2016**, *85*, 248–258. [CrossRef]

32. Varzaneh, S.G.; Abedi, M.; Gharehpetian, G. A new simplified model for assessment of power variation of DFIG based wind farm participating in frequency control system. *Electr. Power Syst. Res.* **2017**, *148*, 220–229. [CrossRef]
33. Hossain, M.E. A non-linear controller based new bridge type fault current limiter for transient stability enhancement of DFIG based Wind Farm. *Electr. Power Syst. Res.* **2017**, *152*, 466–484. [CrossRef]
34. Yao, J.; Li, J.; Guo, L.; Liu, R.; Xu, D. Coordinated control of a hybrid wind farm with PMSG and FSIG during asymmetrical grid fault. *Int. J. Electr. Power Energy Syst.* **2018**, *95*, 287–300. [CrossRef]
35. Li, D.Y.; Li, P.; Cai, W.C.; Song, Y.D.; Chen, H.J. Adaptive Fault Tolerant Control of Wind Turbines with Guaranteed Transient Performance Considering Active Power Control of Wind Farms. *IEEE Trans. Ind. Electron.* **2017**, *65*, 3275–3285. [CrossRef]
36. Chaurasiya, P.K.; Ahmed, S.; Warudkar, V. Comparative analysis of Weibull parameters for wind data measured from met-mast and remote sensing techniques. *Renew. Energy* **2018**, *115*, 1153–1165. [CrossRef]
37. Atighechi, H.; Hu, P.; Ebrahimi, S.; Lu, J.; Wang, G.; Wang, L. An effective load shedding remedial action scheme considering wind farms generation. *Int. J. Electr. Power Energy Syst.* **2018**, *95*, 353–363. [CrossRef]
38. Suganthi, S.; Devaraj, D.; Ramar, K.; Thilagar, S.H. An Improved Differential Evolution algorithm for congestion management in the presence of wind turbine generators. *Renew. Sustain. Energy Rev.* **2018**, *81*, 635–642. [CrossRef]
39. Wan, C.; Xu, Z.; Pinson, P.; Dong, Z.Y.; Wong, K.P. Probabilistic forecasting of wind power generation using extreme learning machine. *IEEE Trans. Power Syst.* **2014**, *29*, 1033–1044. [CrossRef]
40. Nikolić, V.; Motamedi, S.; Shamshirband, S.; Petković, D.; Ch, S.; Arif, M. Extreme learning machine approach for sensorless wind speed estimation. *Mechatronics* **2016**, *34*, 78–83. [CrossRef]
41. Lazarevska, E. Wind speed prediction with extreme learning machine. In Proceedings of the 2016 IEEE 8th International Conference on Intelligent Systems (IS), Sofia, Bulgaria, 4–6 September 2016; pp. 154–159.
42. Wu, S.; Wang, Y.; Cheng, S. Extreme learning machine based wind speed estimation and sensorless control for wind turbine power generation system. *Neurocomputing* **2013**, *102*, 163–175. [CrossRef]
43. Frandsen, S.; Barthelmie, R.; Pryor, S.; Rathmann, O.; Larsen, S.; Højstrup, J.; Thøgersen, M. Analytical modelling of wind speed deficit in large offshore wind farms. *Wind Energy* **2006**, *9*, 39–53. [CrossRef]
44. Pookpunt, S.; Ongsakul, W. Optimal placement of wind turbines within wind farm using binary particle swarm optimization with time-varying acceleration coefficients. *Renew. Energy* **2013**, *55*, 266–276. [CrossRef]
45. Iowa State University of Science and Technology The Iowa Environmental Mesonet (IEM). Available online: http://mesonet.agron.iastate.edu/ (accessed on 5 March 2019).
46. Huang, G.B.; Zhu, Q.Y.; Siew, C.K. Extreme learning machine: A new learning scheme of feedforward neural networks. *Neural Netw.* **2006**, *2*, 985–990.

energies

MDPI

Article

A Flower Pollination Method Based Global Maximum Power Point Tracking Strategy for Point-Absorbing Type Wave Energy Converters

Aqiang Zhao [1], Weimin Wu [1,*], Zuoyao Sun [1], Lixun Zhu [1], Kaiyuan Lu [2,*], Henry Chung [3] and Frede Blaabjerg [2]

1 Department of Electronic Engineering, Shanghai Maritime University, Shanghai 201306, China; zaq18201716171@gmail.com (A.Z.); sunzy205@gmail.com (Z.S.); Lixunzhu@shmtu.edu.cn (L.Z.)
2 Department of Energy Technology, Aalborg University, DK-9220 Aalborg, Denmark; fbl@et.aau.dk
3 Department of Electronic Engineering, City University of Hong Kong, Hongkong 999077, China; eeshc@cityu.edu.hk
* Correspondence: wmwu@shmtu.edu.cn (W.W.); klu@et.aau.dk (K.L.)

Received: 23 January 2019; Accepted: 4 April 2019; Published: 8 April 2019

Abstract: To maximize the generated output power under random waves, the control strategy of maximum power point tracking (MPPT) is incorporated in point-absorbing type wave energy converter (WEC) system. However, due to the influence of mutative wave conditions, the curve of the maximum average power appears in multiple peaks, adding complexity to the tracking process. This paper proposes a new MPPT control technique for a wave power generation system, by using the flower pollination algorithm (FPA) instead of the conventional hill-climbing method. Compared with the hill-climbing method, this method has advantages of achieving a smaller peak-to-average ratio and capturing the more average power under the same sea state. The verification has been carried out through the simulations and the experimental results on a lab test bench.

Keywords: point-absorbing; wave energy converter (WEC); maximum power point tracking (MPPT); flower pollination algorithm (FPA); power take-off (PTO); hill-climbing method

1. Introduction

In recent years, the exhaustion of fossil fuel resources, environmental problems, and growing energy demands have prompted the active study of producing electricity from renewable energy sources. Among renewable energies, ocean wave energy has attracted more attention, due to its high availability and low environmental impact [1]. In addition, ocean wave energy has a quite higher energy density than solar energy and is more stable than wind energy [2–4]. Up to date, several wave energy converters (WEC) concepts have been studied, and the research focuses so far mainly on the mechanical structure, the hydrodynamic aspect and the energy conversion rate of different WECs [5–7].

Compared with larger converter units, the point absorber is one of the most promising solutions in all wave energy conversion technologies, due to its ease of both fabrication and installation [8,9]. Figure 1 shows a typical point absorber WEC, where the wave energy is captured by floating buoys and converted into linear motions. At present, there are several point-absorbing type WECs available, including Ocean Power Technology's Powerbuoy [10] and Wavestar [11]. However, their frequency response tends to be narrow-band, which means that they will perform poorly under the irregular wave conditions. In order to improve the efficiency of point-absorbing type WEC, control strategies for optimizing wave energy capture were usually adopted [12,13], ensuring that the wave power generation system can always capture the maximum power in the real sea state. In addition, the instantaneous output power peak-to-average ratio should be limited in a reasonable range during extracting the maximum average power of wave energy, since it directly affects the cost of electromechanical systems [14].

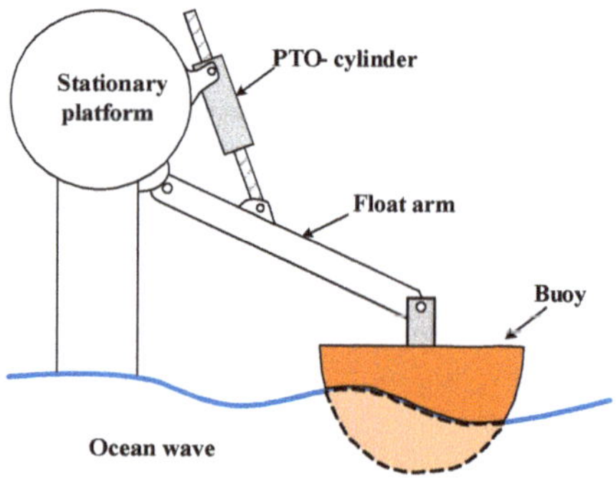

Figure 1. Point-absorber type wave energy converter (WEC).

Currently, various control methods of point-absorber type WEC for optimizing wave energy capture have been proposed, where the implementation complexity, the accuracy and the tracking speed of these techniques are different. These control techniques can be classified into four categories: (1) phase and amplitude control [15–18], (2) complex conjugate control [19–21], (3) model predictive control (MPC) [22–24], and (4) extremum-seeking control [25].

The phase and amplitude control of heaving-buoy WEC constrained by amplitude are described in [15–17], where the phase control is applied to keep the velocity of the buoy in phase with the excitation force of waves, producing a result similar to locking. This method is effective for achieving the maximum power extraction of the point absorber. However, the phase control requires an additional mechanism to be configured to hold and release the buoy. The results show that the phase control suffers from considerable practical challenges, significantly increasing the return on investment periods [18].

For the complex conjugate control [19–21], to maximize the generated power in each sea state, the damping coefficients and the stiffness coefficients of the power take-off (PTO) unit are controlled to be equal to the complex conjugate of the inherent mechanical impedance. In terms of hydrodynamics, the complex conjugate control can achieve the optimal energy extraction, as it aims to obtain resonance between the wave power generation system and the incident waves. However, a substantial amount of reactive power is required, when the resonance frequency drifts. Moreover, realizing the optimal control can lead to excessive motions, and this control technique usually requires the position and velocity of the buoys, which can easily introduce uncertain information. In addition, the application of this method is difficult under random wave conditions.

Another solution is the MPC [22–24], where the model is used to predict the future response of the system. This method can accurately estimate the parameters of the point-absorbing type WEC and search the maximum power point without detecting the dominant frequency of the wave. However, the future value of the wave excitation should be predicted, and the computation burden is increased. Meanwhile, the control system becomes more complex.

The superiority of extreme-seeking based on metaheuristic process in solving nonlinear optimization problems has been recognized. The extremum-seeking is a well-known control strategy that has been used with great success in other forms of renewable energy, such as solar energy and wind energy [26–28]. However, the extremum-seeking technology is rarely used in wave energy. The hill-climbing method is used to implement the MPPT of point-absorber type WEC [25]. This method does not need to predict the wave period of irregular waves, the added impedance, the added mass, and the buoyancy term on the WEC. However, it is found that the curve of the average power has multiple peaks, especially under irregular wave conditions. The hill-climbing method cannot search the global maximum power point among multiple extreme points, due to its monotonous search characteristics. Moreover, the step of the hill-climbing method is fixed, which will greatly affect the search time and search accuracy.

Recently, MPPT researchers witnessed the optimization of swarm optimization and meta-heuristic methods [29]. These methods had been successfully applied in the fields of solar and wind energy [30–32]. The main advantages associated with these methods are (1) starting with a random search, (2) avoiding the convergence to a local minimum, and (3) easy to implement.

In this paper, a new point-absorbing type WEC control strategy based on the flower pollination algorithm (FPA) is proposed for maximum wave power absorption, as also introduced in [33]. The FPA is a new nature-inspired algorithm, based on the characteristics of flowering plants [34] and the FPA had been successfully applied in the photovoltaic field [35,36]. From these literature, it is found that this method has several significant advantages, such as (1) the ability to search global space, (2) being easy to implement, (3) being easy to encode and compile, and (4) the fast convergence. When the control strategy of MPPT based on the FPA is incorporated in a point-absorbing type wave power generation system, it is not necessary to detect the wave frequency, the accurate information of the incoming excitation force, and parameters of the point-absorbing type WEC. This method can obtain the maximum power of the wave energy rapidly, the reduced power peak-to-average ratio, and less overall system costs as well.

The rest of this paper is organized as follows: The analysis model of the point-absorbing type WEC system and power spectrum are first derived in Section 2. Then, the proposed control strategy and its implementation are presented in Section 3. Further, the proposed control strategy is evaluated through simulation in Section 4 and by the experiment in Section 5, compared with the hill-climbing method. Finally, conclusions are drawn in Section 6.

2. Analysis Model of the Point-Absorbing Type WEC and Power Spectrum

2.1. Modeling of Irregular Wave Energy and Heave-Buoy Mechanic Analysis

Ocean waves are typically irregular and consist of multiple regular waves with varying frequencies and amplitudes. To describe a sea state mathematically, the energy spectrum characteristic of ocean waves is constructed by the Bretschneider spectrum which is widely adopted based on representative sea state parameters [37]. The spectrum depends on the significant wave height $H_{1/3}$ and spectrum peak period T_p, as,

$$S(\omega) = \frac{5}{16} H_{1/3}^2 \frac{16\pi^4}{T_\mathrm{p}^4 \omega^5} e^{-\frac{5}{4}\left(\frac{2\pi}{\omega T_\mathrm{p}}\right)^4} \tag{1}$$

The Bretschneider wave energy spectrum of the sea states 2–6 is depicted in Figure 2. The angular frequency ω of the regular waves is mainly concentrated in the range of 0.5 rad/s to 1.2 rad/s. The irregular time-domain excitation can be obtained from the energy spectrum as,

$$A_\mathrm{i}(\omega_\mathrm{i}) = \sqrt{2S(\omega_\mathrm{i})\Delta\omega} \tag{2}$$

$$F_\mathrm{exc}(t) = \sum_{i=1}^{N} A_\mathrm{i} f_\mathrm{exc}(\omega_\mathrm{i}) \cos(\omega_\mathrm{i} t + \varphi_\mathrm{i}) \tag{3}$$

where $A_\mathrm{i}(\omega_\mathrm{i})$ are the wave amplitudes, $\Delta\omega$ represents the wave spectrum discretization step; $f_\mathrm{exc}(\omega_\mathrm{i})$ are the excitation force coefficients, calculated by the hydrodynamics software, such as WAMIT or ANSYS AQWA in the frequency domain, and φ_i are the random phase angles from 0 rad to 2π rad.

Figure 2 shows the energy spectrum obtained by the selected wave period and a significant wave height of the sea state. It can be seen that the available wave energy range is from 0.5 rad/s to 1.2 rad/s, while the wave energy density of other frequencies is quite small and can be ignored when power control is carried out in the wave power generation system. The time-domain expression of the excitation force on the float can be found [38]. An example of an excitation force curve is shown in Figure 3, with a significant wave height ($H_{1/3}$ = 1.2 m) and spectrum peak period (T_p = 7.5 s).

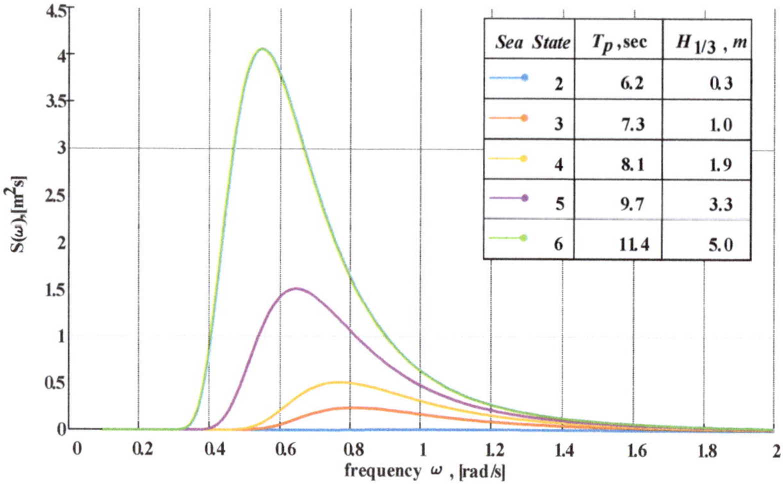

Figure 2. Wave energy spectrum according to Bretschneider model.

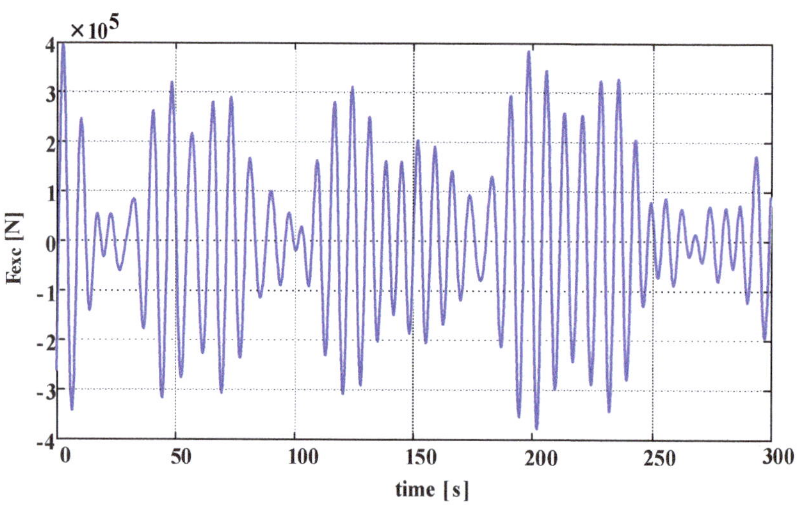

Figure 3. Time-domain excitation force for irregular waves.

In order to analyze the motion and control of WECs in irregular waves, a time-domain model is required [39]. According to the wave–buoy interaction, the analytical force equation is given as,

$$\begin{array}{c}(M + m_{\infty})\ddot{s}(t) + \int_0^t H_{rad}(t-\tau)\dot{s}(\tau)d\tau + Ks(t) = F_{\mathrm{PTO}}(t) + F_{\mathrm{exc}}(t) \\ H_{\mathrm{rad}}(t) = \frac{2}{\pi}\int_0^{\infty} B(\omega)\cos(\omega t)d\omega\end{array} \quad (4)$$

$$m_{\infty} = \lim_{\omega \to \infty} m(\omega) \quad (5)$$

where "." is time derivation operation, M represents the mass of the WEC, the damping ($H_{\mathrm{rad}}(t)$) and added mass (m_{∞}) time-domain components can be determined in (4) and (5) from its frequency-domain parameters ($B(\omega)$ and $m(\omega)$) [39], $s(t)$ is the displacement of the buoy, K is spring stiffness, and F_{exc} is the excitation wave force.

2.2. Analysis of Power Curve of Wave Power Generation System Under Irregular Wave Condition

Based on the existing literature, in order to better understand the hydrodynamic model of wave power generation system, according to Equations (4) and (5), the system can be equivalent to an *RLC* circuit given in Figure 4.

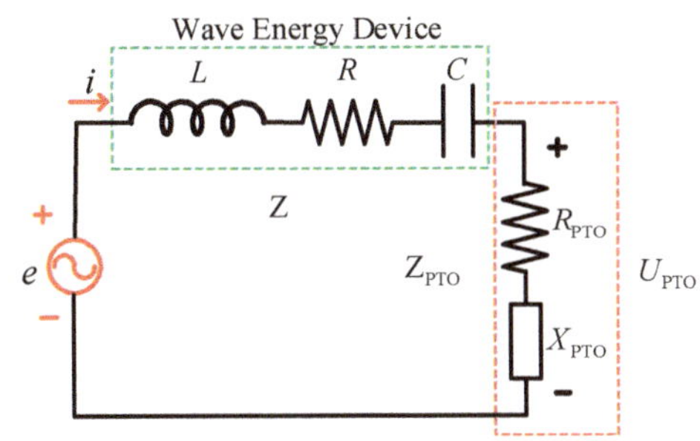

Figure 4. Equivalent electric model of the WEC.

Figure 4 shows the equivalent electric model of the WEC, where the wave excitation force F_{exc} corresponds to the supply voltage e, the velocity of buoy $\dot{s}$ represents the current i, the mass term $(M + m_\infty)$ stands for the inductance term (L) of the WEC impedance, the spring term (K) coefficient stands for the capacitive (C), and the real component of the radiation damping $\Re(H_{rad}(\omega))$ represents the resistive (R). The force applied by the PTO corresponds to the load voltage of U_{PTO}.

Let

$$Z(\omega) = R(\omega) + jX(\omega) = \Re(H_{rad}(\omega)) + j[\omega(M + m_\infty) - \frac{K}{\omega}] \tag{6}$$

where $Z(\omega)$ represents the inherent mechanical impedance of the wave energy device, which is equal to the ratio of force to velocity and only related to waves and buoys, and $X(\omega)$ represents the reactance of the wave energy device. From (6), it can be seen that $Z(\omega)$ is frequency dependent, which means that each frequency corresponds to a different optimal value of $Z(\omega)$. Thus, a question is raised on how to specify PTO impedance for irregular oceans with mixed frequencies.

The mechanical properties of a permanent magnet synchronous machine can be expressed by mechanical impedance,

$$Z_{PTO}(\omega) = R_{PTO}(\omega) + jX_{PTO}(\omega) \tag{7}$$

where $Z_{PTO}(\omega)$ is termed the equivalent mechanical impedance of the permanent magnet synchronous motor, the resistive force $R_{PTO}(\omega)$ provided by the permanent magnet synchronous motor is equal to the radiation damping term, and the imaginary permanent magnet synchronous motor component $X_{PTO}(\omega)$ is responsible for ensuring the resonant operating conditions [40].

When the exciting force of the wave power generation system is under irregular wave conditions, the voltage e can be expressed as follows by Equation (3),

$$e(t) = \sum_{n=1}^{N} E_n \cos(\omega_n t + \alpha_n) \tag{8}$$

where ω_n, α_n are the wave frequency and impedance angle of the nth wave frequency of the irregular wave, respectively. A phasor method is used on each frequency to derive the current and PTO load voltage of the circuit model at each frequency.

Based on Equations (7) and (8), the equivalent current and voltage in Figure 4 can be expressed as,

$$\dot{I}_{\rm n} = \frac{E_{\rm n}\cos(\omega_{\rm n}t+\alpha_{\rm n})}{(R_{\rm n}+R_{\rm PTO})+j[X_{\rm n}(\omega_{\rm n})-\frac{1}{\omega_{\rm n}X_{\rm PTO}}]} \tag{9}$$

$$\dot{U}_{\rm PTO\_n} = \dot{E}_{\rm n} - \dot{I}_{\rm n}\cdot Z_{\rm n}\angle\alpha_{\rm n} \tag{10}$$

where $R_{\rm n}$, $Z_{\rm n}$ are the resistance and impedance of the n$^{\rm th}$ wave frequency of the irregular wave, respectively.

Based on (9) and (10), $U_{\rm PTO}$ and i can be expressed as, PTO-n

$$U_{\rm PTO}(t) = \sum_{n=1}^{N} U_{PTO-{\rm n}}\cos(\omega_{\rm n}t+\varphi_{\rm n}) \tag{11}$$

$$i(t) = \sum_{n=1}^{N} I_{\rm n}\cos(\omega_{\rm n}t+\theta_{\rm n}) \tag{12}$$

where $\varphi_{\rm n}$, $\theta_{\rm n}$ are the phase shift angles at the n$^{\rm th}$ wave frequency of irregular waves. The produced average power can be written using (11) and (12) as,

$$P_{\rm PTO} = \frac{1}{T}\int_0^T U_{PTO}(t)\cdot i(t)dt = \frac{1}{T}\int_0^T\left[\sum_{n=1}^{N} U_{PTO-{\rm n}}\cos(\omega_{\rm n}t+\varphi_{\rm n})\right]\left[\sum_{n=1}^{N} I_{\rm n}\cos(\omega_{\rm n}t+\theta_{\rm n})\right]dt \tag{13}$$

In the circuit model, when the voltage and current of the PTO load are generated by the same frequency wave, the average power obtained can be expressed as,

$$P_{\rm PTO} = \frac{1}{T}\int_0^T\sum_{n=1}^{N} U_{PTO-{\rm n}}I_{\rm n}\cos(\omega_{\rm n}t+\varphi_{\rm n})\cos(\omega_{\rm n}t+\theta_{\rm n}) = \frac{1}{2}\sum_{n=1}^{N} U_{PTO-{\rm n}}I_{\rm n}\cos(\varphi_{\rm n}-\theta_{\rm n}) \tag{14}$$

When the voltage and current of the PTO load are generated by different frequency waves, the average power obtained can be expressed as,

$$P_{\rm PTO} = \frac{1}{T}\int_0^T U_{\rm i}I_{\rm j}\cos(\omega_{\rm i}t+\varphi_{\rm i})\cos(\omega_{\rm j}t+\varphi_{\rm j}) = 0(i\neq j) \tag{15}$$

According to Equations (13)–(15), superimposing the average power at each frequency is the average power captured by the PTO load in the case of irregular waves. After simplification, the average power captured in the frequency domain is as follows,

$$P_{\rm PTO}(\omega) = \sum_{n=1}^{N}\frac{[E_{\rm n}\cos(\omega_{\rm n}t+\alpha_{\rm n})]^2}{8R_{\rm n}}\left[1-1\left|\frac{Z^*(\omega)-Z_{\rm PTO}(\omega)}{Z(\omega)+Z_{\rm PTO}(\omega)}\right|^2\right] \tag{16}$$

Figure 5 depicts the average power and power peak-to-average ratio curves of the system with two control variables (i.e., the damping coefficient ($R_{\rm PTO}$) and the stiffness coefficient ($X_{\rm PTO}$)), which the significant wave height ($H_{1/3}$ = 1.2 m), the spectrum peak period ($T_{\rm p}$ = 7.5 s) and parameters of wave power generation device from references [40]. Figure 5a shows that the curve of function $P_{\rm PTO}$—($R_{\rm PTO}$, $X_{\rm PTO}$) is a multi-extreme point function and the MPP is the extreme value within the range of control variables. Figure 5b shows that the power peak-to-average ratio mapped with $R_{\rm PTO}$ and $X_{\rm PTO}$. Maximum average power is expected to be obtained with a small power peak-to-average ratio. Therefore, searching for the maximum average power and simultaneously reducing the power peak-to-average ratio becomes a new challenge.

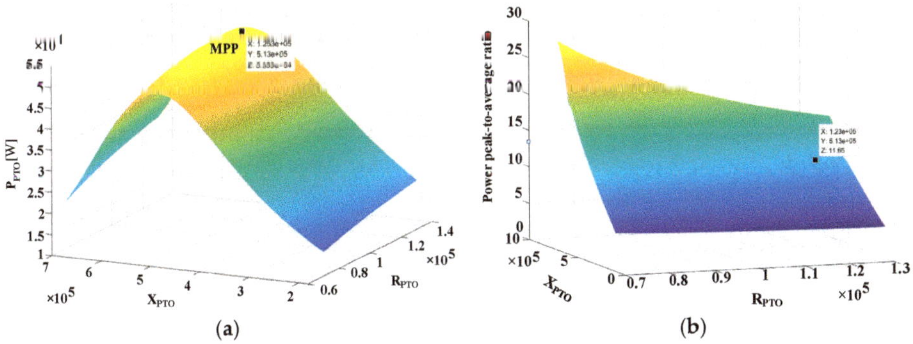

Figure 5. Average power and power peak-to-average ratio of the WEC under different damping and stiffness coefficients, (**a**) average power (**b**) power peak-to-average ratio.

The relationship between the equivalent mechanical impedance of the permanent magnet synchronous motor (i.e., R_{PTO} and X_{PTO}) and the extracted power of the system can be obtained. In Equations (6) and (16), the parameters such as Z, ω, E_n and R_n are determined by wave period and physical characteristics of the sea state. The maximum average power extracted from the ocean is deduced as follows,

$$P_{PTOmax}(\omega) = \sum_{n=1}^{N} \frac{[E_n \cos(\omega_n t + \alpha_n)]^2}{8R_n} \tag{17}$$

where the following conditions must be met as,

$$Z_{PTO}(\omega) = Z^*(\omega) \tag{18}$$

The physical meaning of the Functions (17) and (18) is that when the equivalent mechanical impedance of the permanent magnet synchronous motor matches the inherent mechanical impedance of the point-absorbing type WEC, the oscillation of the buoy resonates with the wave motion, and the power extracted by the system reaches a maximum from the wave. Therefore, the damping coefficient (R_{PTO}) and the stiffness coefficient (X_{PTO}) of the permanent magnet synchronous motor need be controlled. According to Equations (6), (7) and (18), the following conditions should be met as,

$$\begin{array}{c} R_{PTO} = \Re(H_{rad}(\omega)) \\ X_{PTO} = \frac{K}{\omega} - \omega(M + m_\infty) \end{array} \tag{19}$$

However, the added mass and the damping coefficient are nonlinear functions with respect to the wave period [38]. According to the optimal control technique with two control variables, the equivalent mechanical impedance of the permanent magnet synchronous motor must be changed to match the wave frequency to extract the maximum average power. When the wave frequency changes, the control side of the permanent magnet synchronous motor should select the corresponding impedance based on the wave frequency to match the wave frequency. Furthermore, there is still a difference between the hydrodynamic parameters obtained by simulation software, such as ANSYS-AQWA, and the actual hydrodynamic parameters. The WEC control methods, which can adapt to changes in hydrodynamic parameters and working conditions, and automatically to track the MPP, is the key to promoting wave power technique applications.

In this paper, the parameters of the controller can be regulated by the FPA directly and change with hydrodynamic parameters and working conditions automatically, which means that the wave period of irregular waves or the accurate knowledge of the incoming wave excitation force need not be identified anymore.

3. FPA and Its Implementation

3.1. Flower Pollination Algorithm

The FPA developed by Yang [41] is a recent meta-heuristic optimization technique, which is inspired by the nature of flower pollination processing. The pollination refers to a phenomenon of transferring pollen from one species to another. This process helps the emergence of new species of flowers. There are two methods of pollination, (1) cross-pollination, where the pollen transfer occurs between two different species, and pollinators are bees, insects or birds, (2) self-pollination, where pollens come from the same flower, and no pollinator is needed [41]. According to the algorithm tests on the FPA, genetic algorithm and particle swarm optimization, the FPA has the best performance.

As introduced in [42], for the FPA, the pollination method of pollen is determined by the probability switch of P, and $P = 0.8$ works better for most applications. The flower constancy that can be considered as the reproduction probability is proportional to the similarity of two flowers involved. To implement FPA, the characteristics of the pollination process and pollinator behavior are summarized as the following rules:

Rule 1: Cross-pollination is considered to be a global pollination process, where pollinators, such as insects that carry pollen, perform L'evy flight. The first rule can be represented mathematically as,

$$x_{\mathrm{i}}^{\mathrm{t}+1} = x_{\mathrm{i}}^{\mathrm{t}} + L(g_{best} - x_{\mathrm{i}}^{\mathrm{t}}) \tag{20}$$

where $x_{\mathrm{i}}^{\mathrm{t}}$ is the pollen i or solution vector x_{i} at iteration t, g_{best} is the current best solution found in the current population, and L is the L'evy flying step size and is subject to uniform distribution as,

$$L = \frac{\lambda\Gamma(\lambda)\sin(\pi\lambda/2)}{\pi}\frac{1}{S^{1+\lambda}}(S >> S^0 > 0) \tag{21}$$

where $\Gamma(\lambda)$ is the standard gamma function which is valid for large steps S > 0, and S is the L'evy flying step. To ensure faster convergence, $\lambda = 1.5$ works better for most applications.

Rule 2: Self-pollination represents the local pollination process. The characteristic equation for local pollination is obtained as follows,

$$x_{\mathrm{i}}^{\mathrm{t}+1} = x_{\mathrm{i}}^{\mathrm{t}} + \varepsilon(x_{\mathrm{j}}^{\mathrm{t}} - x_{\mathrm{k}}^{\mathrm{t}}) \tag{22}$$

where $x_{\mathrm{j}}^{\mathrm{t}}$ and $x_{\mathrm{k}}^{\mathrm{t}}$ are pollens from different flowers of the same plant species, ε is characterized by the local search in distribution, and $\varepsilon \in [0, 1]$.

The FPA algorithm is well suited to solve nonlinear optimization problems [34]. In addition, this approach works best for MPPT applications, because it is explored globally and exploits locally in a single iteration. Unlike other bio-inspired algorithms, the best part of this algorithm is that the FPA introduces randomness in each iteration by self-pollination. Thus, the FPA is best suitable for maximum energy extraction of the point-absorbing type WEC. Moreover, this method does not rely on the accuracy of the excitation force and the dominant frequency characteristics of the irregular waves.

3.2. FPA Implemented for MPPT

The complete flowchart for the proposed method is pictured in Figure 6. For FPA implemented in MPPT control, it can be explained in four steps:

Step 1: Initialize parameters, including the maximum number of iterations (N), PTO load boundary value (R_{min}, R_{max}, X_{min}, X_{max}), switching probability of pollination (P), standard gamma function parameter λ, PTO load solution vector number (m), and the objective function as shown in (13), where five different PTO loads are considered as the pollen or solution vector as following,

$$x_{5\times2} = [\boldsymbol{R}_{5\times1}, \boldsymbol{X}_{5\times1}] \tag{23}$$

Step 2: Evaluate the fitness, where the pollen's suitability is assessed by using the fitness function. The pollen with high fitness is selected as g_{best}, and then a random number named rand is generated.

Step 3: Start the pollination process, where the condition of self-pollination (if rand > P) or cross-pollination (if rand ≤ P) is determined. Note that all pollens in the population must undergo cross-pollination or self-pollination.

Step 4: Confirm the termination conditions, and continuously continue Step 2 to Step 4 until the maximum power point is captured or the maximum number of iterations appears. Note that the maximum number of iterations is set as 50 in our case.

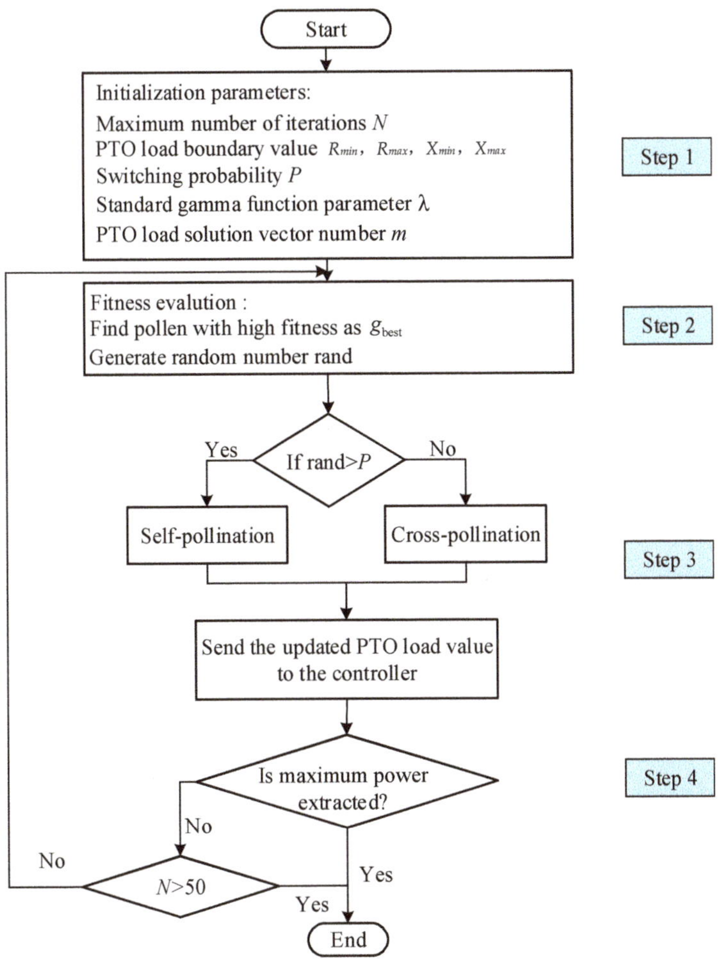

Figure 6. Flow chart of maximum power point tracking (MPPT) control algorithm for point-absorbing type WEC based on flower pollination algorithm (FPA).

4. Simulation Results

In order to validate the proposed MPPT control method based on FPA used in WEC system, a complete model of the WEC system and the controller were constructed. Three cases were studied, where the parameters, the related average power absorption and peak-to-average ratio are listed in Table 1, by using the FPA and the hill-climbing method, respectively.

Table 1. Comparison of average power absorption and peak-to-average ratio in simulation.

Case Number	T_p	$H_{1/3}$	FPA	Hill-Climbing Method
Case 1	7.5 s	1.2 m	P_{PTO} = 53.5 kW *Ratio* = 11.6	P_{PTO} = 51.4 kW *Ratio* = 14.8
Case 2	8 s	1.6 m	P_{PTO} = 103.7 kW *Ratio* = 12.1	P_{PTO} = 98.3 kW *Ratio* = 16.1
Case 3	6.5 s	0.8 m	P_{PTO} = 28.7 kW *Ratio* = 11.3	P_{PTO} = 27.1 kW *Ratio* = 13.8

For irregular waves, the previous analysis was verified by MATLAB/Simulink simulation according to the buoy parameters and sea conditions in Table 2 [40], and only considered the buoy oscillating in heave.

Table 2. Data of the buoy and sea state.

Quantity	Symbol	Value	Units
Wave period	T_p	7.5	[s]
Significant wave height	$H_{1/3}$	1.2	[m]
Buoy mass	M	267040	[Kg]
Added mass	m_∞	156940	[Kg]
Spring stiffness	K	785890	[N/m]
Total buoy damping	B_{rad}	91520	[Kg/s]

In order to detailly analyze the results obtained by the FPA and the hill-climbing method, the average power, the damping coefficient and the stiffness coefficient of Case 1 are depicted in Figures 7 and 8, respectively. It can be observed that the maximum average power is obtained at the 16th iteration and the values of the damping coefficient and the stiffness coefficient finally converges to the best solution. However, the hill-climbing method requires approximately 250 steps to converge and cannot converge to the MPP. Therefore, the damping coefficient and the stiffness coefficient obtained by the hill-climbing method are suboptimum solutions.

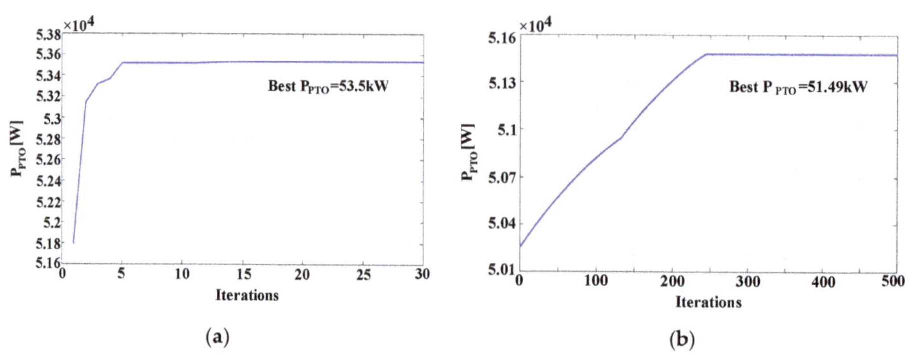

Figure 7. Simulated average power obtained by (**a**) FPA, (**b**) hill-climbing method.

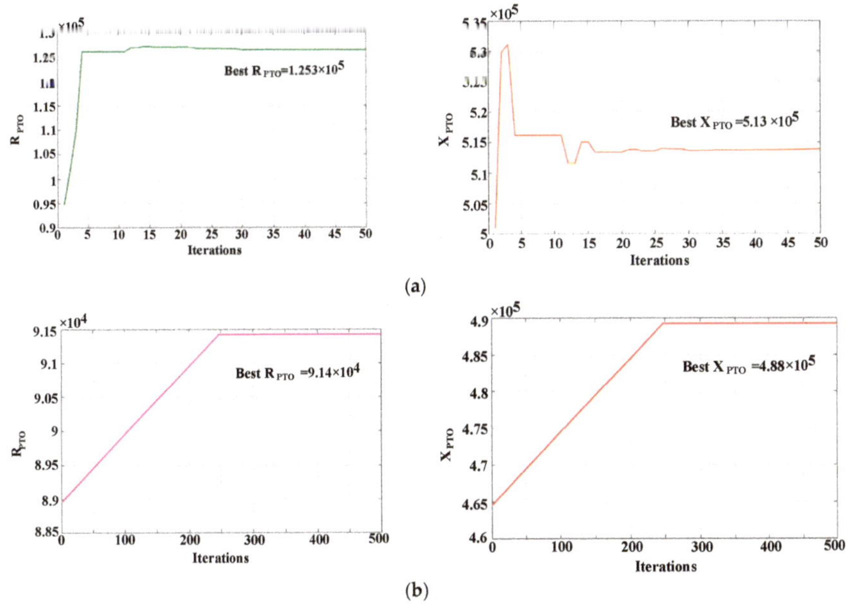

Figure 8. Simulated damping coefficient and stiffness coefficient obtained by (**a**) FPA, (**b**) hill-climbing method.

In order to extract the maximum wave energy, in theory, the excitation force should be in phase with the buoy velocity at any incident wave frequency, where the product of the excitation force and the wave speed is exact extracted power. Thus, the WEC operates at resonant conditions, and the wave power generation system extracts the instantaneous maximum power. This performance can be evaluated from Figure 9, where the extreme-seeking control technique based on FPA has better performance in most areas than the hill-climbing method. The reason that the performance of the hill-climbing method is insufficient is mainly due to the slow convergence rate and easy trapping in a local optimum.

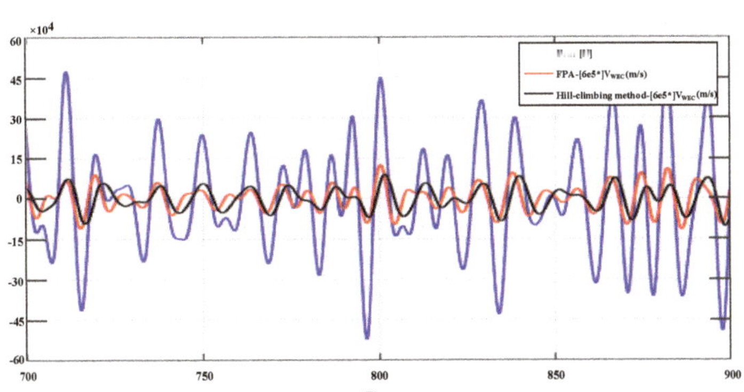

Figure 9. Simulated excitation force and the velocity of the buoy.

Figure 10 presents the simulated instantaneous power and average power absorption when the wave power generation system operates at maximum energy extraction states. As seen from Figure 10, when the wave power generation system works in the optimal state, in order to maximize the average power, the permanent magnet synchronous motor provides energy at some time, and the energy is transferred from the motor side to the float side. In the rest time, the wave provides energy, which is transferred from the buoy side to the motor side. Compared with the hill-climbing method, the MPPT method based on the FPA can capture more average power and reduces power peak to average ratio. The simulated results have proven the advantages of FPA in terms of convergence time, efficiency and accuracy, compared with the hill-climbing method.

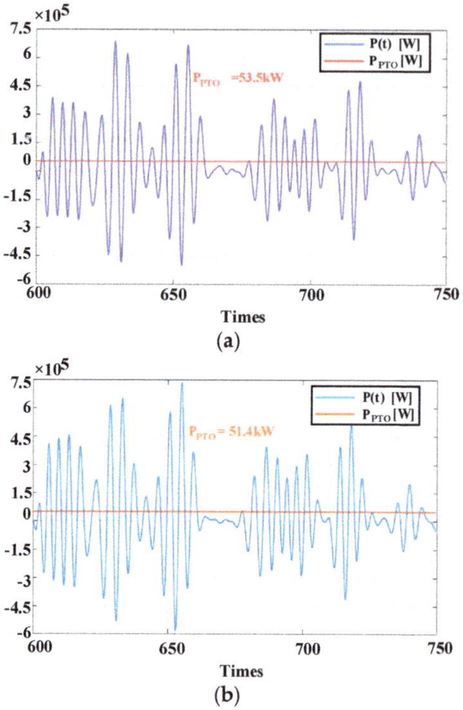

Figure 10. Simulated instantaneous and average power obtained by (**a**) FPA and (**b**) hill-climbing method.

Figure 11 shows the average power (active power) spectrum and the apparent power spectrum captured by the wave power generation system, where the rest power at low frequency is ignored due to the small values. The seventh frequency is the dominant wave frequency of input excitation force wave. More apparent power is extracted in the range near the dominant wave frequency by the hill-climbing method, which means that more reactive power is generated near the dominant wave frequency, leading to a higher peak-to-average ratio. Although the proposed method cannot make the wave power generation system resonate at the dominant wave frequency, the system can capture more power at the other frequencies.

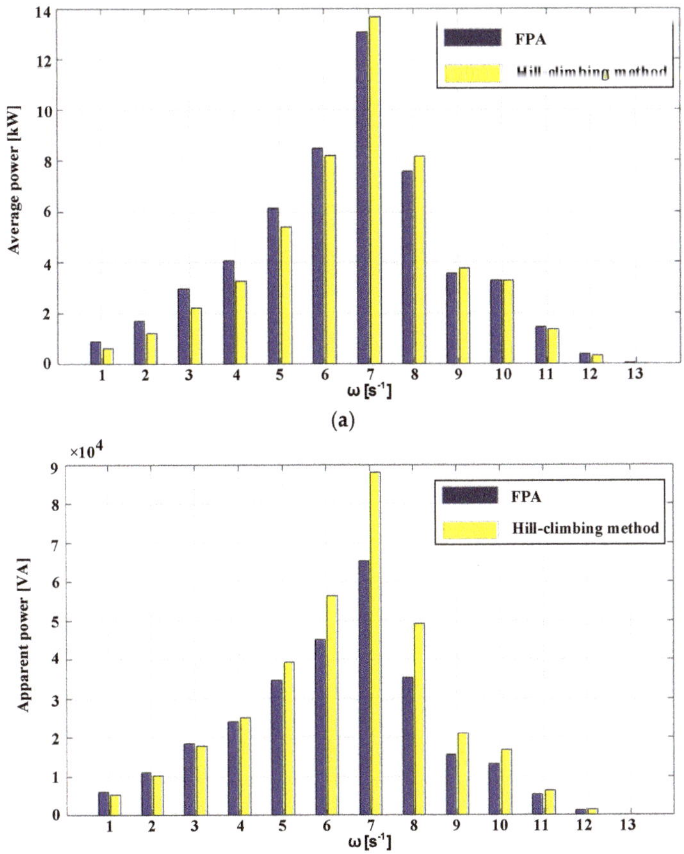

Figure 11. Simulated power spectrum of (**a**) average power, (**b**) apparent power.

Moreover, the robustness of the proposed technique is further validated, when the sea state suddenly changes as shown in Figure 12. After captured by the wave power generation system repeatedly, the average power can be re-converged to the optimal value. Figure 12 shows the optimal average power search curve for the wave power generation system when Sea condition 1 is changed to Sea condition 3 as listed in Table 1. Among them, the sudden change of the sea state occurs in the 24th iteration, and the system detects the change in the 25th iteration, and then re-searches the optimal power point, indicating that the proposed control method has strong adaptability and immunity.

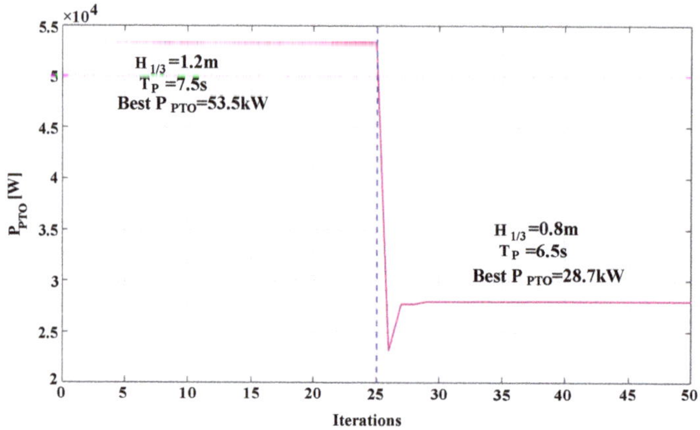

Figure 12. Simulated power search in case of a sudden change in sea state.

5. Experimental Result

In order to study the practicability of the proposed control strategy, a test bench shown in Figure 13 is constructed by using conventional dynamometer system together with two converters, whose parameters are listed in Table 3.

Table 3. Permanent magnet synchronous motor parameters.

Symbol	Parameter	Value
n_p	Pole number	8
P_{rat}	rated power	3000 [W]
V_{rat}	rated voltage	220 [V]
I_{rat}	rated current	12 [A]
n_{rat}	rated speed	1500 r/min
T_{rat}	rated torque	19 [N · m]
T_{max}	Maximum torque	47 [N · m]
f_{PWM}	PWM frequency	5 kHz

In this system, there is a permanent magnet synchronous motor (PMSM) to provide the required load torque for the generator shaft, and a permanent magnet synchronous generator (PMSG) to ensure the required mechanical speed is achieved. The speed and torque controllers are incorporated in the PMSM and PMSG drives, respectively. Due to the logistic limitations of constructing the real point-absorbing WEC prototype, the buoy, the transmission device and the controller of the entire wave power generation system are constructed in the dSPACE real-time system, according to the previous simulation model. The buoy model is operated under the given sea conditions as listed in Table 1. The wave power generation system controller calculates the torque required for power capture, applies this torque to the transmission, and then monitors the buoy velocity. The generator side converter tries to extract the maximum power from the wave. The irregular wave data from the irregular excitation force lookup table are used for the wave mechanical character analysis. The position and speed of the generator are measured by a position sensor. Note that due to the limitations of experimental conditions and protecting the experimental device, the model data is scaled down within a reasonable range.

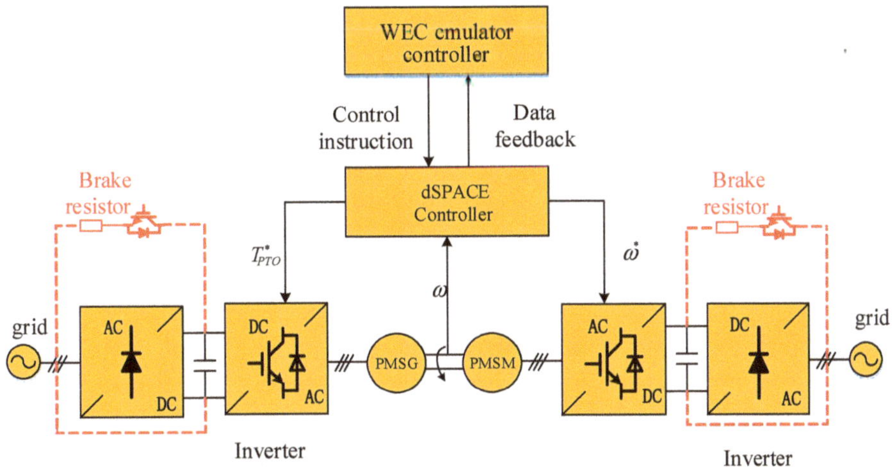

(**a**)

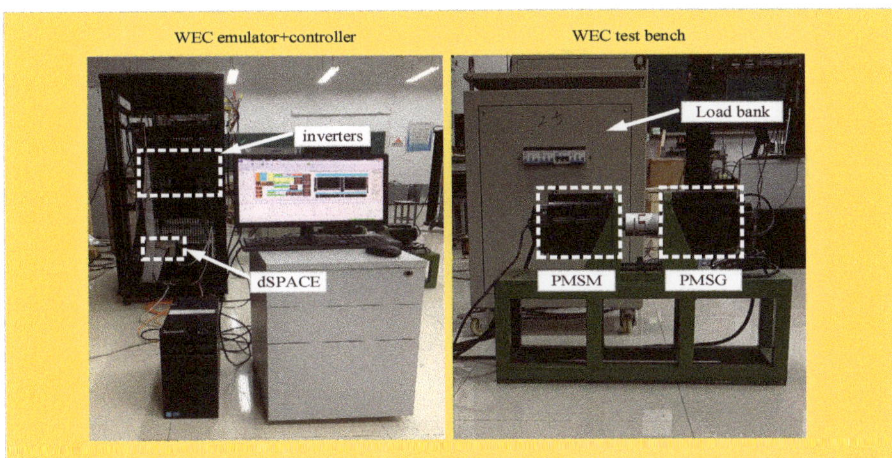

(**b**)

Figure 13. Synchronous test bench, (**a**) overall control block diagram in the laboratory and (**b**) experimental setup.

The point-absorbing WEC performance is analyzed to determine whether the system has met the predefined setup specifications. Figure 14 shows the PMSG speed under Case 1 sea condition. It can be seen that the speed controller has reached the expected performance, although there is a small measurement error caused by the encoder accuracy and switching frequency.

Figure 15 shows the instantaneous mechanical power applied to the shaft of the synchronous test bench by FPA and the Hill-climbing method, respectively. Compared with Figures 10 and 15, it can be seen that the measured instantaneous power is very similar to the simulated result. It is only numerically reduced, due to limitations of experimental conditions.

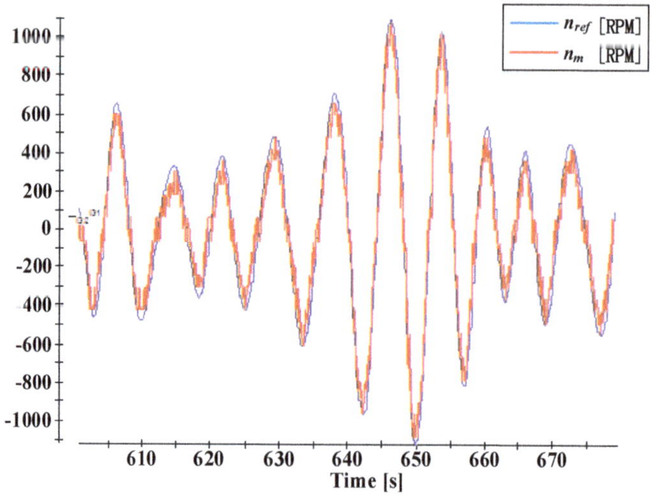

Figure 14. Measured speed vs. the reference speed of permanent magnet synchronous generator (PMSG).

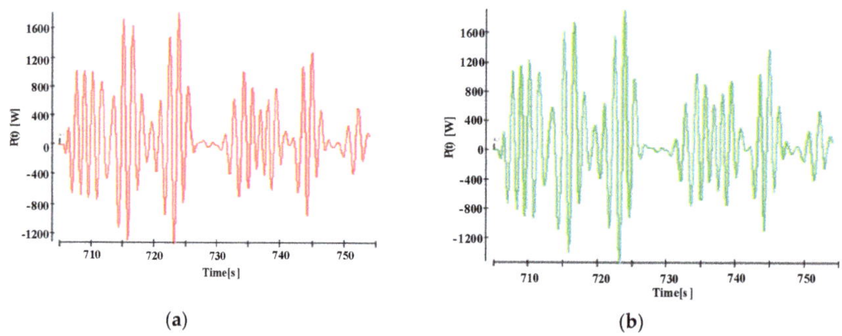

Figure 15. Measured instantaneous mechanical power of the synchronous test bench. (**a**) FPA and (**b**) hill-climbing method.

Figure 16 shows the measured WEC velocity and excitation force. It can be seen that the buoy velocity applied by the FPA is mostly in phase with the excitation force, while there is some phase shifting between the buoy velocity and the excitation force applied by the hill-climbing method. Table 4 lists the average power and the peak-to-average ratio, applied by the FPA and the hill climbing method respectively, under three kinds of sea conditions. It can be seen that the FPA-MPPT is a more suitable technology than the hill-climbing method for point-absorbing type WEC applications since the more average power and the smaller peak-to-average ratio can be achieved.

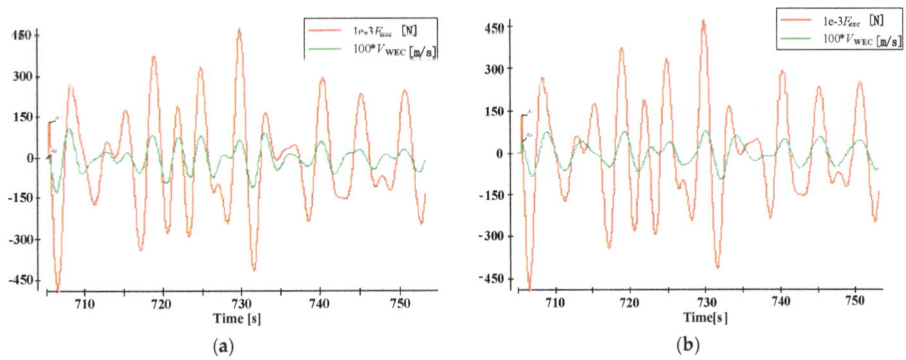

Figure 16. Measured excitation force and the velocity of buoy, (**a**) FPA and (**b**) hill-climbing method

Table 4. Comparison of average power absorption and peak-to-average ratio in experiments.

Case Number	T_p	$H_{1/3}$	FPA	Hill-Climbing Method
Case 1	7.5 s	1.2 m	P_{PTO} = 123.2 $WRatio$ = 12.4	P_{PTO} = 117.8 $WRatio$ = 15.4
Case 2	8 s	1.6 m	P_{PTO} = 243.6 $WRatio$ = 13.5	P_{PTO} = 220.6 $WRatio$ = 17.4
Case 3	6.5 s	0.8 m	P_{PTO} = 61.8 $WRatio$ = 11.8	P_{PTO} = 56.4 $WRatio$ = 14.5

6. Conclusions

In this paper, a new point absorption WEC control strategy is proposed, which is based on the extremum-seeking controller with FPA. The proposed MPPT search algorithm has been compared with the hill-climbing method. The following conclusions can be drawn:

1. Similar to the hill-climbing method, the proposed method is independent of the wave period of irregular waves, the accurate knowledge of the incoming wave excitation force, and the parameters of the point-absorbing type WEC.
2. When using the proposed MPPT control technique, a system starting from a non-optimal state can operate stably at the MPP after several iterations. Moreover, the system can rapidly respond to the MPP deviations caused by ocean wave changes and operate under the new MPP.
3. Compared with the hill-climbing method, the proposed FPA can track the global maximum power under irregular wave conditions and produce faster convergence, better tracking, and higher efficiency. Therefore, the FPA-MPPT is a more suitable technology than the hill-climbing method for point-absorbing type WEC applications.

Author Contributions: A.Z. collected experimental data and worked in the simulation experiments. Z.S. contributed to the theoretical analysis of the FPA. L.Z. optimized the program. All authors analyzed the results and revised the paper.

Funding: This work was funded in part by National Natural Science Foundation of China, grant number 51477094, and in part by the Shanghai Science and Technology Commission, grant number 17040501500.

Conflicts of Interest: The authors declare no conflict of interest.

References

1. Khan, N.; Kalair, A.; Abas, N.; Haider, A. Review of ocean tidal, wave and thermal energy technologies. *Renew. Sustain. Energy Rev.* **2017**, *72*, 590–604. [CrossRef]
2. Uihlein, A.; Magagna, D. Wave and tidal current energy—A review of the current state of research beyond technology. *Renew. Sustain. Energy Rev.* **2016**, *58*, 1070–1081. [CrossRef]
3. Astariz, S.; Iglesias, G. The economics of wave energy: A review. *Renew. Sustain. Energy Rev.* **2015**, *45*, 397–408. [CrossRef]
4. Zheng, C.W.; Li, C.Y.; Pan, J.; Liu, M.Y.; Xia, L.L. An overview of global ocean wind energy resource evaluations. *Renew. Sustain. Energy Rev.* **2016**, *53*, 1240–1251. [CrossRef]
5. Ozkop, E.; Altas, I.H. Control, power and electrical components in wave energy conversion systems: A review of the technologies. *Renew. Sustain. Energy Rev.* **2017**, *67*, 106–115. [CrossRef]
6. Gaspar, J.F.; Calvário, M.; Kamarlouei, M.; Soares, C.G. Design tradeoffs of an oil-hydraulic Power Take-Off for Wave Energy Converters. *Renew. Energy* **2018**, *129*, 245–259. [CrossRef]
7. Wen, Y.; Wang, W.; Liu, H.; Mao, L.; Mi, H.; Wang, W.; Zhang, G. A Shape Optimization Method of a Specified Point Absorber Wave Energy Converter for the South China Sea. *Energies* **2018**, *11*, 2645. [CrossRef]
8. Kolios, A.; Di Maio, L.F.; Wang, L.; Cui, L.; Sheng, Q. Reliability assessment of point-absorber wave energy converters. *Ocean Eng.* **2018**, *163*, 40–50. [CrossRef]
9. Meng, F.; Ding, B.; Cazzolato, B.; Arjomandi, M. Modal analysis of a submerged spherical point absorber with asymmetric mass distribution. *Renew. Energy.* **2019**, *130*, 223–237. [CrossRef]
10. OPT—Ocean Power Technology. Available online: http://oceanpowertechnologies.com/ (accessed on 20 December 2018).
11. Wavestar. Available online: http://wavestarenergy.com/ (accessed on 20 December 2018).
12. Tom, N.M.; Madhi, F.; Yeung, R.W. Balancing power absorption and structural loading for an asymmetric heave wave-energy converter in regular waves. In *ASME 2016 35th International Conference on Ocean, Offshore and Arctic Engineering*; American Society of Mechanical Engineers: Busan, Korea, 19–24 June 2016.
13. Ringwood, J.V.; Bacelli, G.; Fusco, F. Energy-maximizing control of wave-energy converters: The development of control system technology to optimize their operation. *IEEE Control Syst.* **2014**, *34*, 30–55.
14. Tedeschi, E.; Molinas, M. Tunable control strategy for wave energy converters with limited power takeoff rating. *IEEE Trans. Ind. Electron.* **2012**, *59*, 3838–3846. [CrossRef]
15. Park, J.S.; Gu, B.G.; Kim, J.R.; Cho, I.H.; Jeong, I.; Lee, J. Active Phase Control for Maximum Power Point Tracking of a Linear Wave Generator. *IEEE Trans. Power Electron.* **2017**, *32*, 7651–7662. [CrossRef]
16. Têtu, A.; Ferri, F.; Kramer, M.; Todalshaug, J. Physical and Mathematical Modeling of a Wave Energy Converter Equipped with a Negative Spring Mechanism for Phase Control. *Energies* **2018**, *11*, 2362. [CrossRef]
17. Liu, C.; Yang, Q.; Bao, G. Latching control using optimal control method for a raft-type wave energy converter. *Ships Offshore Struct.* **2018**, *13*, 138–154. [CrossRef]
18. Temiz, I.; Leijon, J.; Ekergård, B.; Boström, C. Economic aspects of latching control for a wave energy converter with a direct drive linear generator power take-off. *Renew. Energy.* **2018**, *128*, 57–67. [CrossRef]
19. Cantarellas, A.M.; Remon, D.; Rodriguez, P. Adaptive vector control of wave energy converters. *IEEE Trans. Ind. Appl.* **2017**, *53*, 2382–2391. [CrossRef]
20. Song, J.; Abdelkhalik, O.; Robinett, R.; Bacelli, G.; Wilson, D.; Korde, U. Multi-resonant feedback control of heave wave energy converters. *Ocean Eng.* **2016**, *127*, 269–278. [CrossRef]
21. Abdelkhalik, O.; Zou, S.; Robinett, R.D.; Bacelli, G.; Wilson, D.G.; Coe, R.; Korde, U. Multiresonant Feedback Control of a Three-Degree-of-Freedom Wave Energy Converter. *IEEE Trans. Sustain. Energy* **2017**, *8*, 1518–1527. [CrossRef]
22. Anderlini, E.; Forehand, D.I.; Stansell, P.; Xiao, Q.; Abusara, M. Control of a point absorber using reinforcement learning. *IEEE Trans. Sustain. Energy* **2016**, *7*, 1681–1690. [CrossRef]
23. Anderlini, E.; Forehand, D.I.; Bannon, E.; Abusara, M. Control of a realistic wave energy converter model using least-squares policy iteration. *IEEE Trans. Sustain. Energy* **2017**, *8*, 1618–1628. [CrossRef]
24. Xu, Q.L.; Li, Y.; Lin, Z.L. An improved boundary element method for modelling a self-reacting point absorber wave energy converter. *Acta Mech. Sin.* **2018**, *34*, 1015–1034. [CrossRef]
25. Xiao, X.; Huang, X.; Kang, Q. A hill-climbing-method-based maximum-power-point-tracking strategy for direct-drive wave energy converters. *IEEE Trans. Ind. Electron.* **2016**, *63*, 257–267. [CrossRef]

26. Li, X.; Wen, H.; Jiang, L.; Xiao, W.; Du, Y.; Zhao, C. An improved mppt method for pv system with fast-converging speed and zero oscillation. *IEEE Trans. Ind. Appl.* **2016**, *52*, 5051–5064. [CrossRef]
27. Ghasemi, M.A.; Foroushani, H.M.; Parniani, M. Partial shading detection and smooth maximum power point tracking of PV arrays under PSC. *IEEE Trans. Power Electron.* **2016**, *31*, 6281–6292. [CrossRef]
28. Chen, Z.; Yin, M.; Zou, Y.; Meng, K.; Dong, Z. Maximum wind energy extraction for variable speed wind turbines with slow dynamic behavior. *IEEE Trans. Power Syst.* **2017**, *32*, 3321–3322. [CrossRef]
29. Yang, X.S.; Deb, S.; Fong, S.; He, X.; Zhao, Y. Swarm intelligence: Today and tomorrow. In Proceedings of the 2016 3rd International Conference on Soft Computing & Machine Intelligence (ISCMI), Dubai, United Arab Emirates, 23–25 November 2016; pp. 219–223.
30. Li, H.; Yang, D.; Su, W.; Lü, J.; Yu, X. An Overall Distribution Particle Swarm Optimization MPPT Algorithm for Photovoltaic System Under Partial Shading. *IEEE Trans. Ind. Electron.* **2019**, *66*, 265–275. [CrossRef]
31. Mohanty, S.; Subudhi, B.; Ray, P.K. A new MPPT design using grey wolf optimization technique for photovoltaic system under partial shading conditions. *IEEE Trans. Sustain. Energy* **2016**, *7*, 181–188. [CrossRef]
32. Li, X.; Wen, H.; Hu, Y.; Jiang, L.; Xiao, W. Modified beta algorithm for gmppt and partial shading detection in photovoltaic systems. *IEEE Trans. Power Electron.* **2018**, *33*, 2172–2186. [CrossRef]
33. Sun, Z.; Zhao, A.; Zhu, L.; Lu, K.; Wu, W.; Blaabjerg, F. Extremum-seeking Control of Wave Energy Converters using Two-objective Flower Pollination Algorithm. In Proceedings of the 2018 IEEE International Power Electronics and Application Conference and Exposition (PEAC), Shenzhen, China, 4–7 November 2018; pp. 1–5.
34. Yang, X.S.; Karamanoglu, M.; He, X. Multi-objective flower algorithm for optimization. *Procedia Comput. Sci.* **2013**, *18*, 861–868. [CrossRef]
35. Ram, J.P.; Rajasekar, N. A new global maximum power point tracking technique for solar photovoltaic (PV) system under partial shading conditions (PSC). *Energy* **2017**, *118*, 512–525.
36. Pei, T.; Hao, X.; Gu, Q. A Novel Global Maximum Power Point Tracking Strategy Based on Modified Flower Pollination Algorithm for Photovoltaic Systems under Non-Uniform Irradiation and Temperature Conditions. *Energies* **2018**, *11*, 2708. [CrossRef]
37. Michel, W.H. Sea spectra revisited. *Mar Technol.* **1999**, *36*, 211–227.
38. Ansari, K.A.; Khan, N.U. Computation of environmental forces for use in offshore vessel dynamic analysis. *Int. J. Energy Res.* **1989**, 545–553. [CrossRef]
39. Taghipour, R.; Perez, T.; Moan, T. Hybrid frequency–time domain models for dynamic response analysis of marine structures. *Ocean Eng.* **2008**, *35*, 685–705. [CrossRef]
40. Tedeschi, E.; Carraro, M.; Molinas, M.; Mattavelli, P. Effect of control strategies and power take-off efficiency on the power capture from sea waves. *IEEE Trans. Energy Convers.* **2011**, *26*, 1088–1098. [CrossRef]
41. Yang, X.S.; Karamanoglu, M.; He, X. Flower pollination algorithm: A novel approach for multi-objective optimization. *Eng. Optim.* **2014**, *46*, 1222–1237. [CrossRef]
42. Ram, J.P.; Rajasekar, N. A novel flower pollination based global maximum power point method for solar maximum power point tracking. *IEEE Trans. Power Electron.* **2017**, *32*, 8486–8499.

energies

MDPI

Article

Performance Optimization of a Kirsten–Boeing Turbine by A Metamodel Based on Neural Networks Coupled with CFD

Jan-Philipp Küppers [1,*], Jens Metzger [2], Jürgen Jensen [2] and Tamara Reinicke [1]

1 Chair of Product Development, Universität Siegen, Paul-Bonatz-Str. 9-11, 57076 Siegen, Germany; Tamara.Reinicke@Uni-Siegen.de

2 Research Institute for Water and Environment (fwu), Department of Hydraulic and Coastal Engineering, Universität Siegen, Paul-Bonatz-Str. 9-11, 57076 Siegen, Germany; jens.metzger@uni-siegen.de (J.M.); juergen.jensen@uni-siegen.de (J.J.)

* Correspondence: Jan-Philipp.Kueppers@Uni-Siegen.de; Tel.: +49-(0)-271-740-2730

Received: 13 March 2019; Accepted: 26 April 2019; Published: 9 May 2019

Abstract: The supply of energy is sustainable only if it is predominantly based on renewable or regenerative energies. For this reason, the use of micro-hydropower plants on rivers and streams is considered recently. This is a particular challenge for the preservation of ecologically permeable streams, so that no dams or similar structures can be considered. While the axial turbine design has prevailed in wind power, there is still no consensus for the generation of energy in free water flow conditions. In this work, an existing prototype of an unusual vertical axis Kirsten–Boeing turbine was investigated. A multivariate optimization process was created, in which all important machine parameters were checked and improved. By using neural networks as a metamodel coupled with flow simulations in ANSYS CFX, a broadly applicable optimization strategy is presented that yielded a blade design that is 36% more efficient than its predecessor in experiments. During the process, it was shown how to set up a complex sliding mesh problem with ANSYS expressions while evaluating a free surface problem.

Keywords: Kirsten–Boeing; vertical axis turbine; optimization; neural nets; *Tensorflow*; ANSYS CFX; metamodeling

---

1. Introduction

The Kirsten–Boeing concept first appeared in the late 1920s as a proposed engine for airships [1]. Since then, parts of its functionality have found their way into modern marine propulsion with the Voith–Schneider [2] propulsion. Apart from that, the use as an omnidirectional turbine for ocean and river currents has not been investigated yet. Compared to other traditional vertical axis turbine concepts, such as Savonius or Darrieus, it combines some advantages of a lift- and a drag-driven rotor in one machine. In addition to being independent of the direction of flow, the turbine does not need start-up aid, but at the same time achieves a higher efficiency than pure drag-driven rotors. It is also conceivable that the turbine is used as a hybrid propulsion system for shipping. Port tugs could navigate with it and charge the batteries when at a standstill. In addition, compared to a Darrieus turbine, it may be more environmentally friendly and less destructive to the underwater world as the Kirsten–Boeing turbine rotates relatively slowly. However, these two considerations are not part of this study.

The basic functional principles and the best possible parameter combinations were determined by Computational Fluid Dynamics (CFD). A particular challenge here was the highly asymmetric character of the entire machine. In the numerical model, neither symmetries nor periodic properties of the turbine can be exploited. In addition, stationary solutions have been found to be unreliable due to the massive influence of transient effects and the extensive blade interaction during one revolution, thus time averaged results of transient CFD simulations have to be used. Finally, a fundamental feature of the turbine makes it difficult to model—each blade has its own axis of rotation in addition to the central machine axis (cycloidal propeller, see Figure 1), which requires the use of custom functions for correct kinematics in ANSYS.

In the end, a final model was generated from the collected experience and parameter combinations, which Was then subjected to a final optimization step. A response surface method was used to determine the optimum blade-profile shape (design of experiments) for the five- and four-bladed machines.

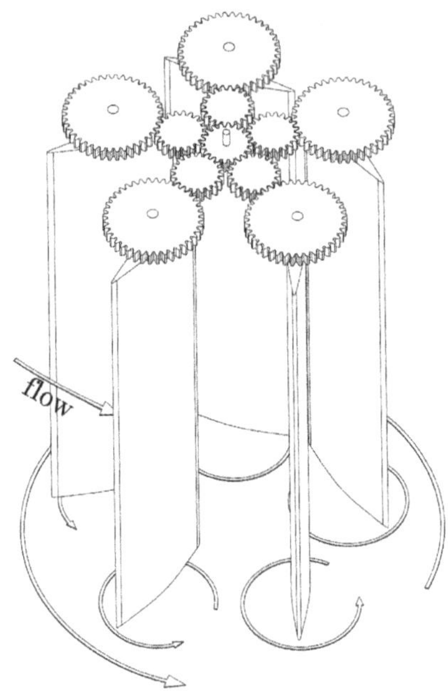

Figure 1. Sketch of the basic turbine components. The blades revolve around themselves while the machine is turning as well. The central gear is fixed in place.

2. The Kirsten–Boeing Turbine

As a cycloidal turbine, the Kirsten–Boeing turbine has the peculiarity that, in addition to the rotation of the machine, the blades also rotate about their own axis at half angular velocity [1]. In this case, the movement is generated by a planetary gearset. Each blade requires two gears to operate the additional rotation. The existing prototype has five blades. A sketch of how it works is shown in Figure 2a. In dangerous situations, it is possible to change the blade pitch (phase angle) in such a way that the turbine comes to a complete halt (indifferent blade position). This can also be used to adjust the machine to any direction of incoming flow (Figure 2b). Due to the vertical construction, most machine components can be installed above the water surface. Furthermore, no additional infrastructure is required for operation, since the turbine operates in free flow conditions, such as in rivers or ocean currents.

(**a**) (**b**)

Figure 2. The turbine can be adjusted to the flow direction. (**a**) Kinematics of the planetary gearbox with a static central gear. The transmission ratio is 2:1. (**b**) Phase angle affected by the central gear rotation.

An important measure for all turbines is the tip speed ratio λ, which defines the ratio of circumferential speed u to flow velocity v_∞.

$$\lambda = \frac{u}{v_\infty} \tag{1}$$

Due to the fixed kinematics, however, the Kirsten–Boeing turbine is unable to rotate much faster than the surrounding fluid flows. At the twelve o'clock position, the blade is always perpendicular to the flow for a short time, no matter what relative speed between the blade and flow. This leads to low possible tip speed ratios, which in turn requires large blade widths [3]. Those blades cause considerable drag forces in the flow direction, very high material costs and momentum losses, which occur because a particularly large counter-torque is exerted on the outflowing fluid. These losses occur in low λ-machines because they achieve their power through high torque and low speed [4] (Equation (15)). According to Gasch, this leads to a reduction in efficiency c_p of almost 30% at $\lambda = 1$ to $c_{p_{max}} = 0.42$. Furthermore, it is to be presumed that, due to the high machine solidity, large separation phenomena occur on the leeward side, which make the blade interaction very complex. Low tip speed machines are thus generally considered economically uninteresting.

3. Validation of the Simulation Environment

Before detailed parameter studies can be carried out, it must be ensured that ANSYS CFX can reliably reproduce the general machine behavior and power output. For this, the complete experimental environment of the prototype is modeled virtually.

3.1. Experimental Environment and Measuring Equipment

The experiments were carried out in the laboratory of Research Institute for Water and Environment at the University of Siegen (fwu). With a maximum possible volume flow of $Q = 150\,\mathrm{L\,s^{-1}}$, flow velocities of approximately $v = 0.65\,\mathrm{m\,s^{-1}}$ are achieved. The laboratory water channel has twice the width of the turbine (0.5 m) and a depth of 0.7 m. The length is 25 m, but was replaced in the simulation by corresponding constraints (see Section 3.5). The task was to find the absolute mechanical power output in relation to the tip speed ratio. To determine mechanical power output P_{mech}, the braking torque of a generator was used, which kept the turbine at a constant speed. Consequently, the rotational speed can be adjusted via a variable resistor. The generator itself is freely rotatably mounted so that it translates the turbines torque into a force that lifts a weight on a scale via the rope force F_{rope} (see Figures 3 and A1). From the measured force F_{rope} and the known lever arm l, the acting moment M can be concluded. The product of the turbine rotational speed and torque was the sought after mechanical power.

$$P_{mech} = M \cdot \omega = F_{rope} \cdot l \cdot \omega \tag{2}$$

The characteristic curve of the machine can be created along by adjusting the variable resistor. The phase angle (angle of rotation of each blade about its own axis) remains constant. All measurements were repeated around 300 times over a period of five seconds and averaged afterwards.

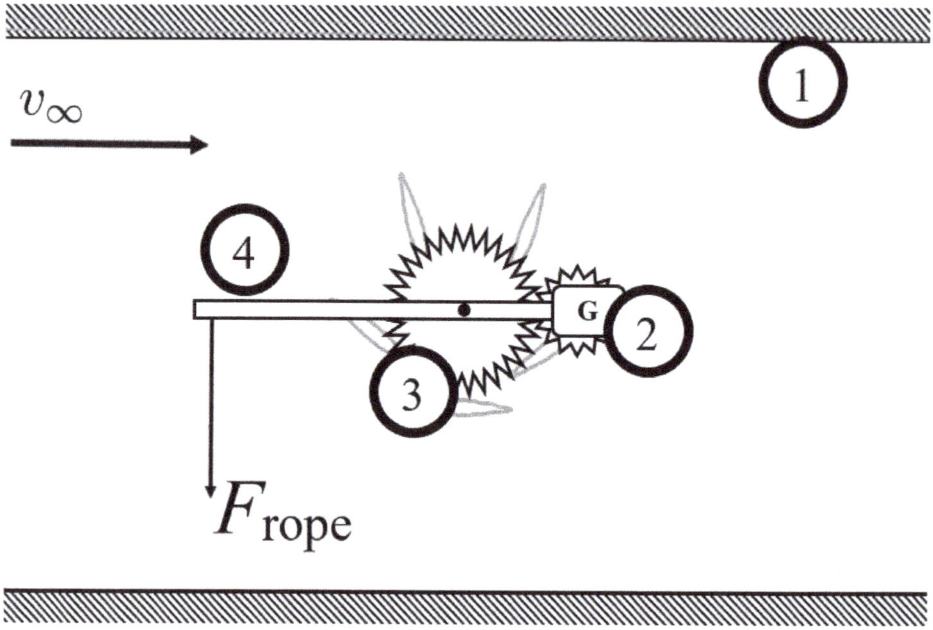

Figure 3. Functional sketch of the measuring setup: (1) duct wall; (2) generator with pinion; (3) turbine with main gear; and (4) free-floating rod.

3.2. Structure of the Simulation

The simulation was modeled based on the experimental setup as closely as possible, therefore a multi-phase simulation with air and water was generated. The data for the blade geometry came from the original CAD files of the turbine. It is an axis-symmetric blade with straight flanks and rounded tips (see Figure 4). The geometry creation was followed by the meshing process and the actual setup, which are described below.

Figure 4. Blade profile of the first prototype [5].

3.3. Mesh Independent Solution

The discretization of the fluid domain was carried out in the ANSYS Mesher and allowed a consistent quality of meshing, also for variable geometries and main dimensions. The meshing structure consisted essentially of quad-elements and a prism layer near the wall with a first element offset of 2.5×10^{-5} m, which was supposed to resolve the fluid-dynamic boundary layer. The mesh consisted of up to 1,000,000 unstructured hexahedral elements. Since the gap between the blades and the channel floor was neglected (approximately 2 mm), the entire structure could be meshed using a sweep algorithm. At critical points, such as the blade tips, or at domain transitions, the element density was further increased by size functions. In an earlier mesh convergence study, this quality of mesh resolution has proven to be sufficient to generate reliable results in an acceptable time (see Figures 5 and 6).

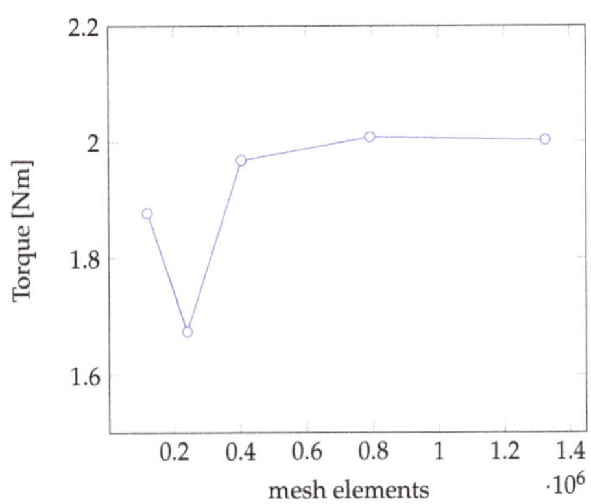

Figure 5. Mesh convergence study for the channel flow.

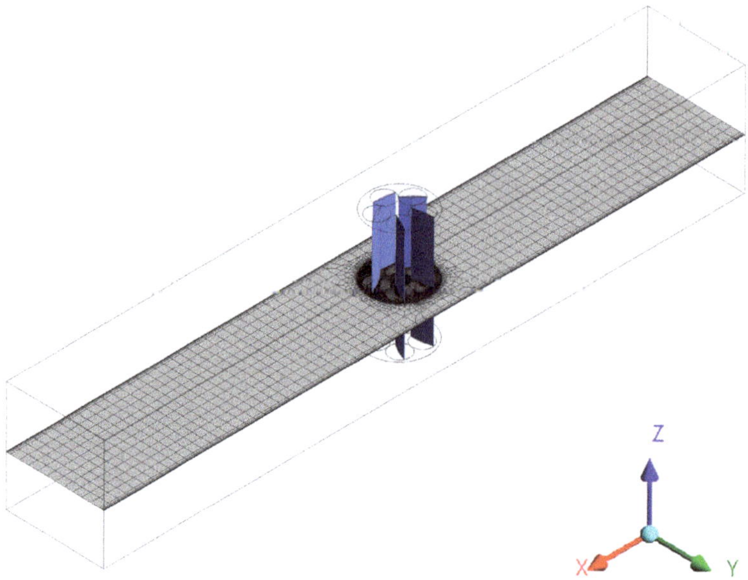

Figure 6. Isometric view of the fluid domain with cross section through the meshing structure.

3.4. Setup–Mesh Movement

The movement of the airfoils was specified by a *moving mesh* approach a no time consuming recreation of the meshing was not necessary in each time step. As the inner cylindrical domain rotates at the angular velocity ω, the blades also move around their own axis at the velocity $\omega/2$. This behavior could be generated in ANSYS by a *subdomain*, which assigns a new position to every single mesh node through an *expression* (function). ANSYS provides the variables *X Coordinate, Y Coordinate* and *Z Coordinate*, which can now be moved relative to the node position in the last time step with Δt. In the following, these variables are summarized in the mesh node vector $\vec{n}_i$. The central domain is rotated by φ around the Z-axis with the rotation matrix $\boldsymbol{R}_z$:

$$\boldsymbol{R}_z = \begin{pmatrix} \cos\alpha & -\sin\alpha & 0 \\ sin\alpha & cos\alpha & 0 \\ 0 & 0 & 1 \end{pmatrix} \tag{3}$$

$$\varphi = \omega \cdot \Delta t \tag{4}$$

$$\vec{n}_i^{\,t+1} = \boldsymbol{R}_z(\varphi)\vec{n}_i^{\,t} \tag{5}$$

For the blade nodes $\vec{n}_{i,s}$ $(s = 0...5)$ a transformation is needed in which the absolute position of the individual blade rotation axis $\vec{p}_s = (x_s, y_s, 0)$ for two time steps is calculated. For the calculation in the ANSYS environment, the expression *"atstep" (accumulated time step)* is used. It outputs the current number of the CFX iteration.

$$\varphi^{t+1} = atstep \cdot \omega \cdot \Delta t \tag{6}$$

$$\varphi^{t} = (atstep - 1) \cdot \omega \cdot \Delta t \tag{7}$$

The position of each blade axis must also be calculated via expressions. The known blade positions before the simulation ($t = 0$) yields:

$$\vec{p}_s^{\,t+1} = \mathbf{R}_z(\varphi^{t+1})\vec{p}^{\,s,t=0} \tag{8}$$

$$\vec{p}_s^{\,t} = \mathbf{R}_z(\varphi^{t})\vec{p}^{\,s,t=0} \tag{9}$$

$$\vec{n}_{i,s}^{\,t+1} = \mathbf{R}_z\left(\frac{\varphi}{2}\right)\left(\vec{n}_{i,s}^{\,t} - \vec{p}_s^{\,t}\right) + \vec{p}_s^{\,t+1} \tag{10}$$

In the event that the phase angle is $\neq 0$, the blade nodes must be initialized with an additional rotation angle φ_{phase}. To ensure the stability of the simulation, the additional rotation should be distributed over several time steps. ANSYS offers the use of an IF function depending on the global time step *atstep*. With the simultaneous conversion to radians for ten iterations follows:

$$\varphi_{phase} = \begin{cases} 0 & ,atstep > 10 \\ \frac{2\pi}{360} \cdot \frac{phaseangle}{10} & ,atstep \leq 10 \end{cases} \tag{11}$$

The relative rotation angle φ can be calculated for a constant angular velocity with $\varphi = \omega \cdot dt$.

3.5. Setup—Domain Properties

From the material library in ANSYS CFX, *Water* and *Air at* 25 °C were selected at a reference pressure of 101 kPa. Since gravity plays a role, the *Buoyancy Model* was activated and the gravitational constant was set to $-9.81\,\text{m/s}^2$. The reference density for air was $1.2\,\text{kg/m}^3$, the *Ref. Location* in the global coordinate system was zero. For multi-phase flow, the *Homogeneous Model* with the standard *Free Surface* option was used, meaning the two different phases shared a single velocity field. *Heat Transfer* was set to 10 °C for simplicity, since no significant temperature changes were expected. The surface tension was taken into account by the *Surface Tension Coefficient* of $0.072\,\text{N}\,\text{m}^{-1}$ in the *Continuum Surface Force* model.

First, a 10-meter-long piece of flow channel was simulated, but, despite the rough side walls, the velocity field was very homogeneous. Since this visually did not seem compatible with reality, and the modeling of the complex inflow into the channel was also not expedient, the velocity field was detected locally by direct measurements. For this purpose, a universal current meter of the type *Ott c2* was used 2 m in front of the turbine. From velocities recorded at a 5 × 5 grid, a function (expression) was generated by using quadratic polynomial regression which, depending on the spatial position (y, z), assigned different entry velocities to the inlet. The x-axis was pointed in the flow direction. The fast core flow, which can be seen in Figure 7, led to a significant change in the turbine performance compared to an erroneously assumed homogeneous velocity. For the entry velocity v_1, the function is:

$$v_1(y,z) = -3.84y^2 + 0.63yz - 1.76z^2 - 0.14y + 1.2z + 0.51 \tag{12}$$

The determined velocity field was also used to initialize the fluid domain. The air/water state (1, water; 0, air) was initialized with the measurement results for the mean water level height h_1, h_2 in front of and behind the turbine, using the following expression:

$$VolFr = \begin{cases} \begin{cases} 1, z < h_1 \\ 0, z > h_1 \end{cases} & ,x < 0 \\ \begin{cases} 1, z < h_2 \\ 0, z > h_2 \end{cases} & ,x \geq 0 \end{cases} \tag{13}$$

Since only a short area in front of and behind the turbine was simulated but the buoyancy-option was active, the hydrostatic pressure at the boundary conditions had to be taken into account. The outlet had a linearly increasing pressure gradient with the depth z. The equation was also used to initialize the whole domain:

$$P_{Profile}(..) = VolFr \cdot \begin{cases} \rho g(h_1 - z) & ,x > 0 \\ \rho g(h_2 - z) & ,x \geq 0 \end{cases} \tag{14}$$

As a solver, CFX was used with the classic URANS-based SST (Shear Stress Transport) turbulence model [6], which is now the industrial standard and allows reliable results in models affected by strong flow separation. The dimensionless wall distance $Y+$ was < 1, and the *Wall Function* option was set *Automatic* by default. The blade surface was defined as hydraulically smooth, while, for the channel walls, a *Sand Grain Roughness* of 0.005 m was assumed. The resulting Reynolds number was 43,000 $\pm$ 1000, while the Froude number for the open channel flow was 0.28.

For transient time integration, an implicit second-order Euler method was used. Each time step was considered converged if the RMS residuals (*Root Mean Square*) for momentum and continuity were below 1×10^{-4} Between 1 and 15 *Coefficient Loops* were required. The time step was chosen such that the mesh elements between two domains *(sliding interface)* in a time step advanced a maximum of half an element further. For $\Delta t = 2\pi/\omega N$, besides the angular velocity ω, only the element number N in the circumferential direction (blade domain) must be taken into account. On average, the time step was 0.005 s while the total simulation time was about 5 s. Between the static and rotating domains, a transient CFX *rotor–stator* interface (General Grid Interface) was created to pass over the fluid properties correctly.

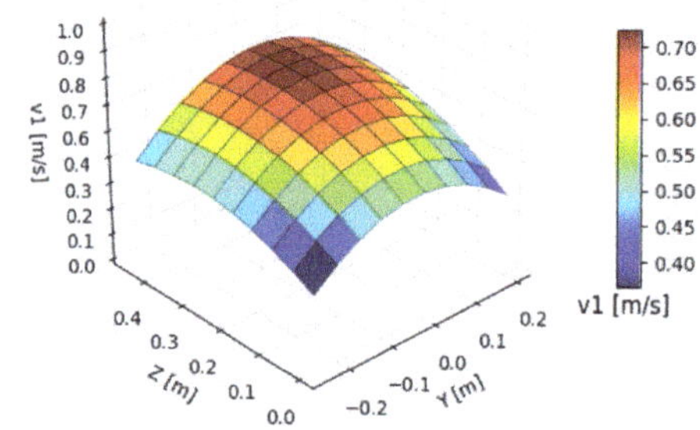

Figure 7. Measured velocity field in front of the wheel extrapolated from 5 $\times$ 5 support points.

3.6. Setup—Performance and Efficiency

The characteristic curve was evaluated at operating points of constant angular velocity. At these operating points, the performance of the turbine was calculated using the following equation:

$$P_T = \omega \cdot M_T \tag{15}$$

The final value of the power was averaged over the last revolution of the turbine. ANSYS has the option to apply various operators to variables using *Transient Statistics*. With *<variable>.Trnavg*, the arithmetic mean is calculated starting from a specified time step, which can be output directly as a solution variable in the parameter manager. The efficiency of the turbine results from the ratio of the shaft power to the kinetic energy present in the fluid P_F and can be determined by the following equation:

$$c_p = \frac{P_T}{P_F} = \frac{\omega \cdot M_T}{\frac{1}{2}\rho A v_\infty^3} \tag{16}$$

The area A is the "harvesting area" of the machine and v_∞ is the mean velocity of the surrounding fluid. The harvesting area was calculated by projecting all blades onto a surface in the flow direction while turning the blades a full revolution. The result was a rectangle with the height of the individual blade and a width that was the doubled distance from the outer most blade tip to the machine centre. All simulations presented in this paper that were to be validated by a physical experiment only outputted the absolute power of the machine. Since the efficiency in the channel flow depends strongly on the ratio of wheel diameter to channel diameter, there would be no generally valid or meaningful value for efficiency. The results of the optimization are given relative to a reference power instead.

3.7. Results of the Simulation

The simulation results are close to the averaged real measurement (see Figures 8 and 9) and the machine behavior could be simulated properly. Due to the low absolute power, however, small interfering factors gained in importance within the experiment. Friction in the bearings, flow around the blade top and bottom (tip losses), blade surface roughness, etc. led to losses that were not shown in the simulation. The deviation from the blue curve at high powers may possibly be explained by the inefficiency of the RANS turbulence model used. At low Reynolds numbers, much of the boundary layer can be laminar, which is not exactly reflected by the classic fully turbulent model. In the real laminar boundary layer, disturbances often led to earlier flow separation compared to a turbulent boundary layer, which increased blade performance and led to problems with the correct calculation of the deep stall [7]. In general, the performance was overestimated, but, since the general machine behavior was well represented, the simulation was still suitable for further optimization of the basic design parameters.

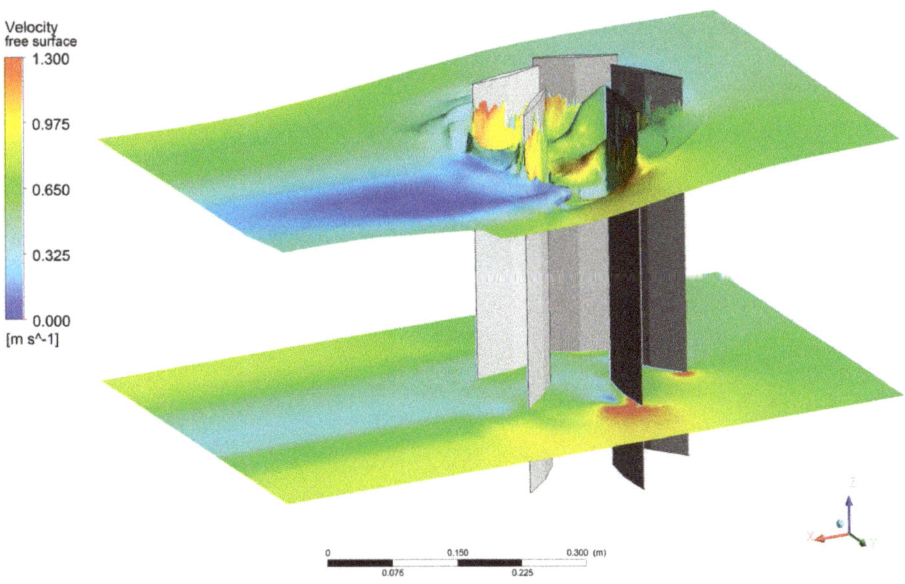

Figure 8. Qualitative flow velocity visualization on free surface plot.

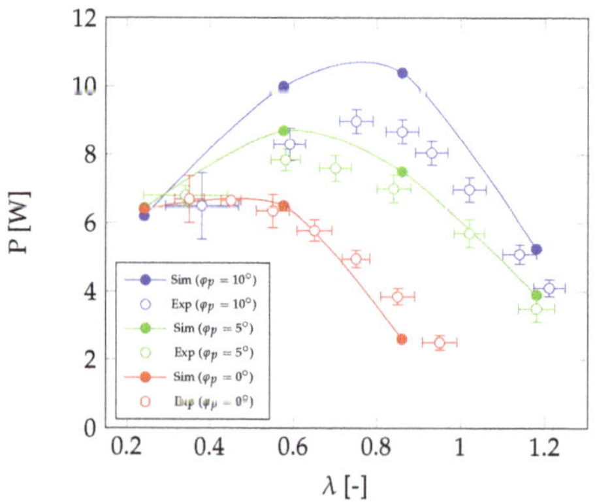

Figure 9. Comparison of performance between simulation and experiment depending on the tip speed ratio. The error bars represent the standard deviation.

4. Study of the Basic Design Parameters

Currently, there is little experience or conventions for the construction of a Kirsten–Boeing turbine. This study evaluated some basic parameters in terms of their efficiency through 2D flow simulations.

- The optimum number of blades
- Improved blade geometry
- Influence of the blade width

In modern airfoil optimization techniques, correct parameterization is crucial. The method must be able to describe a comprehensive selection of wings with minimal use of geometry parameters. The use of Béziersplines [8] in combination with response surface methods [9] or a *Design of Experiments* has proven to be particularly efficient. The direct optimization by *Evolutionary Algorithms*, which also benefit from small parameter sets, is also conceivable .

4.1. Parametric Geometry Creation

A Bézierspline of degree n is defined by $n+1$ support points ($\vec{P}_i$ with ($i = 0...n$). In this case, the entire profile consists of two splines ($n = 4$) on the long flanks, which are connected by two splines ($n = 2$) at the ends. The so-called Bernstein polynomial is [8]:

$$\begin{pmatrix} x(t) \\ y(t) \end{pmatrix} = \sum_{i=0}^{n} B_{i,n}(t) \cdot \begin{pmatrix} P_{x,i} \\ P_{y,i} \end{pmatrix} \tag{17}$$

where the polynomial $B_{i,n}(t)$ is defined as:

$$B_{i,n}(t) = \binom{n}{i} t^i (1-t)^{n-i} \tag{18}$$

Using the parameter $t = 0 \ldots 1$, any number of point coordinates can now be generated for later use in CAD tools. Because of the symmetry, only two out of four curves need to be defined. The fixed blade width b enforces the position of the outer control points. The number of degrees of freedom continues to be reduced by continuity conditions at the transition between the curves. Since c_1-*continuity* is provided, this leads to the complete definition of the spline at the leading edge. Its center control point (P_1^2) lies at the intersection of the tangents with the first points of the two main lines ($\rightarrow c_1$). The control point coordinates are shown in Tables 1 and 2. b is a precalculated width variable to make sure the whole blade has the same desired chord length of 0.09 m every time. A total of seven free parameters are available. An exemplary profile is shown in Figure 10.

Table 1. Definition of the control points of the upper main spline $n = 4$

i	0	1	2	3	4
$P_{x,i}$	$-b/2$	free	0	free	$b/2$
$P_{y,i}$	free	free	free	free	free

Table 2. Definition of the control points of the left short spline $n = 2$.

i	0	1	2
$P^2_{x,i}$	$P_{x,0}$	$\times$	$P_{x,0}$
$P^2_{y,i}$	$P_{y,0}$	$\times$	$-P_{y,0}$

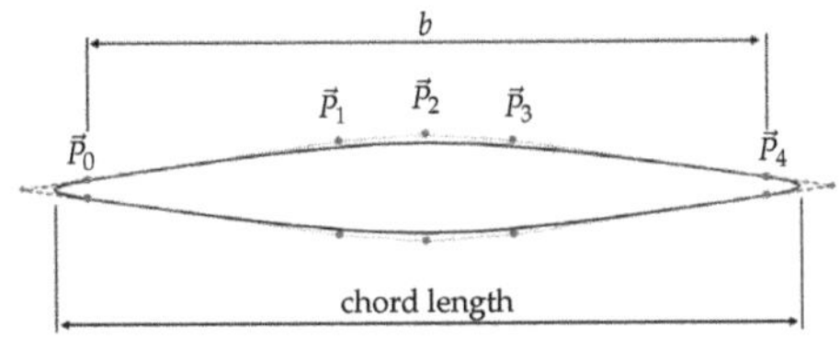

Figure 10. Parameterized blade with Bézierspline control points.

The geometry was parameterized in MS Excel (see Figure 11) and constructed from Béziersplines; the free parameters were initially intuitively assigned. The generated coordinate points can be passed to ANSYS via scripting and converted to true CAD geometry by 3D-Curves in the *Designmodeler* (by extrusion). In addition, conditions were created in order to be able to easily modify or replace the blade shape on-the-fly. With a machine diameter of 0.5 m, an outer fluid domain with an extent of 10 m was created in each direction, which was sufficient to dissipate any transient effects within the domain.

Figure 11. MS Excel geometry representation with interface walls.

4.2. Setup—Parameter Study

The setup in CFX was the same as the validation calculation, except that the free surface was neglected and the channel wall was removed. After all, the final product should be optimized for free flow. The model was subsequently calculated with only one element over the height. As a result of the significant reduction in the number of elements and complexity, a large part of the calculation time could be saved while increasing the element quality in the plane as well. However, it should be noted that the losses due to the free surface were now neglected and the quasi-2D model generally highly overestimated the actual performance. Nonetheless, the quasi-2D simulation provided meaningful, relative views of different blade configurations, movements and shapes.

To match the flow speeds of real operational locations, the design flow rate v_∞ was increased to $1\,\mathrm{m\,s^{-1}}$. The default blade profile used a scaled-up version of the first prototype. The performance increased accordingly.

4.3. Optimal Number of Blades

The number of blades significantly determines the material requirements and the complexity of the machine. Each additional blade requires two additional gears in the transmission, which in turn can lead to increased losses due to friction. A high number of blades provides more blade area, but at the same time the influence on each other is stronger. The most efficient option was then found. To ensure a fair comparison, the blade profiles were scaled so that the outer diameter of the turbine remained constant (see Figure 12).

Figure 13 shows the comparison between three, four and five blades with different tip speed ratios λ. It is striking that the number of blades had a negligible impact on the performance achieved. The high coefficient of performance was due to the nature of the 2D simulations and was not to be considered realistic. For comparison, three additional points in the known 3D configuration were simulated. These showed a drastic reduction in overall efficiency. Real life performance was estimated to be even lower, since in this case the turbine was only simulated in a very wide open channel and the flow could not pass underneath it. Consequently, tip losses were still neglected. It can be noted, however, that fewer blades resulted in larger amplitudes in the torque created (see Figure 14), which may increase the load on the bearings. For example, the maximum fluctuation of the torque on the machine axis with five blades was observed to be 11–18% and for four blades 15–25%. However, the range also depended strongly on the blade shape. The fundamental frequency in the torque curve f_M resulting from rotational speed and number of blades n was:

$$f_M = \frac{\omega}{2\pi} \cdot n[Hz] \tag{19}$$

In reality, the effect on the structure was attenuated due to the inertia of the blades and the surrounding fluid. The consequence is that, for the later prototypes, four blades were considered, since they represent a good compromise among achieved power, machine complexity and torque fluctuations. First, however, the existing five-bladed prototype was further developed, as its blades could be easily swapped.

Figure 12. Different number of blades with the same outer diameter.

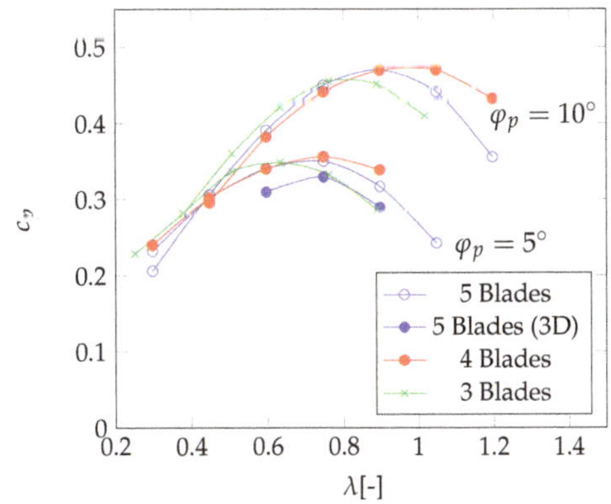

Figure 13. Three, four and five blades compared at different phase angles. Additionally, the simulation was carried out once again in the 3D configuration at $\varphi_p = 10°$.

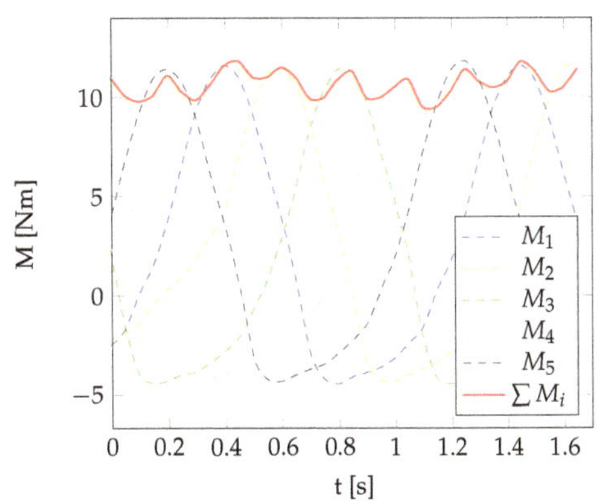

Figure 14. Exemplary torque curve for five blades $M_i = 1.5$ at $\lambda = 0.75$ and $\varphi_p = 10°$.

4.4. Optimization of the Blade Profile

Since the turbine is not a conventional purely lift-driven rotor and the blades have to work from both sides, it is not possible to use conventional profile shapes (such as NACA, etc.). Rather, the optimal profile must be found directly in interaction with all other blades. It is not possible to draw any conclusions on the mutual influence of the blades in the back flow region when considering just a single blade (for example, using classic blade element momentum (BEM) theory). Consequently, different blade profiles were simulated and compared as a whole system of five blades. Due to the characteristics of the machine, only axis- or point-symmetrical profiles could be considered. It was likely that thick profiles with a high tolerance to flow separation at high angles of attack would prevail. While the blades rotate in the fluid, they repeatedly increase the angle of attack until the blade is perpendicular to the fluid flow. Thus, the most efficient blade is probably the one that can produce lift forces the longest.

All blade shapes were first used at the optimum operating point of the reference blade. This reference operating point was set at a phase angle of 10° and a tip speed ratio of $\lambda = 0.9$. Five blades were used. Since factors such as the free water surface, blade tip losses and losses due to friction were not taken into account, the blade interaction could be seen undisturbed. The result is given in ascending order of the achieved performance (see Table 3).

Table 3. Performance comparison of different blade profiles.

No.	Profile	max. Power [W]	Relative to Reference (4)
1	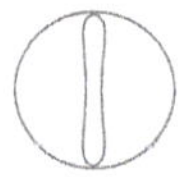	74.8	1.17
2	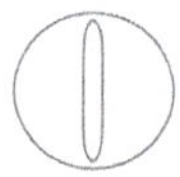	74.4	1.16
3	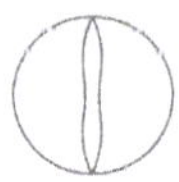	69.3	1.08
4	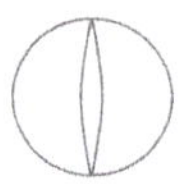	64.1	1.00
5		44.6	0.70
6	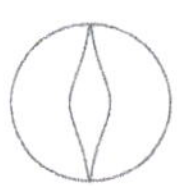	43.3	0.68
7	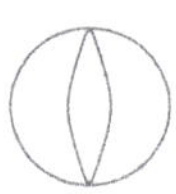	26.0	0.41

It can be seen that blades with a blunt blade tip tended to be ahead in the simulations. The areas with strong surface curvature increase the flow velocity at this point significantly. According to Bernoulli, a negative pressure is created here, which moves the blade. This flow deflection at the blade tip ensured a favorable distribution of the pressure, such that the resulting force vector almost never pointed to the central machine axis, but deviated from it as far as possible (in the circumferential direction) and thus generated a large torque (see Figure 15). The considerations regarding the profile curvature lad to a profile that had a pronounced rounding at the blade tip. Since only one side of the blade tip is involved in torque generation, point-symmetric blade profiles were considered.

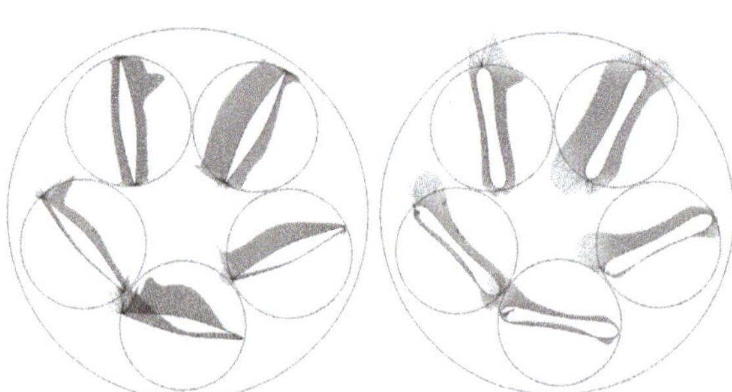

Figure 15. Force distribution on the standard blade (No. 4) (**left**) compared to the best variant (No. 1) (**right**).

4.5. Influence of the Blade Width

In addition to the profile shape, the absolute width also provides an important parameter. Based on a fixed outer diameter of the machine, there is a maximum allowable blade width where no overlapping occurs. A narrower blade can move further outwards—creating a larger available lever arm. To validate this idea and preserve similitude in flow behavior, the blade profiles were scaled by a simple factor (see Figure 16).

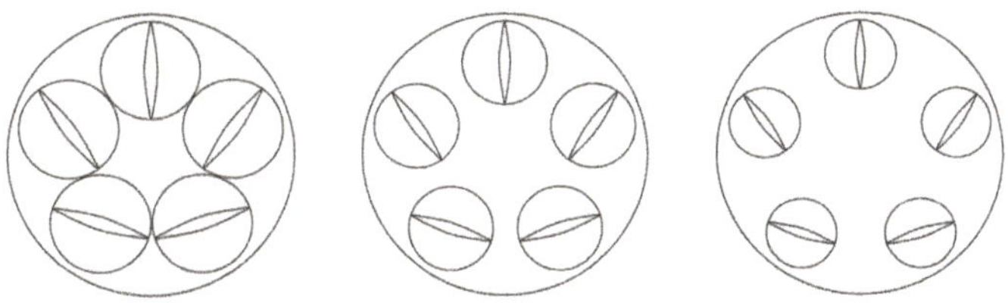

Figure 16. Blade width factors 1 (**left**), 0.85 (**middle**) and 0.7 (**right**).

However, as shown in Figure 17, the narrower blades did not reach the performance level of the unscaled blades. Obviously, the effective surface area ess significantly involved in the generation of the total torque. For low tip speed ratios ($\lambda < 1$), it is generally assumed that the necessary machine surface area to capture all the flow energy is $> 100\%$ of the given flow area [3]. Consequently, the widest possible blade shape was ahead.

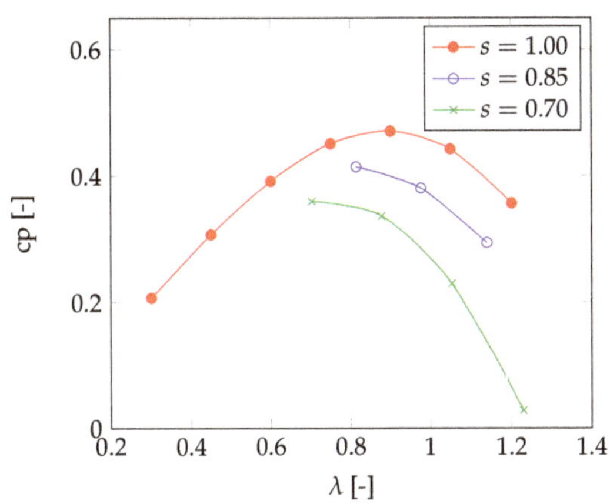

Figure 17. Characteristic curves for different blade widths.

5. Metamodel Based Optimization of The Blade Shape

The preliminary investigations showed that a point symmetrical blade shape could lead to higher performances. However, as the optimal parameter configuration was not known for the time being, an automatic optimization process was developed (see Figure 18). The reference profile was still the one used in the validation calculation (see Section 3). The basis for this optimization process was a Multilayer Perceptron (MLP) [10], which is part of the family of deep learning methods. A classic, fully connected *feed forward* neural network with several layers between input and output was created (see Figure 19). *Feed forward* here means that the flow of information takes place in one direction only. The net was trained with the usual *backpropagation* algorithm [11]. As MLPs are increasingly used in the fields of classification, image recognition and approximation, various programming interfaces (APIs) exist that can speed up the training process using a GPU. In this case, *TensorFlow* [12] was used, controlled by Python scripts.

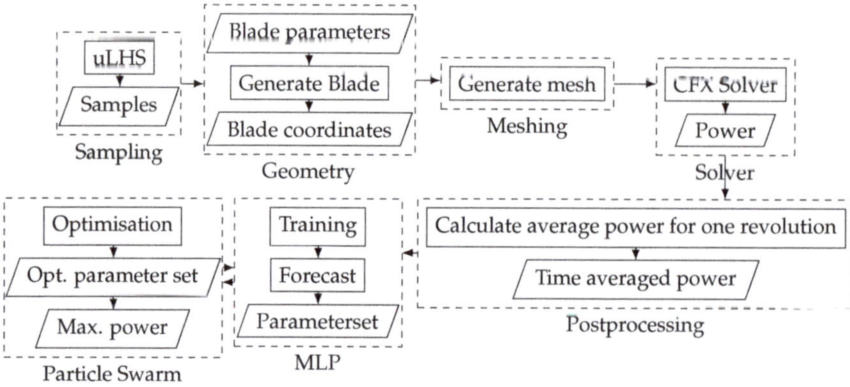

Figure 18. Optimization process.

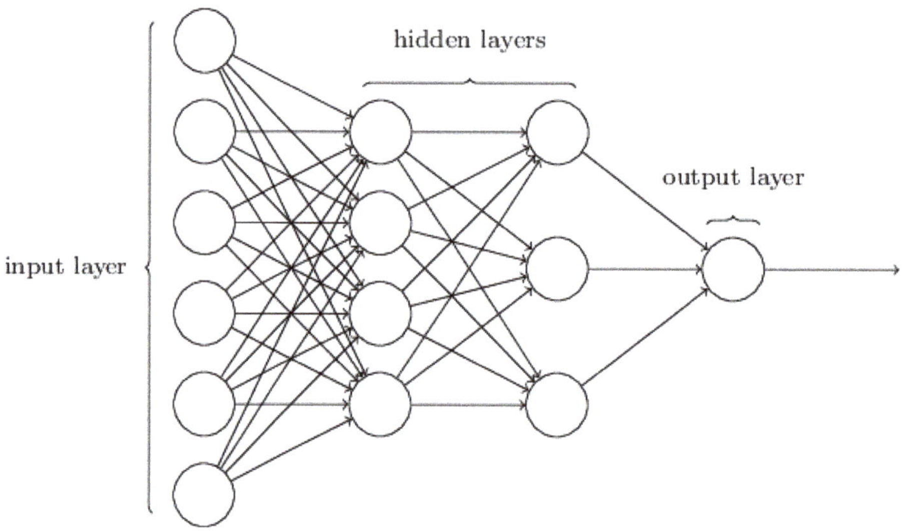

Figure 19. Possible network topology of an MLP [10].

5.1. Geometry Creation

The structure of the geometry works in much the same way as described in Section 4.1, with the difference that the lower spline is mirrored, creating a point symmetrical blade shape (see Figure 20). To ensure that all possible blade shapes were feasible (no self-intersection), the parameter ranges were manually adjusted and normalized between 0 and 1. There were still seven free parameters preserved. For this investigation, the scale of the smaller prototype based on Section 3 was used again, as it provided the opportunity to manufacture the optimal blade by 3D printing and to experimentally validate it in the laboratory channel.

The network was trained with the data of the flow simulation. For this purpose, 90 different blade configurations were generated and calculated. The resulting metamodel could make predictions in the parameter space known to it. These predictions were used to find the desired maximum within the parameter space, leading to the maximum power, by means of a suitable optimization algorithm.

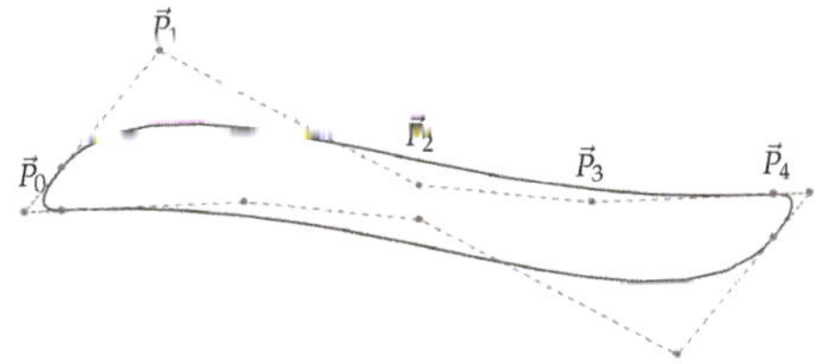

Figure 20. Point symmetrical blade.

Sampling used Latin Hypercube Sampling (LHS) because, unlike Monte Carlo sampling, it has no local clusters (*Sampling Artifacts*) and generally results in better quality results [13]. The samples were evenly distributed in parameter space (*Uniform-LHS*). Finally, all created designs were numbered consecutively and saved as point coordinates. ANSYS could then script-import all designs serially and convert them into 3D geometry.

5.2. Meshing and Solver

The next steps were the same as in the previous calculations and the simplified quasi-2D setup for the free flow remained unchanged. The operating point to be calculated was still at a phase angle of 10° and tip speed ratio $\lambda = 0.9$. After completion of all calculations, the results were split into a training and validation set (80% training, 20% validation), as it must be checked whether the model was sufficiently generalizing and not just memorizing the input data (by over-fitting the model).

5.3. Multilayer Perceptron with Tensorflow

The topology of the neural network was adjusted manually. The best results were provided by a network with three hidden layers, which used *rectified linear units* (ReLU [14]) as an activation function, each layer having 256 nodes. The default optimization algorithm was *Stochastic gradient descent* (SGD [11]). As cost function L to be minimized, in contrast to the frequently used quadratic deviation $L(a) = a^2$, the "Huber loss" function has proven to be particularly effective [15]. Here, a is the difference between prediction and the actual result.

$$L_\delta = \begin{cases} \frac{1}{2}a^2 & , |a| < \delta \\ \delta\left(|a| - \frac{1}{2}\delta\right) & , |a| \geq \delta \end{cases} \tag{20}$$

where δ was set to 0.3 in this case. δ defines the transition at which the piecewise defined cost function changes from the square error to the absolute error. The adjusted loss function reacted less strongly to outliers in the data and led to a better approximation in the validation set. After about 25,000 iterations, there was no further improvement in the loss function of the training or validation set and the calculation was aborted. To measure the quality of the metamodel, a similar method as proposed by *OptiSlang* [16] was used. Instead of cross-validating and recalculating the metamodel at each input value, however, the formula was applied to the validation set only. The OptiSlang-defined *Coefficient of Prognosis* (CoP) helped to interpret the values of the cost function.

$$CoP = 1 - \frac{SS_E^{pred}}{SS_T} \tag{21}$$

where SS_E^{pred} is the sum of the squares from the error of the prediction, which is divided by the total least squares of the actual solutions to the mean SS_T. The result is a percentage that describes the predictive quality of the model. In this case, the model reached a quality of about 75%. This method is generally applicable to all regression methods and makes it easy to compare different metamodels. In contrast to the coefficient of determination r^2, not all interpolation points that contributed to the construction of the metamodel were taken into account, so that the validity of the prediction quality was much more reliable. Figure 21 shows an example of the resulting response surface of the metamodel for parameter $p2$ and $p5$, while all other parameters were set to 0.5. The area showed a recognizable optimum for this pairwise parameter combination. To find the actual optimum for all seven parameters simultaneously, a suitable optimization algorithm must be used.

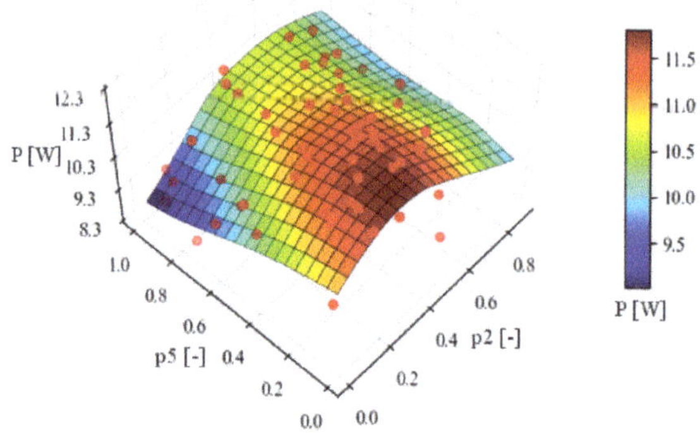

Figure 21. Representation of parameters $p2$ and $p5$ with response surface.

5.4. Particle Swarm Optimization

Since the MLP is a "black box", it is not easy to know after training where the maximum can be found in the parameter space. Therefore, a simple *Particle Swarm Algorithm*(PSA) was programmed to do the search. The algorithm is based on the flock behavior of animals in nature and has three main parameters [17]. In addition to the global best position of all particles, each individual knows his own best position. In addition, there is a factor for inertia. After a random initialization in parameter space, each particle moves randomly into each of the n-dimensions ($n = 7$). In each time step, the speed is manipulated so that each particle moves in the direction of the currently known global or its local maximum. In addition, the weighting between the two attractions is randomly varied during runtime. The inertia factor and the random weighting should ensure that the particles can also get out of local minima. If a particle exceeds the limits of the parameter space, the value of that dimension is randomly reinitialized.

$$x^i_{k+1} = x^i_k + v^i_{k+1} \tag{22}$$

$$v^i_{k+1} = \omega_k v^i_k + c_1 r_1 \left(p^i_k - x^i_k\right) + c_2 r_2 \left(p^g_k - x^i_k\right) \tag{23}$$

x^i_k = particle position
v^i_k = particle velocity
p^i_k = best individual position
p^g_k = best swarm position
ω_k = factor of intertia
c_1, c_2 = weighting
r_1, r_2 = random behavioral change

After fine-tuning the c_1, c_2 and ω_k parameters, a good convergence is achieved after 400 iterations with 100 starting particles. A scatter plot of the individual particle solutions is shown for each iteration in Figure 22.

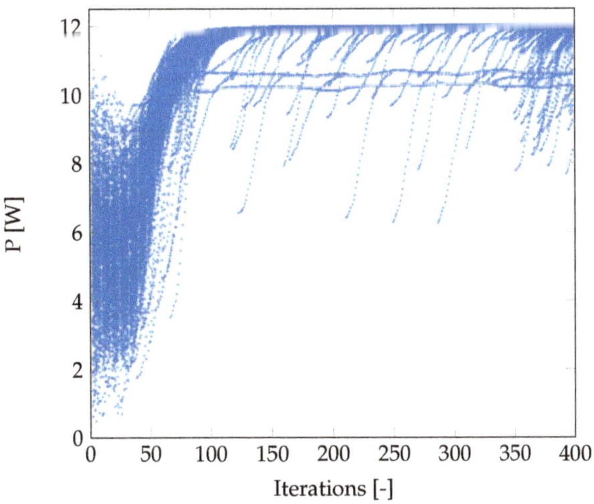

Figure 22. Solutions of individual particles per time step.

Once an ideal solution was found, the coordinates are passed to ANSYS and the performance was validated. The process was repeated iteratively and the new values were added to the old training set. After 10–15 iterations, there was no further improvement in the design and the final blade was found. In the simulated 2D free flow, the machine experienced a massive increase of performance of roughly 50% (see Table 4). Because it was not possible in this study to validate the simulation results of the free flow, the optimized airfoils were simulated again as in Section 3 inside the laboratory channel. The channel flow resulted in a simulative performance increase of 32.9%, as shown in Table 5. Thanks to the aerodynamically optimized airfoil shape, the turbine could draw additional power from the much better used lift forces.

Table 4. Airfoils in comparison from the 2D simulation: Original (1); best sampling (2); and final optimization (3).

No.	Profile	max Power [W]	Relative to Reference
1	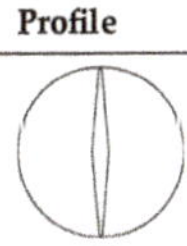	8.13	1.00
2	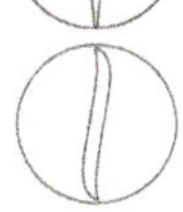	12.01	1.48
3	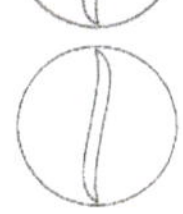	12.25	1.51

Table 5. Airfoils in comparison in the simulated laboratory channel with free water surface: Original (1); and final optimization (2).

No.	Profile	max Power [W]	Relative to Reference
1	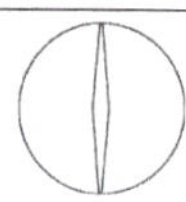	10.22	1.00
2	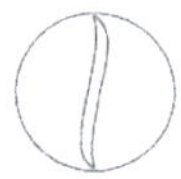	13.59	1.33

For validation of the simulation results, the original blades on the prototype (see Section 3) were replaced by additively manufactured, optimized blades (see Figure 23). The predicted maximum power increase was also observed experimentally in the confined flow of the channel. A maximum performance improvement of 36.5% for the five bladed prototype was achieved (see Figure 24). While the absolute performance values in the simulation were once again overestimated, the relative increase in performance fit very well with the predictions.

Figure 23. 3D printed blades installed on the prototype.

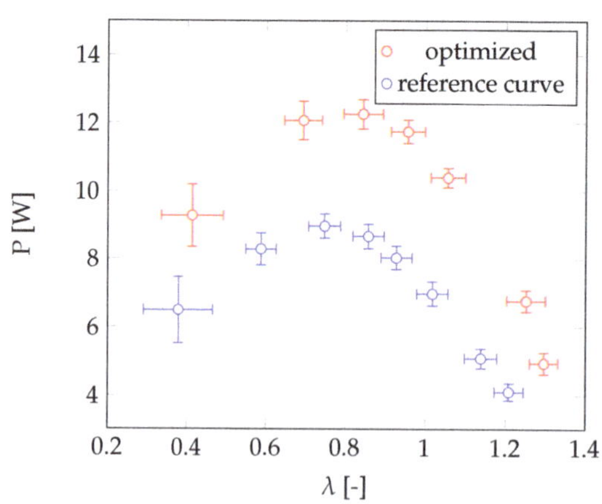

Figure 24. Characteristic curve determined experimentally in the channel at the optimum phase angle of 10°.

6. Design of the Four Bladed Prototype

Based on the new design rules and determined blade profiles, the optimization process of the final prototype was started. The requirements were a turbine with outer diameter of 1.2 m with a vertical length of at least 1.5 m. Based on the previous findings, the turbines blade number was set to four, while each blade had the maximum possible width. The blades were completely submerged, and, at the optimum operating point, they moved approximately with the flow rate of the water at $\lambda = 0.9$. The whole optimization process proceeded again as described in Section 5. Only minor changes were made to cope with the changed machine dimensions.

6.1. Sampling

Since the optimization of the blade shape had previously only been done for a small five-blade machine, it was required to repeat the same steps for the large four-blade variant. The default profile for the new optimization was the previously determined best variant, because it could be assumed that the new optimum was close to the old one in parameter space. The individual parameter ranges were normalized again between 0 and 1 and the actual values were adjusted so that only functioning blade designs were created (for example, no overlapping of the splines; see Figure 25). When sampling the new parameter variations, the old, upscaled design was placed exactly in the middle of the new parameter space so that the average value (0.5) for all free variables resulted in the original design. In total, 102 unique blade designs were generated using the known uniform Latin Hypercube Sampling (uLHS).

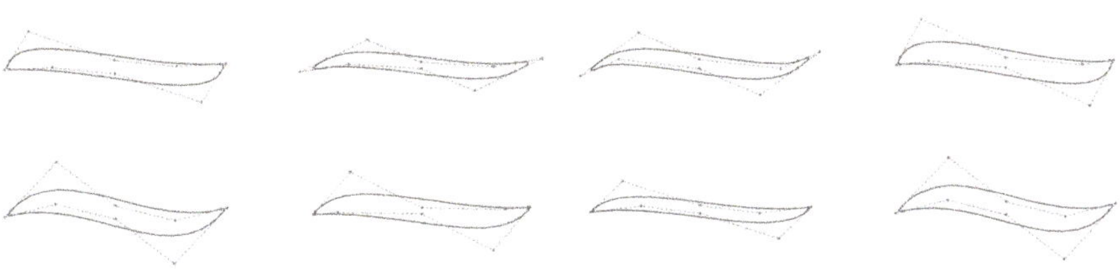

Figure 25. Visualization of the first eight samples.

6.2. Meshing and Solver

The quasi-2D model now consisted of approximately 500,000 unstructured tetrahedral elements (see Figure 26). To approximate the actual environmental conditions at the site of future measurements and experiments, the flow velocity v_∞ at the inlet was increased to $1.4\,\mathrm{m\,s^{-1}}$. The total torque was determined for a virtual blade length of 1.5 m.

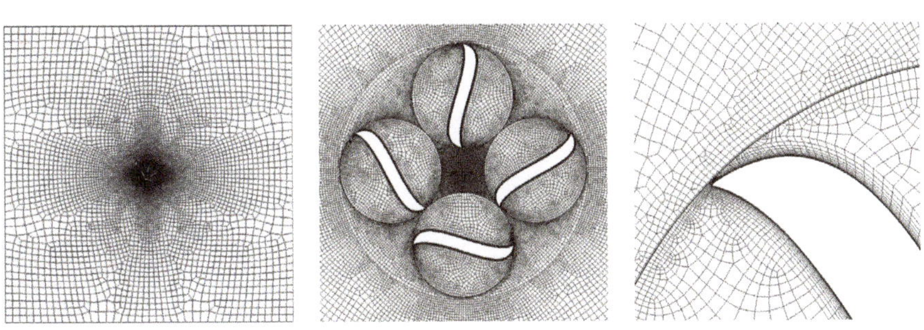

Figure 26. Global domain (**left**), mesh refinement on the interfaces (**middle**) and inflation layer on a blade tip (**right**).

6.3. Results of the Optimization

After calculating all 102 samples, the averaged torque of each was passed to the MLP. The topology remained unchanged. Finally, when the evaluation of the loss function showed no further improvement, the training of the neural network was stopped. The quality of the metamodel was again described by the CoP value (see Section 5.3) and was now 79.1%. Next, the *Particle Swarm Algorithm* generated different suggestions for new parameter configurations that were validated in CFX (Figure 27). Figure 28 shows the initialization and movement of particles in parameter space in two dimensions. Since the search in the background actually worked in seven dimensions, the particles were apparently able to leave the response surface and find an optimum near 1800 W. In contrast to the algorithm described in Section 5.4, the particles were not randomly reinitialized when the permissible range was exceeded, but they bounced in the opposite direction at twice the speed. However, this change had mainly advantages in the visualization and did not add any significant advantage in the speed of convergence. The last result of each search was computed in CFX, added to the set of training data, and the neural network was trained again on several hundred iterations. After the last five variations were very close to each other, the optimization process was stopped (see Figure 28).

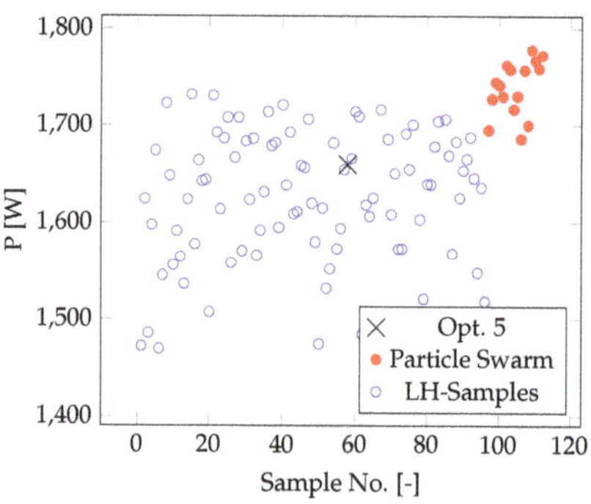

Figure 27. Scatter plot of the power achieved for all samples and the designs generated by the PSA.

Figure 28. PSA finds a maximum on the response surface visualized in two dimensions.

The final blade design achieved a performance of 1772.5 W, which was 6.8% more efficient than the optimum of the design actually developed for the five-blade prototype (1659.7 W). Notable was the increased curvature of the profile top, as shown in Figure 29. Highly curved profiles are used in aircraft designs for slow-moving aircraft, which require a high lift coefficient. The increased lift could probably also be converted into additional energy in the turbine. The large leading edge radius on the front also provided an improved flow separation behavior [18], thus the blade had more time to generate lift during its continuous turning.

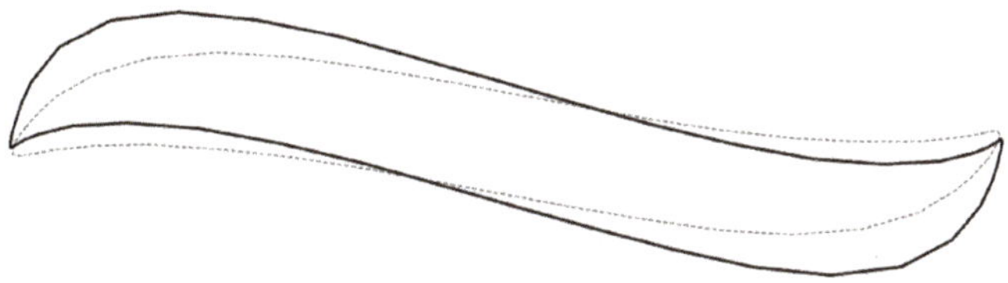

Figure 29. Final blade design compared to the old optimum (dashed).

Figure 30 shows that only the front of the turbine that is facing the flow was involved in the actual power generation. On the back side of the machine, the blade pitch in combination with the passage through the detached flow led to a significantly negative torque contribution. This undesired effect is a fundamental issue. A low tip speed ratio in vertical axis machines led to a large blade width that blocked the flow through the machine itself and created larger areas of detached flow. The effect worsened for the Kirsten–Boeing turbine design when the angle of attack was tweaked by the φ_{phase} parameter. Because of the global nature of that specific mechanism, all blades were turned either clock- or anticlockwise at once, which resulted in an increased angle of attack on one side of the machine, while the angle on the other side decreased. The negative angle of attack yielded a negative torque contribution. However, since less energy was converted on the leeward side due to the large area of detached flow, there was nevertheless a net advantage for the adjustment of the blade angle.

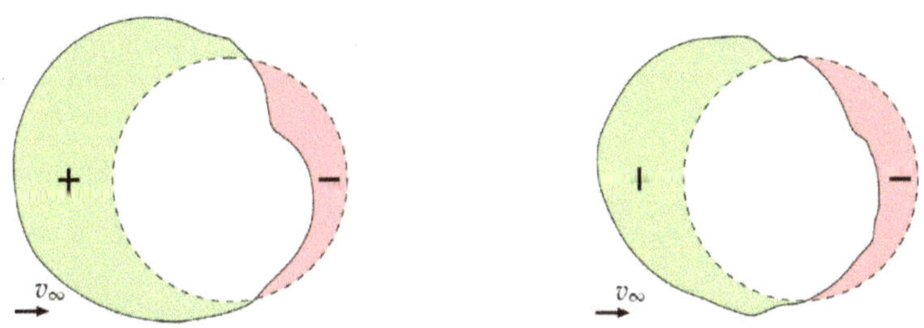

(**a**) Four blade optimizied design (**b**) Five blade unoptimizied design

Figure 30. Qualitative torque distribution around the circumference. The optimized blade profile leads to an increased torque generation windward while keeping the losses leeward roughly the same.

7. Conclusions

To solve a variety of optimization problems related to the Kirsten–Boeing turbine, comprehensive 3D flow simulations were made in ANSYS CFX. The tests included a general study of all parameters of the turbine, such as the number of blades, blade width and the optimization of the blade profiles themselves. It was found that the number of blades only plays a subordinate role in terms of performance. However, due to the low tip speed ratio, the blade width should be designed as large as possible. Through this realization, the structure of the turbine has changed fundamentally, since now four instead of five blades can be used. In addition to a slightly higher power output, this design also results in significantly reduced material consumption since fewer blades also mean fewer gears and bearings in the gearbox.

After calculating some intuitively hand-made blade designs, it was found that blade tip curvature has a significant impact on the more efficient use of lift forces. As part of these considerations, a point symmetric blade design was set up parametrically to serve as a blueprint for a large number of similar blades. The possibility to freely modify the blade shape allowed identifying and optimizing relevant parameters by means of statistical experimental design. From several hundred randomly generated blade designs calculated in CFX, neural networks were used to generate a metamodel. Within the higher-dimensional response surface given by the model, it was possible to find an optimum with a simple *Particle Swarm Algorithm*, which is superior to all previous blade designs. Overall, a substantial and validated performance increase of nearly 36% was achieved in the laboratory channel. The optimization process shown here can also easily be applied to similar problems, be it the optimization of individual profiles or entire machines.

The machine efficiency could not be validated in free flow conditions. However, the performed 3D simulations and the experiments in the laboratory channel hinted at an efficiency of 25–30% (see Section 4.3). This means that the Kirsten–Boeing turbine is positioned between the two other vertical axis turbines, Savonius ($c_{p,max} = 0.15$) and Darrieus ($c_{p,max} = 0.4$ [19]), in terms of efficiency. In terms of complexity, it is above the alternatives, which in addition to the high production costs also means difficulties in scaling to large diameters. When the overall machine diameter is increased, the blade weight increases significantly faster than with a comparable Darrieus turbine. It has been shown nevertheless that CFD is a suitable tool for simulating and optimizing even complex turbomachines where transient effects are decisive. However, due to the fundamentally limited efficiency of the design itself, the economical use of a Kirsten–Boeing turbine remains unlikely.

Author Contributions: Conceptualization, J.-P.K.; Data curation, J.M.; Formal analysis, J.-P.K.; Funding acquisition, J.M. and J.J.; Investigation, J.-P.K. and J.M.; Methodology, J.-P.K.; Project administration, J.J. and T.R.; Supervision, J.J. and T.R.; Visualization, J.-P.K.; Writing—original draft, J.-P.K.; and Writing—review and editing, J.-P.K., J.M. and T.R.

Funding: The presented research is part of the project "StECon-Infra", funded by the European Regional Development Fund (ERDF), project funding number: ERDF-0800555. Scientific supervision by the LeitmarktAgentur.NRW.

Ministerium für Wirtschaft, Energie, Industrie, Mittelstand und Handwerk des Landes Nordrhein-Westfalen

EFRE.NRW Investitionen in Wachstum und Beschäftigung

EUROPEAN UNION Investing in our Future European Regional Development Fund

Conflicts of Interest: The authors declare no conflict of interest.

Appendix A

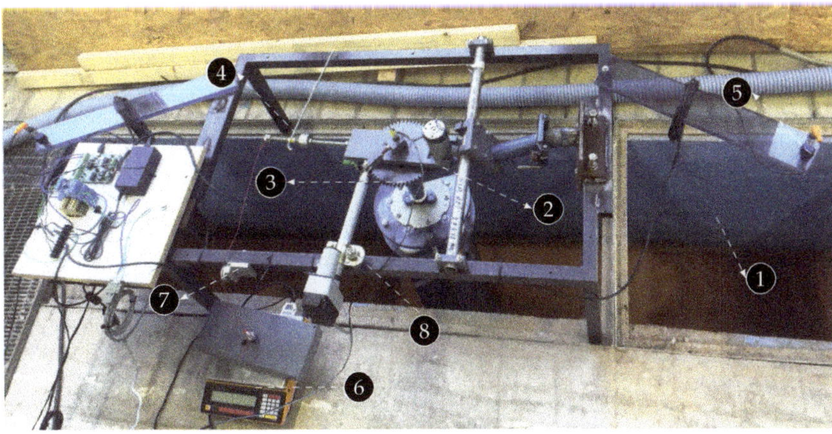

Figure A1. Presentation of the measuring equipment with placeholder blades: (1) duct wall; (2) generator with pinion; (3) turbine with connected main gear; (4) free-floating rod connected to the generator; (5) ultrasonic sensor for the water level; (6) scale with weight; (7) rigid rope which is guided over a pulley; and (8) linear motor for the phase adjustment. It should be noted that the generator only serves as a brake to make the mechanical power output calculable.

References

1. Sachse, H. *Kirsten-Boeing Propeller*; Technical Report; National Advisory Committee for Aeronautics: Washington, DC, USA, 1926.
2. Voith. Schaufelrad mit beweglichen Schaufeln. Available online: https://patents.google.com/patent/DE500340C/zh (accessed on 13 March 2019).
3. Miller, R.H. *Energietechnik und Kraftmaschinen 6, Techniklexikon*; Rowohlt Taschenbuch-Verlag: Hamburg, Germany, 1972.
4. Gasch, R.; Twele, J.; Bade, P.; Conrad, W.; Heilmann, C.; Kaiser, K.; Kortenkamp, R.; Kühn, M.; Langreder, W.; Liersch, J.; et al. *Auslegung von Windturbinen nach Betz und Schmitz*; Vieweg+Teubner Verlag: Wiesbaden, Germany, 2005.
5. Jensen, J.; Wieland, J.; Schmidt, J.; Metzger, J.; Stiller, H.L. *BMWi-Projekt StEwaKorad*; Technical Report; Lehrstuhl für Hydromechanik, Binnen- und Küstenwasserbau: Siegen, Germany, 2016.
6. Menter, F. Two-Equation Eddy-Viscosity Turbulence Models for Engineering Applications. *AIAA J.* **1994**, *32*, 1598–1605. [CrossRef]
7. Lanzafame, R.; Mauro, S.; Messina, M. 2D CFD Modeling of H-Darrieus Wind Turbines Using a Transition Turbulence Model. *Energy Procedia* **2014**, *45*, 131–140. [CrossRef]
8. Bezier, P. *Numerical Control; Mathematics and Applications*; John Wiley and Sons Ltd.: Hoboken, NJ, USA, 1972.
9. Box, G.; Draper, N. *Empirical Model-building and Response Surfaces*; John Wiley and Sons Inc.: Hoboken, NJ, USA, 1987.
10. LeDell, E. Mulitlayer Perceptron. Available online: https://github.com/ledell/sldm4-h2o/blob/master/sldm4-deeplearning-h2o.Rmd (accessed on 13 March 2019).
11. Nielsen, M. How the Backpropagation Algorithm Works. Available online: http://neuralnetworksanddeeplearning.com/chap2.html (accessed on 13 March 2019).
12. Google. An Open-Source Machine Learning Framework for Everyone. Available online: https://www.tensorflow.org (accessed on 13 March 2019).
13. Chrisman, L. LATIN-hypercube vs. Monte-Carlo-Sampling. Available online: http://www.lumina.com/blog/latin-hypercube-vs.-monte-carlo-sampling (accessed on 13 March 2019).
14. Nair, V.; Hinton, G. *Rectified Linear Units Improve Restricted Boltzmann Machines*; Technical Report; ICML: Toronto, ON, Canada, 2010.

15. Huber, P.J. Robust Estimation of a Location Parameter. *Ann. Stat.* **1964**, *35*, 73–101. [CrossRef]
16. Most, T.; Will, J. *Sensitivity Analysis using the Metamodel of Optimal Prognosis*; Technical Report; Dynardo: Weimar, Germany, 2011.
17. Kennedy, J.; Eberhart, R. Particle Swarm Optimization. Available online: https://ieeexplore.ieee.org/document/488968 (accessed on 13 March 2019).
18. Schneider, H. Profil: Charakteristische Geometrische Größen. Available online: http://wiki.rc-network.de/index.php/Profil:_Charakteristische_geometrische_Gr%C3%B6%C3%9Fen (accessed on 13 March 2019).
19. Bayeul-lainé, A.C.; Bois, G. Unsteady simulation of flow in micro vertical axis wind turbine. In Proceedings of the 21st International Symposium on Transport Phenomena, Kaohsiung City, Taiwan, 2–5 November 2010.

energies

MDPI

Article

Proposal of a Novel Semi-Submersible Floating Wind Turbine Platform Composed of Inclined Columns and Multi-Segmented Mooring Lines

Zhenqing Liu [1], Qingsong Zhou [1,*], Yuangang Tu [1], Wei Wang [2] and Xugang Hua [3]

1 School of Civil Engineering and Mechanics, Huazhong University of Science and Technology, Wuhan 430074, China; liuzhenqing@hust.edu.cn (Z.L.); tyg@hust.edu.cn (Y.T.)
2 Department of Architecture and Building Engineering, Tokyo Institute of Technology, Yokohama, Kanagawa 1528550, Japan; wang.w.aj@m.titech.ac.jp
3 Key Laboratory for Bridge and Wind Engineering of Hunan Province, College of Civil Engineering, Hunan University, Changsha 410082, China; cexghua@hnu.edu.cn
* Correspondence: zhouqingsong@hust.edu.cn

Received: 12 April 2019; Accepted: 8 May 2019; Published: 12 May 2019

Abstract: The semi-submersible floating offshore wind turbine has been studied in detail due to its good stability. However, the occurrence of typhoons are very frequent in China's offshore area, putting forward a higher requirement for the stability of the floating wind turbine system. By changing the connection mode of the mooring line as well as the structural form of the platform based on the original OC4 model, two groups of models were examined by an in-house developed code named as the Analysis Tool of Floating Wind Turbine (AFWT). The influence of the arrangement of the mooring lines and the inclination angle of the upper columns on the motion response were clarified. It was found that the surge motion of the platform would be obviously decreased by decreasing the length of the upper segments of the mooring lines, while the heave motion of the platform would be significantly decreased as increasing the inclined angle of the columns. Therefore, a new model integrating the optimized multi-segmented mooring lines and the optimized inclined columns was proposed. The examinations showed that compared with the response motions of the original OC4 semi-submersible model, the proposed model could reduce both the surge and heave motions of the platform effectively.

Keywords: FOWT; multi-segmented mooring line; inclined columns; semi-submersible; AFWT

1. Introduction

Wind energy continues to receive more and more attention due to its renewable and non-polluting advantages. From the beginning of the 21st-Century, wind power has been continuously developed around the world, making crucial contributions to dealing with global warming [1–3]. The on-shore wind energy has developed rapidly in the recent decades. However, on-shore wind power accounts for a large amount of land resources, and in some countries, such as China, the on-shore wind energy is always far away from the cities with high demand for electricity. Furthermore, studies about wind turbine noise (WTN) found that WTN has the indirect health effects, such as sleep disturbance and annoyance [4–8], which can be overcome by offshore wind turbines to some extent. In addition, considering the advantages that the offshore wind energy is close to the developed cities and the offshore wind speed is relatively high in Chinese seas, developing offshore wind energy has become an inevitable trend. Moreover, with the increase of water depth in the construction area of offshore wind farms, in order to ensure the operation of the wind turbine and reduce the construction cost, the foundation of the offshore wind turbine develops from the traditional fixed type to the floating type.

Several offshore wind turbine concepts have been studied to explore the feasibility using offshore wind energy. In shallow to medium water depths below 60 m, conventional bottom support monopiles, tripods, and jacket structures can be utilized. However, in deeper waters, floating platforms may be more feasible and economical as support structures [9]. Additionally, there are several widely used floating platforms, including spar type [10–20], tension leg platform (TLP) type [21–23], and semi-submersible type [24–28]. Among these platforms, the semi-submersible foundation tends to rely mainly on the balance of its own gravity and buoyancy to maintain vertical stability and the mooring system to ensure the stability of the system in the other directions. Therefore, the semi-submersible wind turbines have become a popular floating platform type [29].

Many studies have aimed at reducing the motions of the semi-submersible platform. It has not only been studied numerically [30–33], but also through experiments [34–37]. For instance, Zambrano et al. [38] proposed a MiniFloat concept for a semi-submersible platform, capable of supporting three wind turbines. Meanwhile, a coupled motion model, including the floating body and the anchor chain, was established. However, the cost was found to be too high to be adopted in the actual application. Shimada et al. [39] put forward a semi-submersible structure with three wind turbines. They carried out a 1:150 model test and the experimental results indicated that the dynamic response of the platform could be reduced effectively. However, the whole system was extremely large, increasing the difficulties for construction. Robertson et al. [40] proposed the well-known OC4 semi-submersible floating system which was extensively examined [41–43]. This concept has several advantages. Firstly, its static and dynamic stability provides sufficiently low pitch performance enabling the use of commercial offshore wind turbines; second, its design and fabrication methodology allow for onshore assembly of the complete system including the wind turbine. Then, several floating platform concepts similar to the OC4 semi-submersible floating system were proposed, such as WindFloat by Roddier et al. [44], Tri-Floater by Huijs et al. [45], and the fish-farming cage type by Zheng et al. [46].

In the design phase, numerical simulation is an important way examining the system behavior. Jonkman et al. [47] developed a numerical code to calculate the dynamic performances of the floating platforms based on both the potential flow theory and the Morison equations, by which Ormberg et al. [48] examined several floating wind turbine concepts. Then, Karimirad et al. [49] further developed a code and carried out a comprehensive numerical study for the spar floating wind turbines with catenary and tension mooring systems, considering the effects of aerodynamic and hydrodynamic damping under extreme environmental conditions. The National Renewable Energy Laboratory (NREL) developed a code named as FAST (Fatigue, Aerodynamics, Structures, and Turbulence), which was adopted in many researches [50,51]. At the same time, Borg et al. [26] developed a code especially for floating vertical axis wind turbines (VAWTs) and used it for calculating a series of complex simulation load cases.

However, the occurrence of typhoons is very frequent in the China offshore area, putting forward a higher requirement for the stability of the floating wind turbine system. Therefore, based on the original OC4 semi-submersible model, a new semi-submersible floating wind turbine platform is proposed to improve the stability of the system in the present study. The mooring system adopts multi-segmented mooring lines, while the platform model consists of three vertically inclined columns. The motion responses of the platform under different load conditions are calculated by AFWT. The basic calculation theories are introduced in Section 2. Section 3 introduces the main parameters of each part of the original model firstly, and then it will demonstrate the specific parameters of the modified models. In Section 4, AFWT will be validated by comparing with the experiments. After that, the platform motion responses of each model under regular wave and irregular wave conditions will be studied and compared with the original OC4 model.

2. Numerical Calculation Theory

Compared to traditional onshore wind turbines, the floating wind turbines are subject to more complex environmental loads, including aerodynamic loads on the turbine rotor, hydrodynamic loads, and mooring forces on the floating platform. The aerodynamic force is transmitted to the floating platform through the tower, and the mooring system provides restoring force to the platform, which affects the movement of the platform. The calculation theory of the aerodynamics, mooring system, and platform will be briefly introduced below.

2.1. Dynamic Coupling Process

AFWT employed in this paper consists of the models of aerodynamics, hydrodynamics, and structural dynamics. Additionally, the inflow wind, waves, and currents are considered in this paper as the external conditions, where the blade element moment (BEM) theory is used to calculate the aerodynamic loads on the blade, and the hydrodynamic loads are calculated by the potential theory. Figure 1 shows the main modules and their coupling relationship in AFWT. In AFWT, the wind turbine tower and platform are considered as a rigid body, which is calculated by multi-body dynamics method. In the calculation process, the low-frequency response of six degrees of freedom (DOF) is analyzed, and the second-order effects of the platform as well as the wind turbine are ignored. The mooring system uses quasi-static multi-segment catenary theory to calculate the line shape and the forces on the mooring lines.

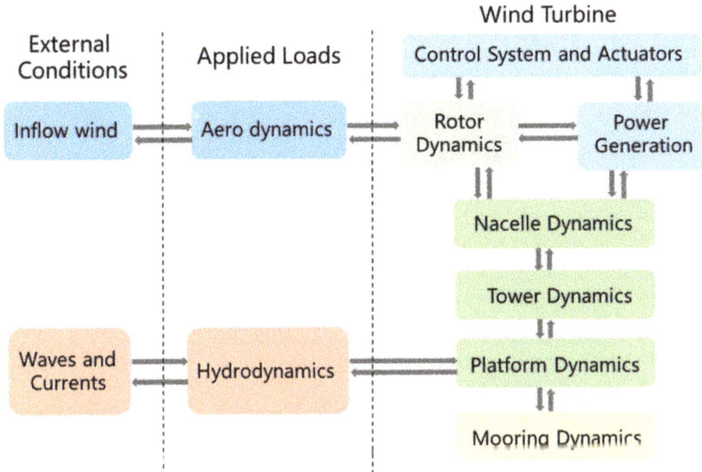

Figure 1. Sketch of the models in AFWT.

2.2. BEM Theory

The blades and towers of the wind turbine are subject to aerodynamic loads, and the BEM theory is used to calculate the aerodynamic loads of the blades. During the calculation process, the upper structure of the wind turbine has been modelled as a rigid body, and the calculated aerodynamic loads are transmitted to the whole structure, which can be calculated by BEM [52,53]. The normal force and the torque on the control volume of the thickness dr are expressed as follows:

$$dT = \frac{1}{2}\rho BC_n V_1^2 (1-a)^2 c \frac{dr}{sin^2\varphi} \tag{1}$$

$$dM = \frac{1}{2}\rho BC_t V_1 (1-a)\omega r(1+a')c \frac{rdr}{sin\varphi cos\varphi} \tag{2}$$

where ρ represents the air density, B is the number of the blades, V_1 is the wind speed, c is the blade chord length, φ represents the relation flow angle, equaling to the blades pitch angle adding the angle of attack, and C_n and C_t are the coefficients for the normal force and thrust force, which can be calculated by the lift coefficient C_l and drag coefficient C_d, respectively.

$$C_n = C_l cos\varphi + C_d sin\varphi \tag{3}$$

$$C_n = C_l cos\varphi + C_d sin\varphi \tag{4}$$

a and a' are the axial and tangential induction factor, respectively. They can be calculated using the expression below:

$$a = \frac{1}{\frac{4sin^2\varphi}{\sigma C_n} + 1},\ a' = \frac{1}{\frac{4sin\varphi cos\varphi}{\sigma C_t} - 1} \tag{5}$$

where σ is the fraction of the annular area in a control volume.

2.3. Mooring System Theory

Quasi-static theory [49] is applied to calculate the line shape as well as the tension of the mooring lines. The analytical equation for a single mooring line between two fixed points is expressed in Equations (6) and (7) and the key parameters are shown in Figure 2.

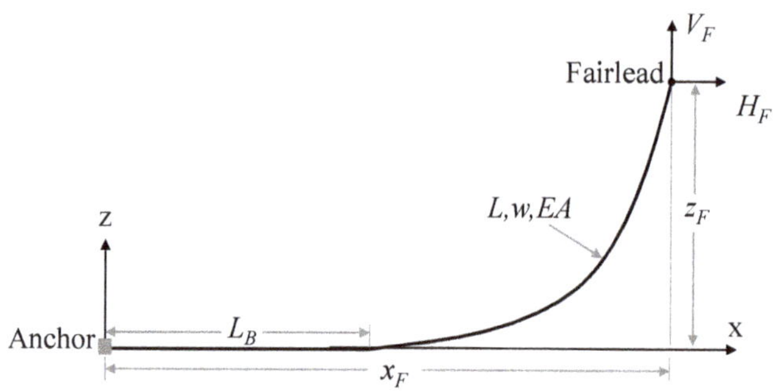

Figure 2. Local coordinate system of the mooring lines.

When a portion of the line lays on the seabed [54,55]:

$$x_F(H_F, V_F) = L_B + \frac{H_F}{w} ln\left[\frac{V_F}{H_F} + \sqrt{1 + \left(\frac{V_F}{H_F}\right)^2}\right] + \frac{H_F L}{EA} + \frac{C_B w}{2EA}\left[-L_B^2 + \left(L_B - \frac{H_F}{C_B w}\right) MAX\left(L_B - \frac{H_F}{C_B w}, 0\right)\right] \tag{6}$$

$$z_F(H_F, V_F) = \frac{H_F}{w}\left[\sqrt{1 + \left(\frac{V_F}{H_F}\right)^2} - \sqrt{1 + \left(\frac{V_F - wL}{H_F}\right)^2}\right] + \frac{1}{EA}\left(V_F L - \frac{wL^2}{2}\right) \tag{7}$$

where x_F and z_F are, respectively, the horizontal and vertical coordinates of the fairlead position relative to the anchor. H_F and V_F are, respectively, the horizontal and vertical components of the effective tension in the mooring line at the fairlead, w represents the mass of the mooring line per unit length, L is the total unstretched mooring line length, and EA represents the sectional stiffness of the mooring line, C_B is the coefficient of static friction between the seabed and the mooring line, $L_B = L - \frac{V_F}{w}$ is the length of catenary relaxation on the seabed. Equations (6) and (7) are solved iteratively in the local coordinate system using methods such as the Newton–Raphson method.

Subsequently, the line shape and tension of the catenary can be calculated using Equations (8)–(11), when a portion of cable is relaxing on the seabed:

$$x(s) = \begin{cases} s \text{ for } 0 \le s \le \left(L_B - \frac{H_F}{C_B w}\right) \\ s + \frac{C_B w}{2EA}\left[s^2 - 2s\left(L_B - \frac{H_F}{C_B w}\right) + \left(L_B - \frac{H_F}{C_B w}\right)\lambda\right] \text{ for } \left(L_B - \frac{H_F}{C_B w}\right) < s \le L_B \\ L_B + \frac{H_F}{w} ln\left[\frac{w(s-L_B)}{H_F} + \sqrt{1+\left(\frac{w(s-L_B)}{H_F}\right)^2}\right] + \frac{H_F s}{EA} + \frac{C_B w}{2EA}\left[\lambda\left(L_B - \frac{H_F}{C_B w}\right) - L_B^2\right] \text{for } L_B < s \le L \end{cases} \tag{8}$$

$$z(s) = \begin{cases} 0 \text{ for } 0 \le s \le L_B \\ \frac{H_F}{w}\left[\sqrt{1+\left(\frac{w(s-L_B)}{H_F}\right)^2} - 1\right] + \frac{w(s-L_B)^2}{2EA} \text{ for } L_B < s \le L \end{cases} \tag{9}$$

$$T_e(s) = \begin{cases} MAX[H_F + C_B w(s-L_B),\ 0] \text{ for } 0 \le s \le L_B \\ \sqrt{H_F^2 + [w((s-L_B))]^2} \text{ for } L_B < s \le L \end{cases} \tag{10}$$

$$\lambda = \begin{cases} L_B - \frac{H_F}{C_B w} \text{ if } \left(L_B - \frac{H_F}{C_B w}\right) > 0 \\ 0 \text{ otherwise} \end{cases} \tag{11}$$

where s is the length from the catenary point to the anchor, x and z are the lengths in the horizontal and vertical directions, respectively, and T_e is the line tension. Similarly, H_A and V_A are the horizontal and vertical components of the effective tension in the mooring line at the anchor, respectively. They can be calculated as $H_A = MAX\ [H_F + C_B w L_B, 0]$ and $V_A = 0$.

The mooring system of the modified models is specially connected by multi-segment catenaries. Thus, to obtain the analytical solution of the multi-segment catenary equations, the coordinate transformation is carried out. Additionally, the analytical model is converted from the local xz coordinate system to the global XYZ coordinate system. In addition, the unknowns for each line are solved in the local xz frame, and the static equilibrium is ascertained by solving for line properties, resulting in zero sum forces on the nodes in the XYZ frame.

In the multi-segmented mooring lines in Figure 3, the coordinates of nodes 1, 3, and 4 are determined. In order to find the coordinates of the connecting node 2 and the force at each point, the force on the node 2 must satisfy the following force balance equation:

$$\{H_{1u}\}_X + \{H_{2l}\}_X + \{H_{3l}\}_X = 0 \tag{12}$$

$$\{H_{1u}\}_Y + \{H_{2l}\}_Y + \{H_{3}l\}_Y = 0 \tag{13}$$

$$V_{1u} + V_{2l} + V_{3l} = F_{ext} \tag{14}$$

where H_{iu}, V_{iu}, H_{il}, and V_{il} are the magnitudes of horizontal ("H") and vertical ("V") forces at the upper node ("$_u$") and lower node "$_l$" of line i, respectively. F_{ext} is the external force at the connection node 2, such as the buoyancy module or the node weight. In this particular application, $F_{ext} = 0$. Then we can determine the line shapes and tensions of each catenary according to Equations (8)–(11). More detailed introduction can be found in the study by Masciola et al. [55].

The force of the sub-mooring line is calculated by the above-mentioned single-line catenary theory under local coordinate system. The numerical value of force in the global coordinate system can be obtained by the coordinate conversion matrix.

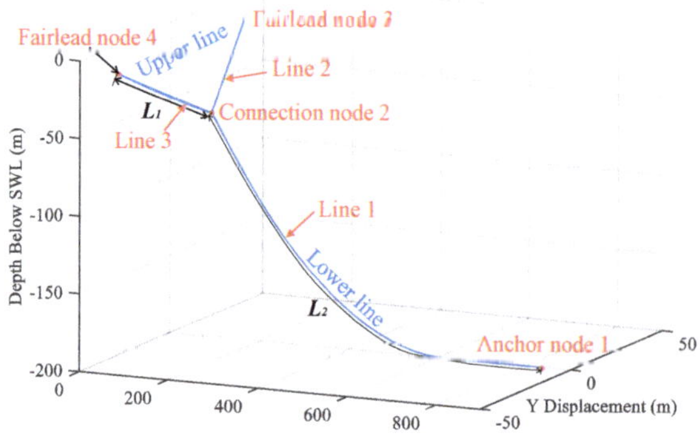

Figure 3. Connection form of the multi-segment mooring lines.

2.4. Hydrodynamic Theory

The research by Jonkman [47,56] suggested two theories for calculating the hydrodynamic loads. One is the Morison's equation and the other is the potential flow theory. Morison's equation is applicable for calculating the hydrodynamic loads on slender cylindrical structures when the effects of diffraction and radiation damping are negligible. In order to model the interaction between the waves and the structures in a more accurate way, the potential flow theory was adopted, which is capable of considering the forces induced by both the diffraction and the radiation. Therefore, in the present study the potential flow theory is applied. By defining a potential function $\phi(x, y)$, the governing equation for ϕ is:

$$\nabla^2\phi = 0 \tag{15}$$

where $\partial\phi/\partial x = u$, $\partial\phi/\partial y = v$, and $\partial\phi/\partial z = w$. u, v, and w are the velocity components of the flow fields in the Cartesian coordinates. After determining the boundary conditions at the free surface, the surface of the structures, and the sea bed, the potential function can be solved and the velocity distribution can be obtained.

2.5. Platform Motion Equation

To study the dynamic responses of a FOWT in the time domain, the platform was considered as a rigid body, with its motion from the equation given as [57]:

$$(M + A)\ddot{X}(t) + C\dot{X}(t) + KX(t) = F^{waves}(t) + F^{wind}(t) + F^{mooring}(t) + F^{vis}(t) \tag{16}$$

where $X(t)$ is the generalized displacement of the platform in the time domain; $\dot{X}(t)$ and $\ddot{X}(t)$ represent the generalized velocity and acceleration, respectively; M is the mass matrix; A and C are the added mass matrix and damping coefficient matrix, respectively, which are caused by the wave radiation; K represents the hydrostatic restoring force matrix. $F^{waves}(t)$ is the incident-wave induced force; $F^{wind}(t)$ is the wind loads acting on the blades and tower, $F^{mooring}(t)$ represents the mooring tension, and $F^{vis}(t)$ is the drag force caused by fluid viscosity.

The wave induced force, F^{waves}, is calculated by:

$$F^{waves} = \int_{-\infty}^{+\infty} \frac{1}{2\pi} W(\omega) \sqrt{2\pi B(\omega)} X_j(\omega, \beta) e^{-i\omega t} d\omega \tag{17}$$

where ω is the frequency of the incident wave; $W(\omega)$ is Fourier transform of a white noise time series with unit variance, $B(\omega)$ is the wave spectrum (P-M spectrum in this paper), $X_j(\omega, \beta)$ is the wave-induced force, and β is the incident wave direction angle.

The flow separation behind the structures will induce the viscous drag force, F_i^{vis}, which is calculated by:

$$F_i^{vis} = \int_0^{d_i} \frac{1}{2}\rho C_{dw} D(v_{w,i} - v_{s,i})\left|v_{w,i} - v_{s,i}\right| dz \tag{18}$$

where C_{dw} is the drag coefficient of water, $v_{w,i}$ is the velocity of water and $v_{s,i}$ is the velocity of structure, and h is the height of the structure.

The added mass matrix, A_{ij}, and damping coefficient matrix, C_{ij}, is calculated as:

$$A_{ij} = \rho_w Re\left\{\oiint \frac{\partial \phi_i}{\partial n}\phi_j dS\right\} \tag{19}$$

$$C_{ij} = \rho_w Im\left\{\oiint \frac{\partial \phi_i}{\partial n}\phi_j dS\right\} \tag{20}$$

where ϕ is the potential function, S is the mean body wetted surface, and ρ_w is the seawater density.

The hydrostatic restoring force matrix K_{ij}, affected by the wetted surface and the hydrostatic pressure, is defined as:

$$K_{33} = \rho_w g \oiint n_3 dS \tag{21}$$

$$K_{34} = \rho_w g \oiint n_3 y_b dS \tag{22}$$

$$K_{35} = -\rho_w g \oiint n_3 x_b dS \tag{23}$$

$$K_{44} = \rho_w g \oiint n_3 y_b^2 dS + \rho_w g \forall z_b - m g z_g \tag{24}$$

$$K_{45} = -\rho_w g \oiint n_3 x_b y_b dS \tag{25}$$

$$K_{46} = -\rho_w g \forall x_b + m g x_g \tag{26}$$

$$K_{55} = \rho_w g \oiint n_3 x_b^2 dS + \rho_w g \forall z_b - m g z_g \tag{27}$$

$$K_{56} = -\rho_w g \forall y_b + m g y_g \tag{28}$$

where all other values of $K_{ij} = 0$, g is the gravity acceleration equaling to 9.8 m s^{-2}. $\forall$ is the average of the submerged volume, defined as:

$$\forall = -\oiint n_1 x dS = -\oiint n_2 y dS = -\oiint n_3 z dS \tag{29}$$

and x_g, y_g, and z_g are the coordinates of the center of gravity of the structure, and x_b, y_b, and z_b are the coordinates of the center of buoyancy of the structure.

3. Models of the Semi-Submersible FOWT

The configuration of the original OC4 semi-submersible model along with the modified models will be introduced, and the load cases will be also summarized in this section.

3.1. Wind Turbine

The NREL offshore 5 MW baseline wind turbine [58] is a conventional three-blade upwind turbine, which is mounted on the top of semi-submersible floating platform. The tower base is located at 10 m high from the still water level (SWL). Table 1 summarizes the key properties for the wind speed and the mass distribution of the wind turbine.

Table 1. Selected properties of the NREL 5MW baseline wind turbine.

Turbine Properties	Value
Rotor configuration	Upwind, 3 Blades
Rotor diameter	126 m
Hub height above SWL	90 m
Cut-in wind speed	3 m/s
Rated wind speed	11.4 m/s
Cut-out wind speed	25 m/s
Rotor mass	110,000 kg
Nacelle mass	240,000 kg
Tower mass	347,460 kg

3.2. Floating Platform of Original OC4

The semi-submersible floating platform in OC4 consists of three main offset columns along with a central column used to support the wind turbine, and a series of diagonally intersecting and horizontal bracing components. The base columns are connected to the bottom of the upper columns to prevent the platform from generating excessive heave motions. The center column of the platform has a diameter of 6.5 m and a length of 30 m. The upper columns of the three main offset columns have a diameter of 12 m and a length of 26 m. Similarly, the base columns have a diameter of 24 m and a length of 6 m, which are connected by 1.6 m diameter cross-bracings [59–61]. Table 2 summarizes the main configurations of the platform. Meanwhile, Figure 4 illustrates the structure of the original OC4 model.

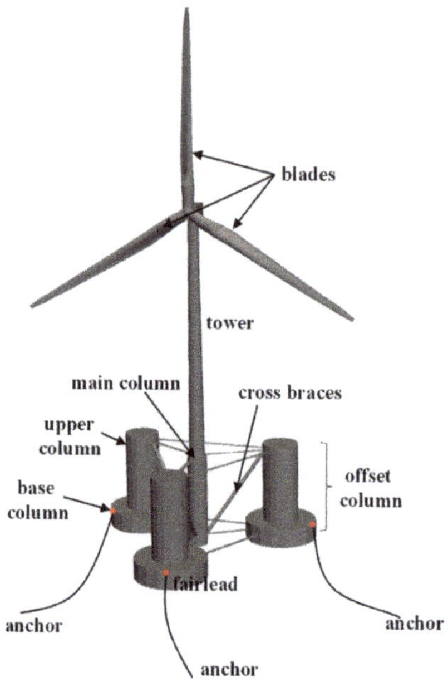

Figure 4. Sketch of the original OC4 model.

Table 2. Platform properties in OC4.

Items	Value
Depth to platform base below SWL (total draft)	20.0 m
Elevation to platform top (tower base) above SWL	10.0 m
Platform mass, including ballast	13,473,000 kg
Displaced volume	13,986.8 m^3
Center of mass (CM) location below SWL	14.4 m
Platform roll inertia about CM	8.011×10^9 kg m^2
Platform pitch inertia about CM	8.011×10^9 kg m^2
Platform yaw inertia about platform centerline	1.391×10^{10} kg m^2

3.3. Mooring System of Original OC4

The platform is moored by three mooring lines which are 120 degrees apart from each other. In the mooring system, the fairleads of the mooring lines are situated at the top of the base columns, 14.0 m deep below the SWL, and 40.87 m from the centerline of the platform. Table 3 shows the main characteristics of the mooring line.

Table 3. Key properties of mooring lines in OC4.

Mooring System Properties	Value
Depth to anchors below SWL (water depth)	200 m
Depth to fairleads below SWL	14 m
Radius to anchors from platform centerline	837.6 m
Radius to fairleads from platform centerline	40.87 m
Unstretched mooring line length	835.5 m
Mooring line diameter	0.0766 m
Equivalent mooring line mass density	113.35 kg/m
Equivalent mooring line mass in water	108.63 kg/m
Equivalent mooring line extensional stiffness	753.6×10^6 N
Seabed drag coefficient	1.0

3.4. Modifications of the Original OC4 Model

Two groups of modified models based on the original OC4 model will be examined. The first group applies multi-segmented mooring lines and the second group applies inclined columns. Figure 3 shows the mooring system adopted in the first group. Here, the mooring line is divided into three segments. Lines 2 and 3 are the upper lines from the connection node to the fairlead node, while line 1 is the lower line from the anchor to the connection node. Models 2, 3, and 4, as listed in Table 4, are generated by changing the length of the upper mooring line L_1. Furthermore, to ensure that the total weight of the mooring lines is a constant, the density of the upper line is changed to be 55.05 kg/m. The second group (Model 5 and Model 6), as listed in Table 4, is generated by changing the vertical upper columns in OC4 to be inclined upper columns. The tilt angles, θ, of the upper columns of Model 5 and Model 6 are 15° and 30°, respectively. The tilt angle is applied with respect to the base columns, and is positive for clockwise rotation with respect to the main column. Finally, Model 7, as an integration of Model 4 and Model 6, is proposed. Table 4 summarizes the sketch of the models and the key geometrical parameters of each model. It should be pointed out that Model 5 and Model 6 only change the tilt angle of the upper columns. The length, as well as the diameter, of the upper columns is unchanged. The length of the upper braces becomes longer but the weight of the upper braces is set to be a constant.

Table 4. Models examined in the present study.

Name	Sketch of the Models	L_1 (m)	θ (°)
Model 1 (original OC4 model)	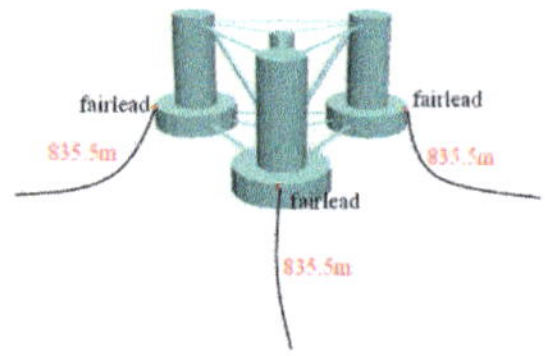	0	0
Model 2	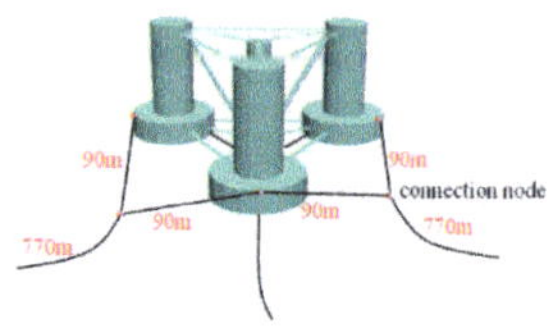	90	0
Model 3	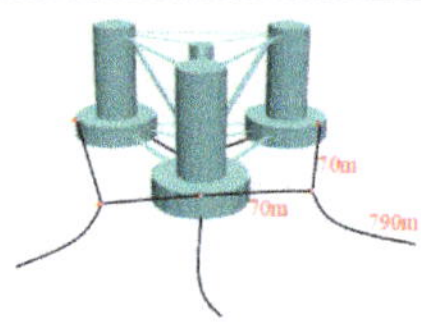	70	0
Model 4	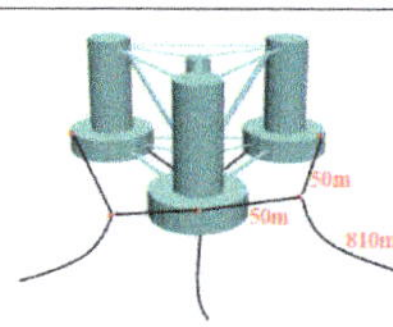	50	0
Model 5	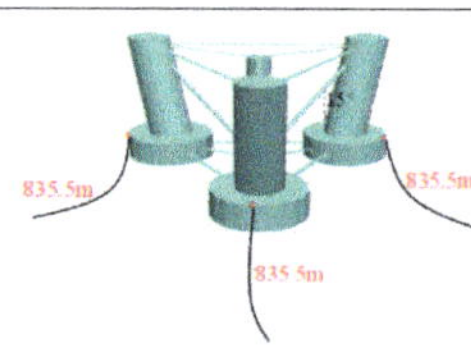	0	15
Model 6	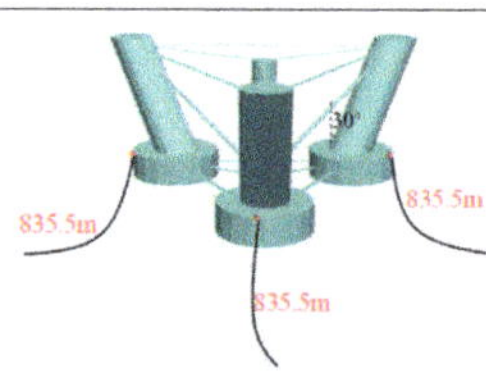	0	30
Model 7	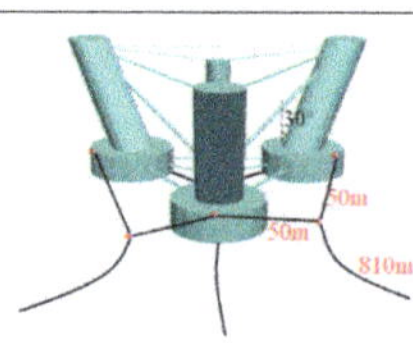	50	30

Mooring line pretensions for the different models are listed in Table 5. It can be seen from Table 5 that the pretension of Line 1 in the original Model 1 is 1.11×10^6 N. From Models 1–4, the pretension of Line 1 is gradually increased. The pretensions of Lines 2 and 3 are found to be very sensitive to the change of the length of the upper segment lines. The pretensions of Lines 2 and 3 in Model 4 are about twice as large as those in Model 1. Figure 5 shows the restoring force as a function of the displacement in each direction to calculate the mooring stiffness, which is listed in Table 6. The increase of the mooring stiffness in the surge direction from Model 1 to Model 4 is clear; however, there is almost no change for the stiffness in heave and pitch directions. The natural periods of each model determined by free decay tests are listed in Table 7, from which we can find that the natural period of the surge direction is decreased as the model changes from Model 1 to Model 4, due to the increase of the surge stiffness of the system. On the other hand, for the natural period in the heave direction, a clear increase can be found as change the model from Model 1 to Model 5 and Model 6, due to the fact that the more inclined columns increase the waterplane area which, in turn, increases the added mass of the system.

Table 5. Mooring line pretension of the different models.

	Line 1 (N)	Line 2 (N)	Line 3 (N)
Model 1	1.11×10^6	/	/
Model 2	1.19×10^6	6.68×10^5	6.68×10^5
Model 3	1.24×10^6	7.94×10^5	7.94×10^5
Model 4	1.32×10^6	1.25×10^6	1.25×10^6
Model 5	1.11×10^6	/	/
Model 6	1.11×10^6	/	/
Model 7	1.32×10^6	1.25×10^6	1.25×10^6

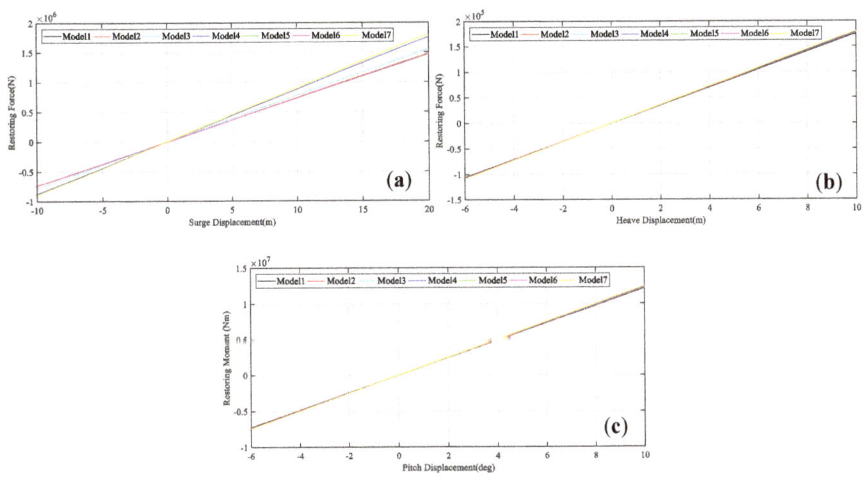

Figure 5. Restoring forces as a function of the displacement for determining the stiffness of the system in (**a**) surge direction; (**b**) heave direction; and (**c**) pitch direction.

Table 6. Mooring stiffness of each model.

	Surge (N/m)	Heave (N/m)	Pitch (Nm/deg)
Model 1	73,047	17,534	1.213×10^6
Model 2	73,676	17,681	1.219×10^6
Model 3	77,551	17,776	1.228×10^6
Model 4	87,832	17,912	1.239×10^6
Model 5	73,302	17,641	1.217×10^6
Model 6	72,986	17,682	1.223×10^6
Model 7	89,901	18,025	1.245×10^6

Table 7. Natural period of each model.

	Model 1	Model 2	Model 3	Model 4	Model 5	Model 6	Model 7
Surge(s)	109.4	108.9	106.2	98.7	109.2	109.4	97.6
Heave(s)	17.4	17.4	17.4	17.4	18.4	19.5	19.4
Pitch(s)	26.0	26.0	26.0	25.9	26.0	26.0	26.1

3.5. Load Cases

To study the performance of the models, the dynamic motions of these 7 models at different wave conditions are calculated. Table 8 lists the main parameters of load cases, where H is the wave height and T is the period for regular waves. Additionally, Hs is the significant wave height and Tp is the peak spectral period for irregular waves. More specifically, wind and waves are in the surge direction. Linear Airy wave theory is used in the generation of regular waves, and irregular waves are described by the Pierson Moskowitz spectrum.

Table 8. Load case definitions.

Load Case	Wave Condition	H or Hs (m)	T or Tp (s)	Wind Condition
LC1	Regular wave	2.44	6.2	Steady, uniform, V_{hub} = 8 m/s
LC2	Regular wave	5.32	8.4	Steady, uniform, V_{hub} = 8 m/s
LC3	Regular wave	7.56	10.6	Steady, uniform, V_{hub} = 8 m/s
LC4	Irregular wave	3.5	7.2	Steady, uniform, V_{hub} = 8 m/s
LC5	Irregular wave	5.5	9.4	Steady, uniform, V_{hub} = 8 m/s
LC6	Irregular wave	7.5	11.6	Steady, uniform, V_{hub} = 8 m/s

4. Results

In this section, the motion responses of the models under different load conditions are calculated by AFWT. The rigid body platform consists of six DOFs, including surge, sway, heave, roll, pitch, and yaw. In general, the surge, heave, and pitch responses of the platform are relatively apparent motions of the whole system. As such, only the surge, heave, and pitch motions of the platform will be studied. Firstly, AFWT is validated. After that, the code is used in calculating the motion of the models in different load conditions.

4.1. Validation of the Code

Before calculating the response of the platform, AFWT is first verified numerically. A free-decay test is carried out on the floating platform without wind or waves. Although there is no wind and the water is still, the radiated waves produced by the initial motion of the platform will still generate hydrodynamic loads on the platform and the mooring system. The details of the settings of the verification case can be found in the study of Liu et al. [62]. The motions in surge, heave and pitch directions in the time domain obtained by FAST and those by AFWT are shown in Figure 6, where satisfactory agreement is achieved. The platform's surge natural period is 109.4 s, the heave natural period is 17.4 s, and the pitch natural period is 26.0 s as determined by Fast Fourier Transformation (FFT).

Further verification of the code is compared with the available OC4 FOWT experimental data [63], as shown in Figure 7. The experimental scale model is carried out under the regular wave conditions with a wave height of 0.14 m and a wave period of 1.33 s. The code calculation results are in good agreement with the experimental data.

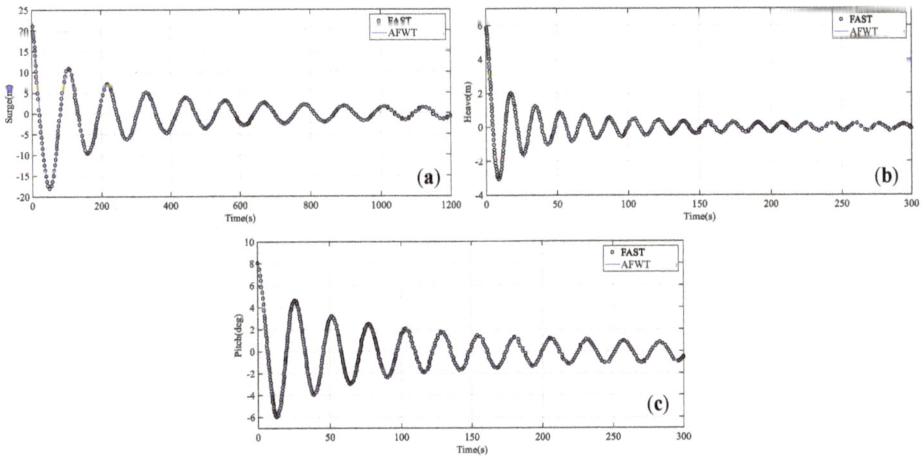

Figure 6. Dynamic response of the original Model 1 under the free decay condition calculated by FAST and AFWT: (**a**) surge decay; (**b**) heave decay; and (**c**) pitch decay.

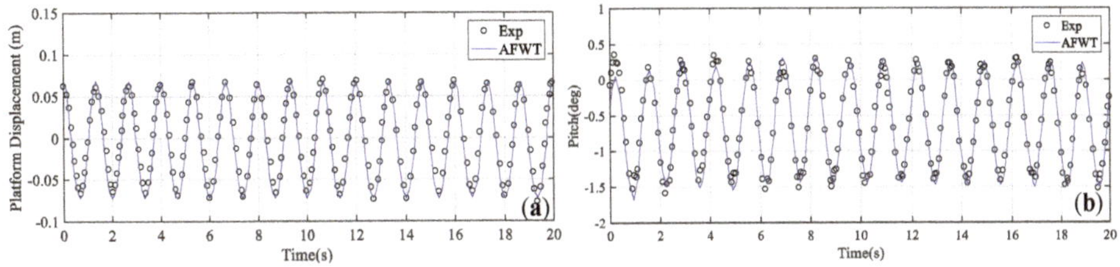

Figure 7. Comparison of the surge and pitch responses between the experiment and AFWT under a regular wave: (**a**) surge response; and (**b**) pitch response.

4.2. Multi-Segmented Mooring Line Models

4.2.1. Regular Waves

Figures 8–10 illustrate the signals of platform motions of the multi-segmented mooring line models (Model 2~4) under regular waves in time domain. As found in Figure 8, the surge motion of the models reaches the stable stage at about 400 s for all of the load conditions. In LC1, the peak value of the surge motion of the Model 1 is estimated to be 9.2 m before the stable stage and about 5.3 m after the stable stage. While, the peak value of the surge motion of the Model 4 is about 6 m before the stable stage and about 3 m after the stable stage. Not only for LC1, but also for LC2 and LC3, the reduction of the surge motion is obvious from Model 1 to Model 4. This demonstrates that the shorter the upper segment lines are, the greater there will be a reduction to the surge of the platform in the regular waves. Similar with the surge motion, the heave motion of the platform is also reduced from Model 1 to Model 4, as shown in Figure 9. However, the magnitude of the reduction is small, indicating that changing the length of the upper segment lines has little effects on the platform heave motion. For the pitch motions illustrated in Figure 10, there is no obvious changes from Model 1 to Model 4.

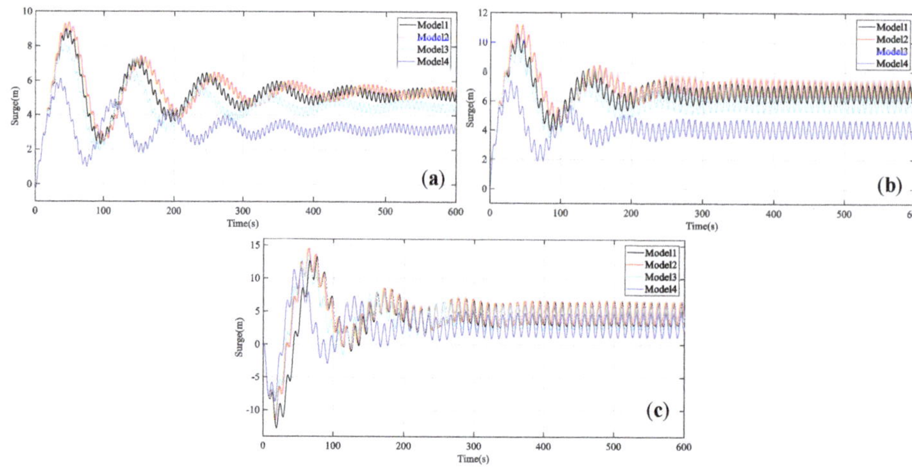

Figure 8. Time domain curves of the surge motion of Models 1~4: (**a**) LC1; (**b**) LC2; and (**c**) LC3.

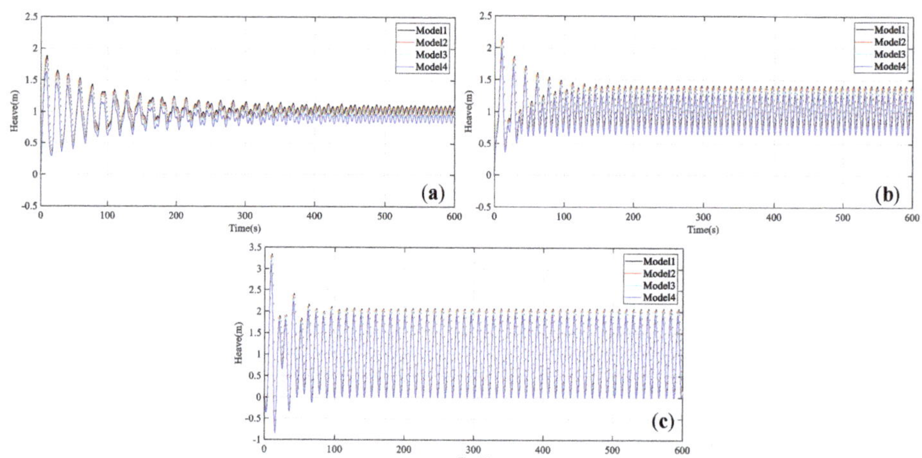

Figure 9. Time domain curves of the heave motion of Models 1~4: (**a**) LC1; (**b**) LC2; and (**c**) LC3.

Due to the computational start-up fluctuation, the representative statistics are calculated based on the 400–600 s signals and the results are listed in Table 9. Most obviously, the mean surge offset decreases from 6.12 m (Model 1) to 3.86 m (Model 4) in LC2, while the mean heave motions just decrease from 1.10 m (Model 1) to 0.96 m (Model 4), and the mean pitch of platform even increases from 1.85° (Model 1) to 2.05° (Model 4). Similarly, the maximum surge motion is also reduced enormously from Model 1 to Model 4, while the maximum heave and pitch motions are found to show a very small change. The stationary motion amplitudes of Models 1–4 are listed in Table 10. It can be seen that the surge amplitude gradually decreases from Model 1 to Model 4. However, the heave and pitch amplitudes are insensitive to the change of the length of the upper segment lines. Therefore, it can be concluded that the multi-segmented mooring line models are feasible to increase the stability of the platform under regular waves, especially for the surge motion. The increase of the stiffness in surge direction of the mooring system due to the reduction of the length of the upper segment lines should be the major reason. In addition, the introduction of multi-segmented lines also has a significant effect on the yaw stiffness of the structure, which can partially explain the reduction of surge motions due to the reduction of surge-yaw coupling.

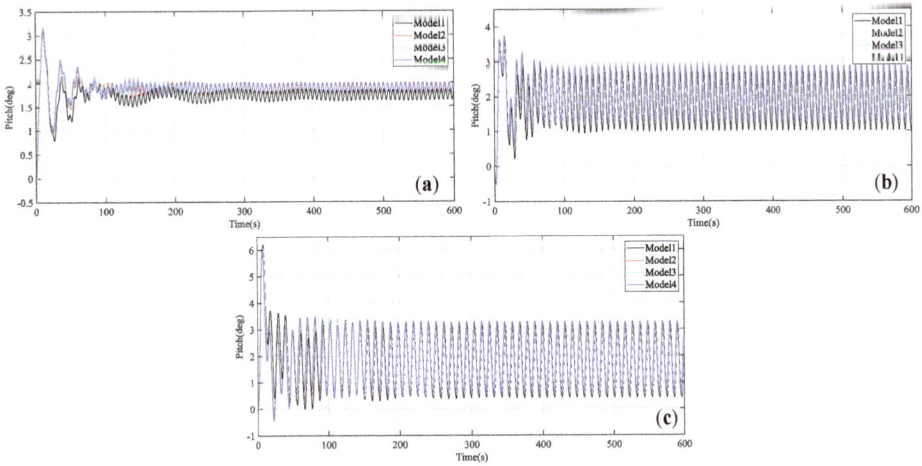

Figure 10. Time domain curves of the pitch motion of Model 1–4: (**a**) LC1; (**b**) LC2; and (**c**) LC3.

Table 9. Statistics of the stationary motion of the multi-segmented mooring line models.

DOF	Load Case	Model 1			Model 2			Model 3			Model 4		
		Max.	Min.	Mean	Max.	Min.	Mean	Max.	Min.	Mean	Max.	Min.	Mean
Surge (m)	LC1	6.03	4.97	5.47	6.04	5.08	5.57	5.28	4.41	4.84	3.76	3.01	3.38
	LC2	7.20	5.05	6.12	7.56	5.52	6.51	6.54	4.61	5.58	4.74	3.03	3.86
	LC3	6.67	2.63	4.66	6.91	3.09	4.89	6.06	2.35	4.09	4.97	1.44	3.11
Heave (m)	LC1	1.16	0.93	1.04	1.13	0.91	1.02	1.10	0.88	0.98	1.01	0.79	0.90
	LC2	1.41	0.79	1.10	1.39	0.77	1.08	1.35	0.74	1.05	1.27	0.65	0.96
	LC3	2.07	0.14	1.10	2.04	0.12	1.08	2.01	0.08	1.05	1.92	0.01	0.96
Pitch (°)	LC1	1.87	1.60	1.74	2.01	1.76	1.88	2.02	1.77	1.89	2.03	1.77	1.90
	LC2	2.69	1.01	1.85	2.88	1.21	2.04	2.89	1.22	2.05	2.90	1.22	2.05
	LC3	3.16	0.40	1.76	3.31	0.51	1.88	3.31	0.52	1.89	3.31	0.53	1.90

Table 10. Amplitudes of the stationary motion of the multi-segmented mooring line models.

DOF	Load Case	Model 1	Model 2	Model 3	Model 4
Surge (m)	LC1	1.06	0.96	0.87	0.75
	LC2	2.15	2.04	1.93	1.71
	LC3	4.04	3.82	3.71	3.53
Heave (m)	LC1	0.23	0.22	0.22	0.22
	LC2	0.62	0.62	0.61	0.62
	LC3	1.93	1.92	1.93	1.91
Pitch (°)	LC1	0.27	0.25	0.25	0.26
	LC2	1.68	1.67	1.67	1.68
	LC3	2.76	2.80	2.79	2.78

4.2.2. Irregular Wave

The motion response of the multi-segmented mooring line models is further studied under irregular wave conditions. Since the time domain curves of the motion responses under irregular waves are random, it is difficult to present the result clearly using the time domain signals. As such, the results are discussed in the frequency-domain obtained by transforming the time-domain signal to the frequency-domain signal through FFT [64]. The total time of the simulations is 1400 s and the time step is 0.1 s. The motion data in the first 400 s is deleted to remove the transient response, then the mean value of the motions in the remaining 1000 s was obtained. Lastly, removing the mean value and FFT was carried out to obtain the spectra of motions in the frequency domain.

As shown in Figure 11, the first peak of the platform surge motion appears at about 0.01 Hz, close to the natural frequency of the platform. And it can be seen in Figure 11 that the surge spectral peak frequency has a slight shift from Model 1 to Model 4, due to the increased surge stiffness. As the wave frequency decreases, two peaks in surge power spectral density (PSD) occur, with one corresponding to the structure frequency and the other corresponding to the wave frequency. In addition, it is interesting that with changing the wave frequency close to the natural frequency of the system, the additional motions with the frequency close to the wave frequency will be enhanced. Also, in the frequency range 0~0.03 Hz, the surge PSD curves of Model 4 are found to be the lowest under LCs 4, 5, and 6, consistent with the result of regular wave in the previous section. However, similar to the findings in the regular wave examinations, there is no obvious effect on the heave and pitch motions with changes in the length of the upper segment lines. Furthermore, there is a trend that, as the wave frequency close to the natural frequency of the system changes, the additional motions with the frequency close to the wave frequency is also obviously found in both heave and pitch motions (Figures 12 and 13).

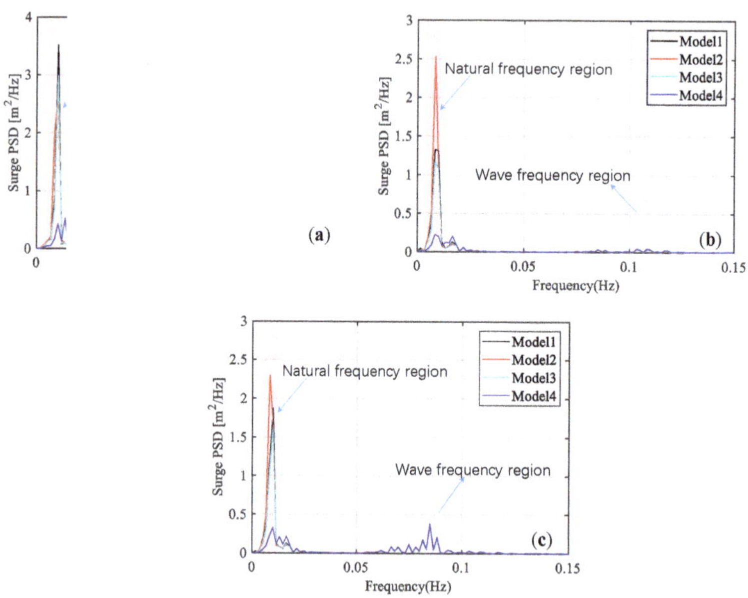

Figure 11. Frequency domain curves of the surge PSD of the multi-segmented mooring line models: (**a**) LC4; (**b**) LC5; and (**c**) LC6.

Table 11 summarizes the first peaks of the surge, heave, and pitch PSD of the multi-segmented mooring line models in the frequency domain along with their corresponding frequencies. The peak frequencies of the platform response are basically the same from Model 1 to Model 4 and exhibit a small difference from LC4 to LC6. The most obvious reduction of the surge motion is found in LC4, where the peak value of the surge PSD is reduced from 3.516 m^2/Hz (Model 1) to 0.520 m^2/Hz (Model 4). For the other load conditions, the surge PSDs also show apparent reductions. However, the reduction of the peak heave and surge PSDs are very limited. The motion standard deviations of the multi-segmented mooring line models under irregular waves are listed in Table 12, where the standard deviation of the surge motion is found to be decreased as decreasing the length of the upper segment lines and the effects seem to be more obvious as increasing the wave frequency.

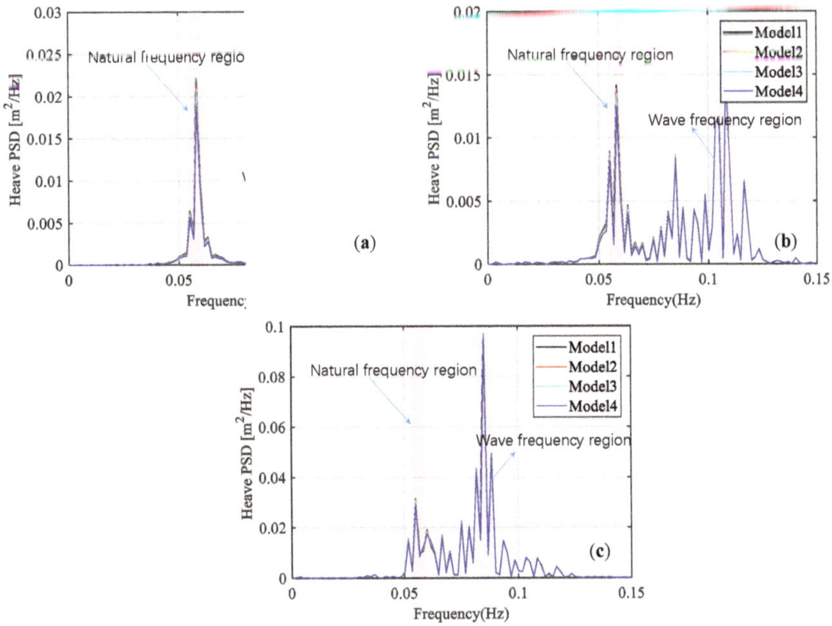

Figure 12. Frequency domain curves of the heave PSD of the multi-segmented mooring line models: (**a**) LC4; (**b**) LC5; and (**c**) LC6.

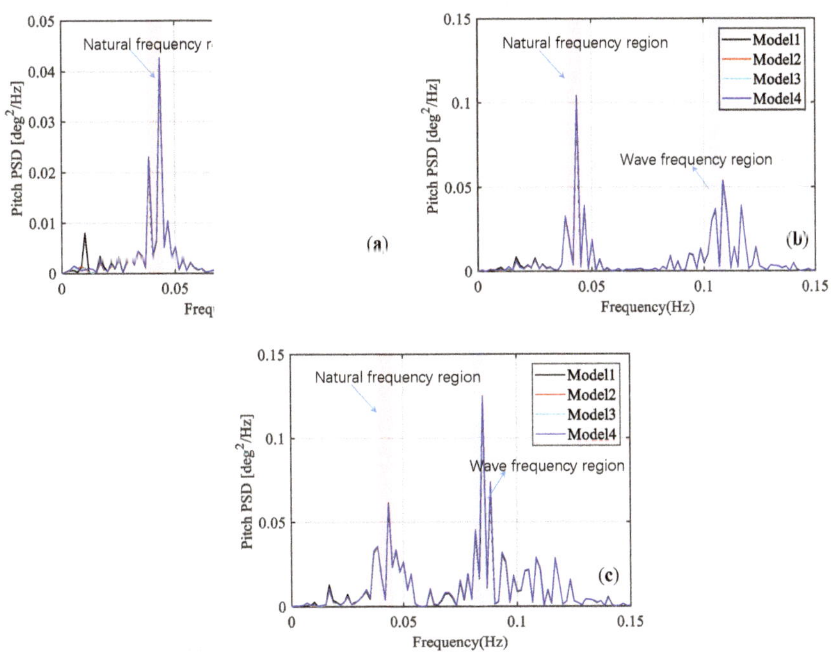

Figure 13. Frequency domain curves of the pitch PSD of the multi-segmented mooring line models: (**a**) LC4; (**b**) LC5; and (**c**) LC6.

Table 11. Surge, heave, and pitch PSDs of the multi-segmented mooring line models in the frequency domain.

DOF	Load Case	Model 1		Model 2		Model 3		Model 4	
		Frequency (Hz)	Peak	Frequency (Hz)	Peak	Frequency (Hz)	Peak	Frequency (Hz)	Peak
Surge (m^2/Hz)	LC4	0.010	3.516	0.010	2.696	0.011	2.985	0.013	0.520
	LC5	0.008	1.325	0.008	2.531	0.010	1.165	0.012	0.223
	LC6	0.010	1.872	0.008	2.295	0.010	1.59	0.012	0.329
Heave (m^2/Hz)	LC4	0.058	0.022	0.058	0.022	0.058	0.021	0.058	0.019
	LC5	0.058	0.014	0.058	0.014	0.058	0.013	0.058	0.012
	LC6	0.085	0.096	0.085	0.095	0.085	0.095	0.085	0.094
Pitch (deg^2/Hz)	LC4	0.043	0.043	0.043	0.043	0.043	0.043	0.043	0.043
	LC5	0.043	0.104	0.043	0.101	0.043	0.104	0.043	0.104
	LC6	0.082	0.125	0.082	0.123	0.082	0.122	0.082	0.122

Table 12. Motion standard deviation of the multi-segmented mooring line models under irregular waves.

Load Case	DOF	Model 1	Model 2	Model 3	Model 4
LC4	Surge (m)	1.276	1.258	1.211	0.707
	Heave (m)	0.086	0.086	0.086	0.086
	Pitch (°)	0.276	0.273	0.272	0.273
LC5	Surge (m)	1.051	1.191	1.015	0.701
	Heave (m)	0.256	0.256	0.256	0.256
	Pitch (°)	0.588	0.590	0.590	0.591
LC6	Surge (m)	1.401	1.529	1.373	1.041
	Heave (m)	0.516	0.516	0.516	0.516
	Pitch (°)	0.744	0.753	0.752	0.753

4.3. *Inclined Upper Column Models*

4.3.1. Regular Waves

The dynamic responses of the inclined upper column models (Models 5 and 6) under regular waves are shown in Figures 14–16. As illustrated in Figures 14 and 16, the surge and the pitch motion of the inclined upper column models are not much different from the original Model 1 under LC1, LC2, and LC3. However, as indicated in Figure 15, the heave motions of the inclined upper column models are much lower than that of the original Model 1. Model 6, whose upper columns are tilted 30 degrees, shows the largest reduction of the heave motions. The mean heave response of Model 6 is even close to 0 m. This means that the heave motion of the platform will decrease as increasing the inclined angle of the upper columns. The increases in added mass in the heave direction and the increase of the waterplane area due to the inclination of the columns should be the reason inducing the decrease of the heave displacement and the overall stability of the concept.

From the statistic point of view, the mean surge and pitch motions of the inclined upper column models are found to be almost the same with the original Model 1, as listed in Table 13. At the same time, the statistics of the surge and pitch motions are also nearly consistent no matter how inclined the upper columns are. It is notable that increasing the inclined angle of the upper columns is very effective to decrease not only the mean heave motions but also the maximum ones. From LC1, LC2, and LC3, it is evident that the maximum of the heave and pitch responses gradually increase. However, the minimum of the heave and pitch responses gradually decrease. It explains that the larger the wave height is, the larger the range of motion responses will be. For Model 1, 5, and 6, as increasing the inclined angle of the upper columns, an obvious reduction of the heave motion amplitudes can only be found for the wave with a high frequency; see Table 14. Slight decreases of the surge and pitch amplitudes can be identified for the wave with low frequency.

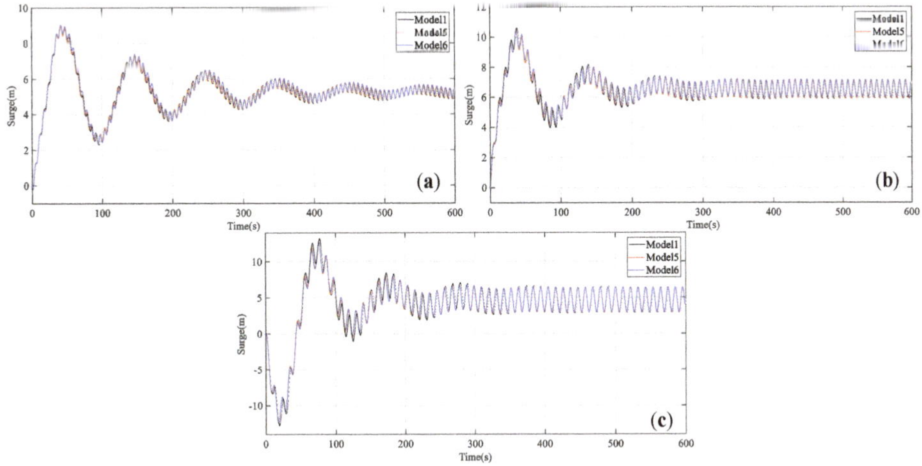

Figure 14. Time domain curves of the surge motion of the inclined upper column models: (**a**) LC1; (**b**) LC2; and (**c**) LC3.

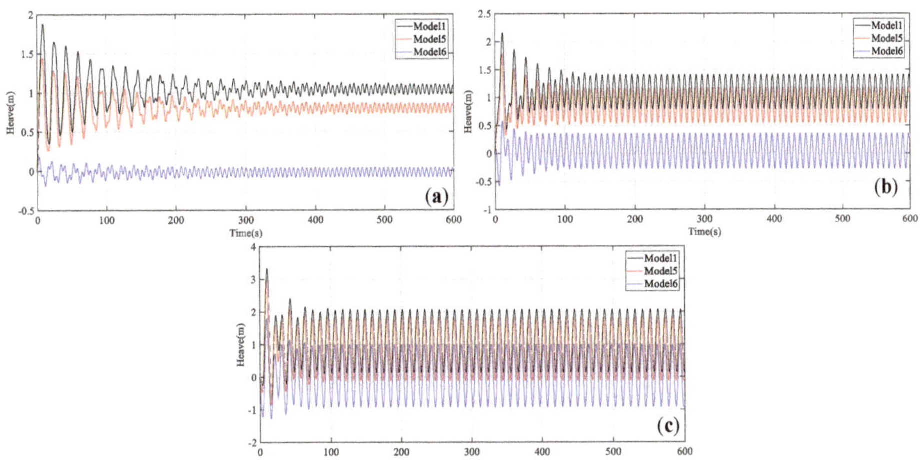

Figure 15. Time domain curves of the heave motion of the inclined upper column models: (**a**) LC1; (**b**) LC2; and (**c**) LC3.

4.3.2. Irregular Waves

To illustrate the effect of the inclined upper column models on the platform motion responses under irregular wave conditions, the models under the irregular waves are further examined, as shown in Figures 17–19. Comparing Model 1, Model 5, and Model 6, it is apparent that only the platform heave PSD is significantly reduced. The surge PSD shows slight reduction, while the pitch PSD increases slightly. This finding is consistent with the results of the regular wave in the previous section. It is also observable that the heave PSD of Model 6 is the smallest, which indicates that the platform whose upper columns are tilted 30° has the best performance for the heave motion under the irregular wave among Models 1, 5, and 6. The most obvious reduction in the heave direction is found to be in LC4. It is evident from Table 15 that the peak heave PSD in Model 6 is about only one tenth that of Model 1. In addition, there is a clear shift of the peak frequency for the heave motion, which is 0.058 Hz in Model 1, 0.55 Hz in Model 5, and 0.52 Hz in Model 6, due to the fact that the effects providing added mass area should dominate the additional stiffness from the increased waterplane area. The motion standard deviations of the inclined upper column models under irregular waves are listed in Table 16, where the decrease of the heave motion as changing the model from Model 1 to Model 5 and Model 6 is obvious when the wave frequency is high; however, this reduction reduces with increasing wave frequency.

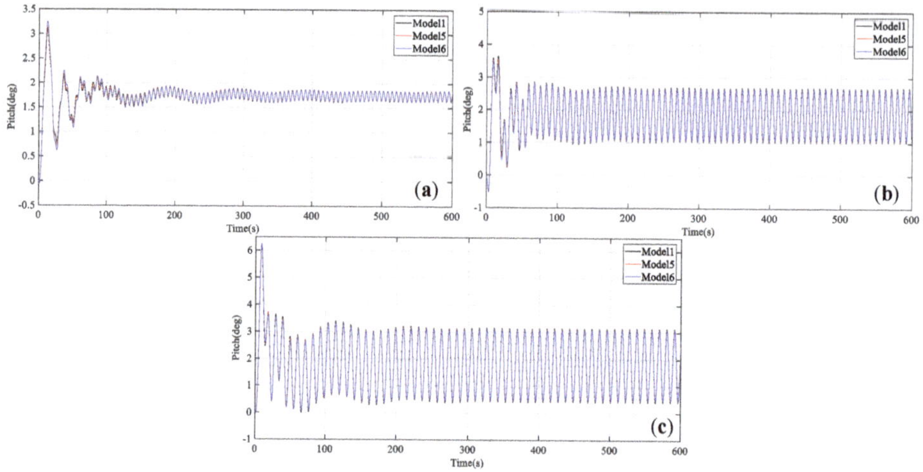

Figure 16. Time domain curves of the pitch motion of the inclined upper column models: (**a**) LC1; (**b**) LC2; and (**c**) LC3.

Table 13. Statistics of the stationary motion of the inclined upper column models.

DOF	Load Case	Model 1			Model 5			Model 6		
		Max.	Min.	Mean	Max.	Min.	Mean	Max.	Min.	Mean
Surge (m)	LC1	6.03	4.97	5.47	6.01	4.99	5.48	6.07	5.01	5.54
	LC2	7.20	5.05	6.12	7.18	5.09	6.13	7.23	5.16	6.19
	LC3	6.67	2.63	4.66	6.64	2.62	4.68	6.69	2.71	4.72
Heave (m)	LC1	1.16	0.93	1.04	0.91	0.71	0.80	0.09	−0.08	0.01
	LC2	1.41	0.79	1.10	1.18	0.55	0.86	0.36	−0.27	0.05
	LC3	2.07	0.14	1.10	1.83	−0.10	0.86	1.01	−0.91	0.05
Pitch (°)	LC1	1.87	1.60	1.74	1.87	1.60	1.74	1.88	1.60	1.74
	LC2	2.69	1.01	1.85	2.67	1.03	1.85	2.66	1.04	1.85
	LC3	3.16	0.40	1.76	3.14	0.42	1.76	3.11	0.45	1.76

Table 14. Amplitudes of the stationary motion of the inclined upper column models.

DOF	Load Case	Model 1	Model 5	Model 6
Surge (m)	LC1	1.06	1.02	1.06
	LC2	2.15	2.09	2.07
	LC3	4.04	4.02	3.98
Heave (m)	LC1	0.23	0.2	0.17
	LC2	0.62	0.63	0.63
	LC3	1.93	1.93	1.92
Pitch (°)	LC1	0.27	0.27	0.28
	LC2	1.68	1.64	1.62
	LC3	2.76	2.72	2.66

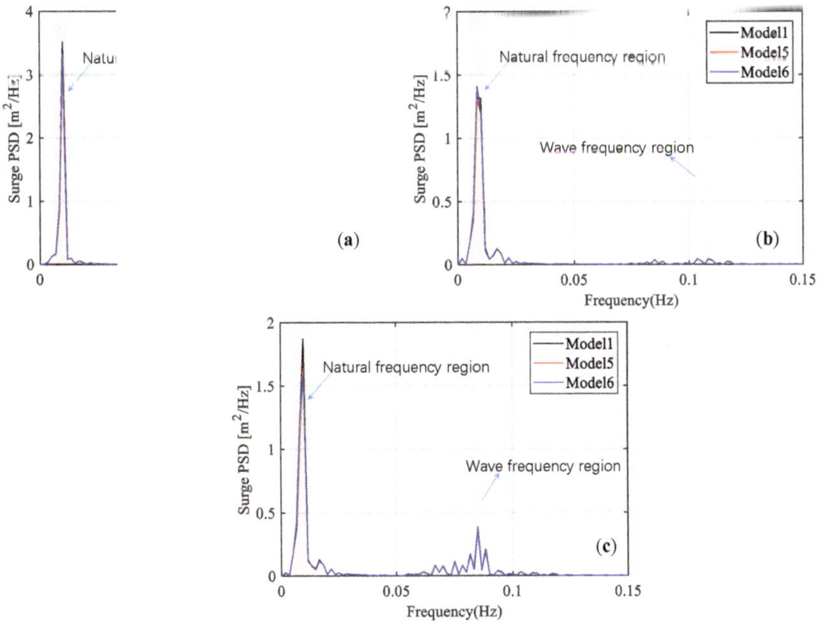

Figure 17. Frequency domain curves of the surge PSD of the inclined upper column models: (**a**) LC4; (**b**) LC5; and (**c**) LC6.

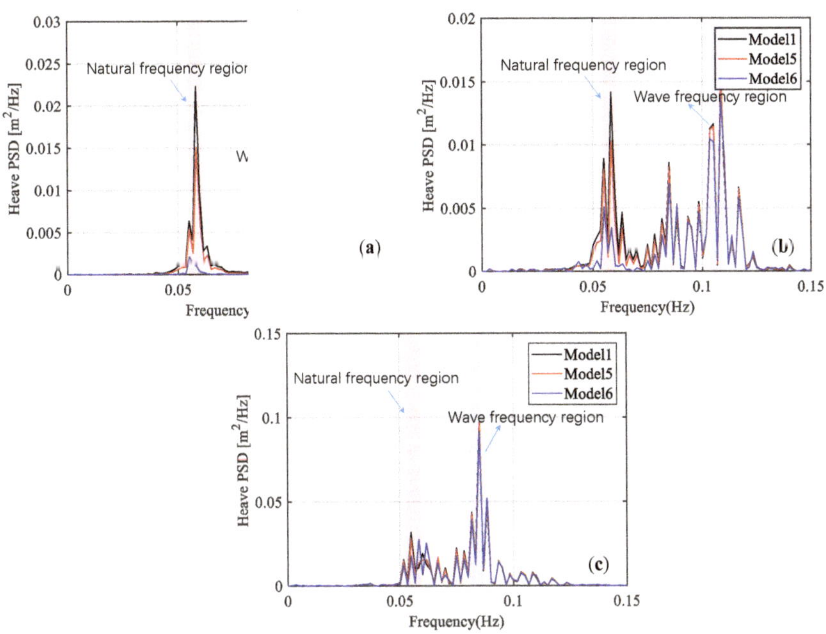

Figure 18. Frequency domain curves of the heave PSD of the inclined upper column models: (**a**) LC4; (**b**) LC5; and (**c**) LC6.

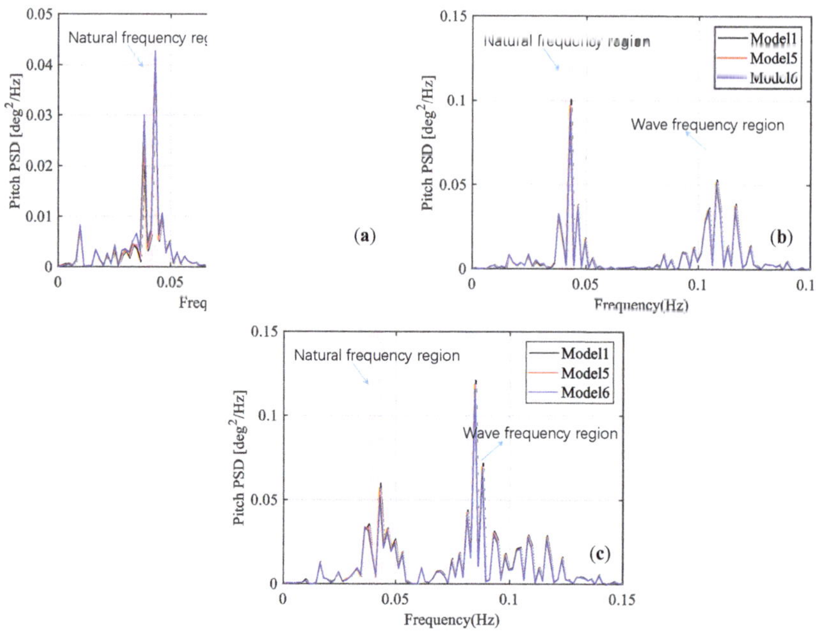

Figure 19. Frequency domain curves of the pitch PSD of the inclined upper column models: (a) LC4; (b) LC5; and (c) LC6.

Table 15. Surge, heave, and pitch PSD statistics of the inclined upper column models in the frequency domain.

DOF	Load Case	Model 1		Model 5		Model 6	
		Frequency (Hz)	Peak	Frequency (Hz)	Peak	Frequency (Hz)	Peak
Surge (m^2/Hz)	LC4	0.010	3.516	0.010	3.329	0.010	3.329
	LC5	0.008	1.325	0.008	1.307	0.008	1.413
	LC6	0.012	1.872	0.012	1.641	0.012	1.574
Heave (m^2/Hz)	LC4	0.058	0.022	0.055	0.015	0.052	0.002
	LC5	0.058	0.014	0.058	0.010	0.112	0.015
	LC6	0.085	0.097	0.085	0.097	0.085	0.092
Pitch (deg^2/Hz)	LC4	0.043	0.043	0.043	0.043	0.043	0.043
	LC5	0.043	0.104	0.043	0.097	0.043	0.092
	LC6	0.082	0.060	0.082	0.057	0.082	0.051

Table 16. Motion standard deviation of the inclined upper column models under irregular waves.

Load Case	DOF	Model 1	Model 5	Model 6
LC4	Surge (m)	1.276	1.244	1.264
	Heave (m)	0.086	0.075	0.042
	Pitch (°)	0.276	0.273	0.269
LC5	Surge (m)	1.051	1.030	1.040
	Heave (m)	0.256	0.257	0.232
	Pitch (°)	0.588	0.578	0.565
LC6	Surge (m)	1.401	1.377	1.381
	Heave (m)	0.516	0.519	0.514
	Pitch (°)	0.744	0.734	0.720

It is interesting that there is no clear frequency shift for the pitch motion, and we think it is probably due to the fact that the added mass in the pitch direction from the increased waterplane area should be very limited compared with the wind turbine blades and the hub, together with the fact that the stiffness in pitch direction is not sensitive to the inclination of the columns, as can be found from Table 15. The natural pitch frequency can be abruptly determined as $\omega = \sqrt{k_\theta / I}$, where k_θ is the stiffness in pitch direction, I is the moment of inertia determined as $I = \int (m + m_a) r^2 dr$, m is the mass and m_a is the added mass. Considering that the distance of the added mass to the center of the floater is much less than that of the wind turbine and the hub, and also I is proportional to the square of r, the natural frequency shift due to the added mass should not be as obvious as that in the heave direction.

4.4. Combination of Models 4 and 6

The multi-segmented mooring line models or the inclined upper column models examined previously have conspicuous impacts on the platform motion response in only one direction, i.e., Model 4 can only greatly reduce the platform surge response, while Model 6 can only greatly reduce the platform heave response. In this section, Model 7, as a combination of Model 4 and Model 6, is proposed, in which the mooring system of Model 7 is arranged in the same way as Model 4, and the upper columns same as those of the platform of Model 6 tilted 30°.

4.4.1. Regular Waves

As illustrated in Figure 20, different from Model 1, the surge motions of Model 4 and Model 7 are the most reduced. Meanwhile, the surge motion of Model 4 and Model 7 are nearly coincident. At the same time, Figure 21 shows that the platform heave motions are gradually reduced, while the heave motions of Model 6 and Model 7 are around the same. Compared to Model 1, the pitch motions of Models 4, 6, and 7 do not change much, see Figure 22. Overall, Model 7 depicts the smallest motions in both the surge and heave directions as compared with the other models under the regular waves.

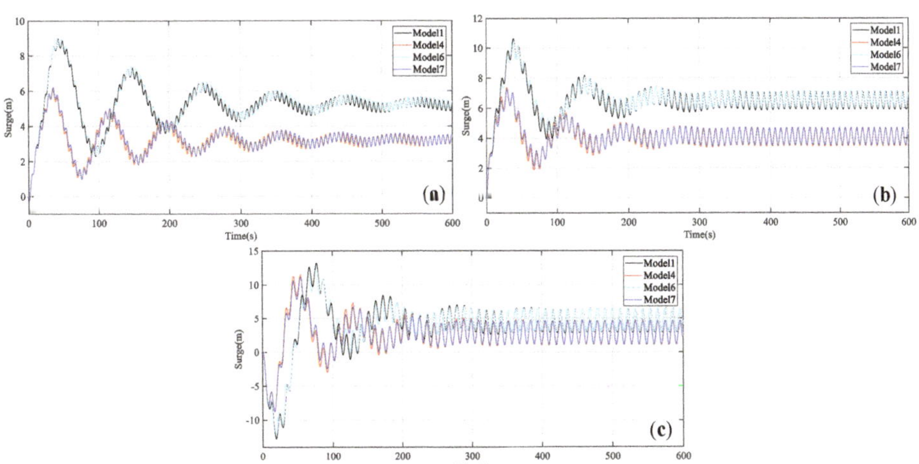

Figure 20. Comparison of the surge motion of Model 1, 4, 6, and 7 in the time domain: (**a**) LC1; (**b**) LC2; and (**c**) LC3.

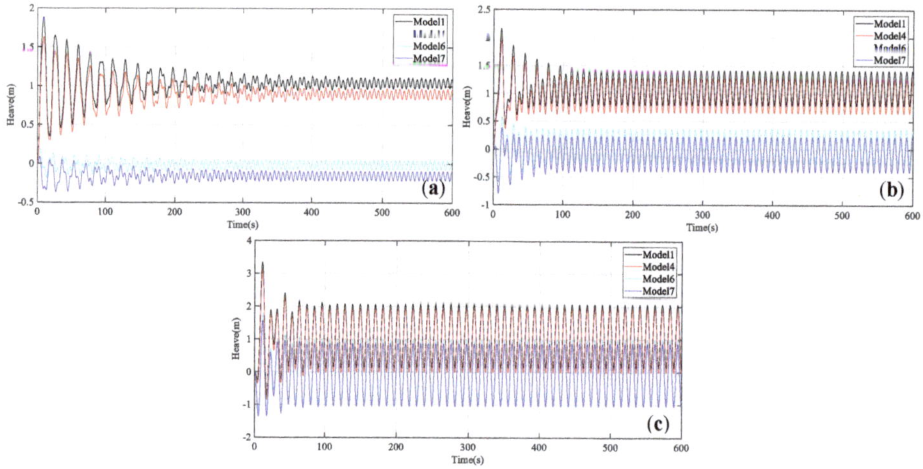

Figure 21. Comparison of the heave motion of Model 1, 4, 6, and 7 in the time domain: (a) LC1; (b) LC2; and (c) LC3.

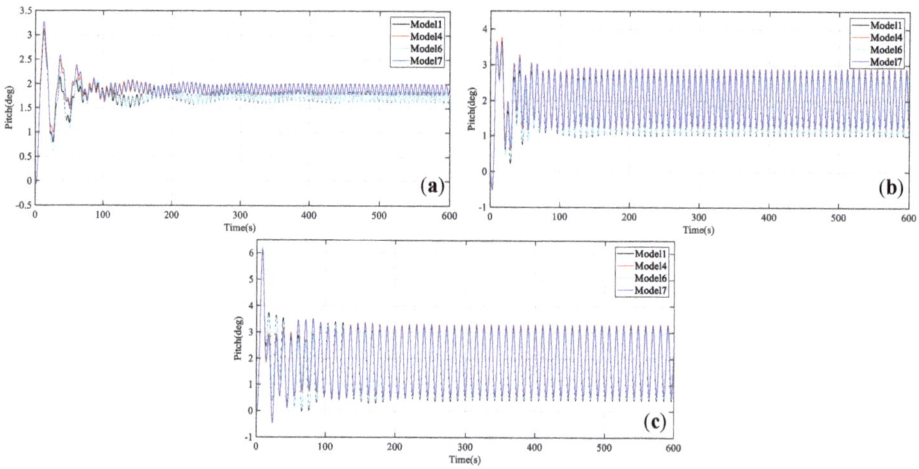

Figure 22. Comparison of the pitch motion of Model 1, 4, 6, and 7 in the time domain: (a) LC1; (b) LC2; and (c) LC3.

Figure 23 shows the maximum, minimum, mean values, and the amplitude of the response motions from 400–600 s. The averaged value of the surge motion of Model 7 is about 0.7 as large as that in Model 1 under different load conditions. Meanwhile, the mean value of heave motion of Model 7 is close to 0, indicating that Model 7 can greatly decline the platform heave motion. As clarified in Figure 23c, there are no significant changes to the average values of pitch motion from the original Model 1 to the modified models. As for the amplitudes of the motion, Model 7 also provides the best performance in both surge and heave directions.

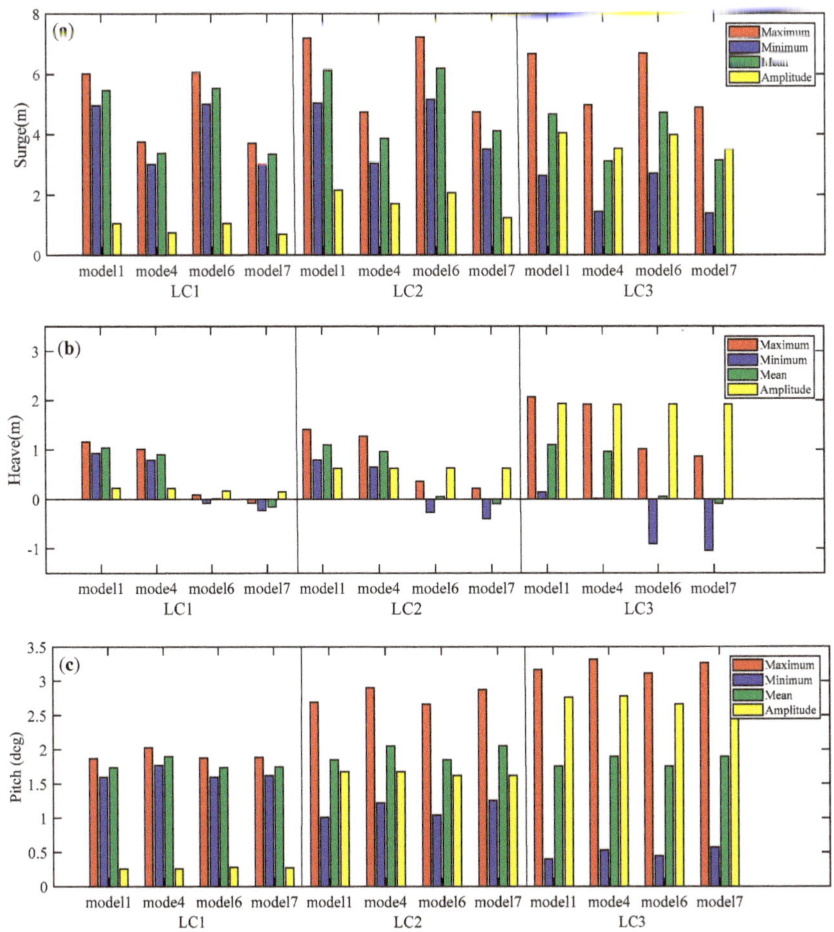

Figure 23. Comparison of the statistics of the stationary motion of Model 1, 4, 6, and 7 from 400–600 s: (**a**) surge; (**b**) heave; and (**c**) pitch.

4.4.2. Irregular Waves

The frequency domain curves of the platform motion PSD of Model 7 in surge, heave, and pitch directions under different irregular waves are plotted in Figures 24–26, respectively. From the figures, it is evident that the peak values of the surge and heave PSD of Model 7 are much smaller than those of the other models. However, the pitch PSD peak values of the models are not much different. Table 17 lists the motion standard deviations of Models 1, 4, 6, and 7 under irregular waves, from which it is clear that the motion standard deviations in both surge and heave directions of Model 7 are minimal compared with the other models. Overall, we can say that Model 7 has a much better performance in comparison with the original OC4 model.

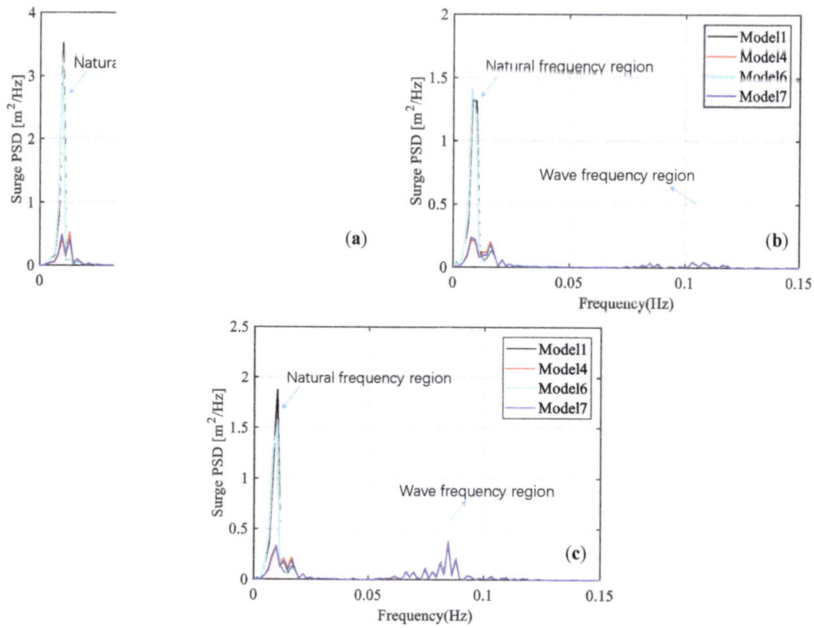

Figure 24. Comparison of the surge motion of Model 1, 4, 6, and 7 in the frequency domain: (**a**) LC4; (**b**) LC5; and (**c**) LC6.

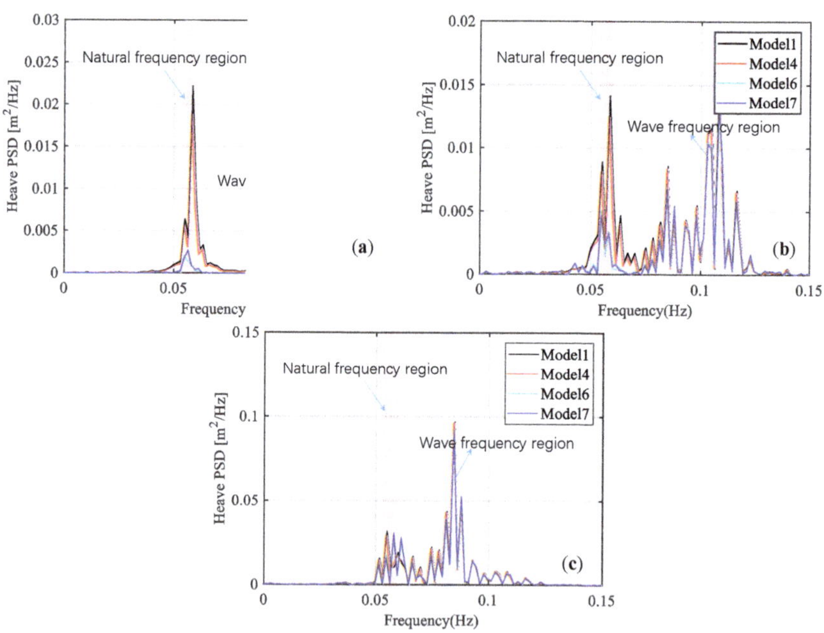

Figure 25. Comparison of the heave motion of Model 1, 4, 6, and 7 in the frequency domain: (**a**) LC4; (**b**) LC5; and (**c**) LC6.

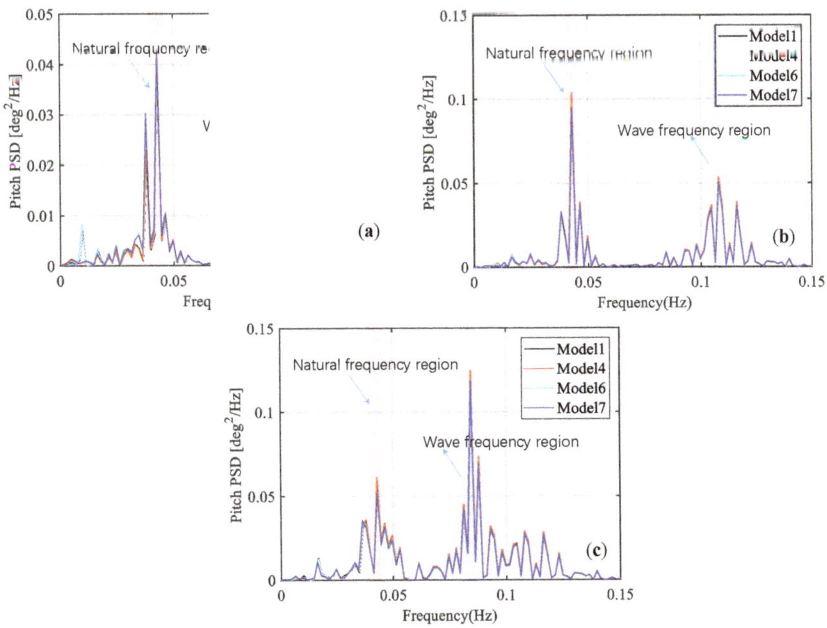

Figure 26. Comparison of the pitch motion of Model 1, 4, 6, and 7 in the frequency domain: (a) LC4; (b) LC5; and (c) LC6.

Table 17. Motion standard deviation of Models 1, 4, 6, and 7 under irregular waves.

Load Case	DOF	Model 1	Model 4	Model 6	Model 7
LC4	Surge (m)	1.276	0.707	1.264	0.706
	Heave (m)	0.086	0.086	0.042	0.042
	Pitch (°)	0.276	0.273	0.269	0.272
LC5	Surge (m)	1.051	0.701	1.040	0.702
	Heave (m)	0.256	0.256	0.232	0.231
	Pitch (°)	0.588	0.591	0.565	0.578
LC6	Surge (m)	1.401	1.041	1.381	1.039
	Heave (m)	0.516	0.516	0.514	0.513
	Pitch (°)	0.744	0.753	0.720	0.744

5. Conclusions

In this study, by changing the connection mode of the mooring line as well as the structural form of the platform based on the original OC4 model, two groups of models were examined by AFWT. The influence of the arrangement of the mooring lines and the inclination angle of the upper columns on the motion response were clarified. Based on the findings in both of the two groups of the models, a new model is proposed. From the study, some important conclusions are drawn, which can be summarized into the following aspects:

1. The numerical results calculated by AFWT are basically consistent with FAST and those in experiment, which verifies the accuracy of AFWT used in this paper.
2. Changing the arrangement of the mooring system can greatly reduce the surge response of the platform, although it will have little effect on the response of heave and pitch. Among Models 2, 3, and 4, the mooring system layout in Model 4 has the most remarkable effect on reducing the

platform surge response by about 30% compared with the original Model 1, indicating that the shorter the upper lines are, the greater the reduction to the surge of the platform will be.

3. Tilting the upper columns of the floating platform can greatly reduce the platform heave response. It is found that the heave motion of the platform will decrease as the inclined angle of the upper column increases. However, it has little effect on the responses in surge and pitch directions.
4. A new model integrating the optimized multi-segmented mooring lines and the optimized inclined columns is proposed. The surge and heave motion responses of this new model are greatly declined. It should be able to increase the stability of the floating platform during operation and improve the working efficiency of the floating wind turbines.

6. Discussion

The proposed structural modifications will also increase the internal tensions in mooring lines and structural elements. This will lead to an increase in the requirement of the structural strength, which may increase the costs of installation, manufacturing, and maintenance of the FOWT. Therefore, additional research considering the economic issues and the service life of the mooring lines as well as the structural elements due to the increases of internal tensions should be carried out. The method of adding dampers to reduce the wind turbine motions is also an important direction.

Author Contributions: Author Contributions: Conceptualization: Q.Z. and Z.L.; methodology: Z.L.; software: Q.Z.; validation: Q.Z. and Z.L.; formal analysis: Q.Z. and Z.L.; investigation: Q.Z. and Y.T.; data curation: Q.Z. and Z.L.; writing—original draft preparation: Q.Z., Y.T. and W.W.; writing—review and editing: Q.Z., Z.L., Y.T., W.W. and X.H.; supervision: Z.L., Q.Z., Y.T., W.W. and X.H.

Funding: The study is funded by National Key Research and Development Plan of China (2016YFE0127900), and the National Natural Science Foundations of China (51608220).

Acknowledgments: The corresponding author sincerely thanks Key Laboratory for Bridge and Wind Engineering of Hunan Province for the support to this research.

Conflicts of Interest: No potential conflict of interest was reported by the authors.

Nomenclature

a	Axial induction factor
a'	Tangential induction factor
B	Number of the blades
$B(\omega)$	Wave spectrum (P-M spectrum)
c	Leaf chord length (m)
C_B	Seabed friction coefficient
C_n	Coefficients for the normal force
C_t	Coefficients for the thrust force
EA	Sectional stiffness of the mooring line (kN)
$\phi(x, y)$	Potential function
$F^{waves}(t)$	Incident-wave induced force (kN)
$tF^{wind}(t)$	Wind loads acting on the blades and tower (kN)
$F^{mooring}(t)$	Mooring tension (kN)
$F^{vis}(t)$	Drag force caused by fluid viscosity (kN)
F_{ext}	External force at the connection node (kN)
H_A	Horizontal component of the effective tension at the anchor (kN)
H_F	Horizontal component of the effective tension at the fairlead (kN)
L	Total unstretched mooring line length (m)
L_B	Unstretched length of cable lying on the seabed (m)
s	Length from the catenary point to the anchor (m)
Te	Mooring line tension (kN)
V_1	Wind speed (m/s)
V_{hub}	Wind speed at the hub (m/s)

V_A	Vertical component of the effective tension at the anchor (kN)
V_F	Vertical component of the effective tension at the fairlead (kN)
φ	Relation flow angle (°)
ω	Frequency(Hz)
w	Mass of the mooring line per unit length (kN/m)
$W(\omega)$	Fourier transform of a white noise time series with unit variance
$X_j(\omega,\beta)$	Wave-induced force array normalized per unit wave amplitude (kN/m)
x_F	Horizontal coordinate of the fairlead position relative to the anchor (m)
z_F	Vertical coordinate of the fairlead position relative to the anchor (m)
AFWT	Analysis tool of floating wind turbine
BEM	Blade element momentum
CM	Center of mass
DOF	Degree of freedom
FAST	Fatigue, Aerodynamics, Structures, and Turbulence
FFT	Fast Fourier Transform
FOWT	Floating offshore wind turbine
HAWT	Horizontal axis wind turbine
LC	Load case
MSQS	Multi-Segmented, Quasi-Static
NREL	National Renewable Energy Laboratory
OC3	Offshore Code Comparison Collaboration
OC4	Offshore Code Comparison Collaboration Continuation
P-M	Pierson Moskowitz
PSD	Power spectral density
RAO	Response amplitude operator
SWL	Still water level
TLP	Tension Leg Platform
WAMIT	Wave Analysis Massachusetts Institute of Technology
WTN	Wind turbine noise
VAWT	Vertical axis wind turbine

References

1. Heronemus, W.E. *Pollution-Free Energy from the Offshore Winds*; Marine Technology Society: Washington, DC, USA, 1972.
2. European Wind Energy Association. *The European Offshore Wind Industry-key Trends and Statistics 1st Half 2013*; European Wind Energy Association: Brussels, Belgium, 2013.
3. Maine International Consulting LLC. *Floating Offshore Wind Foundations: Industry Consortia and Projects in the United States, Europe and Japan*; Maine International Consulting LLC: Bremen, ME, USA, 2013.
4. Gille, L.A.; Marquis-Favre, C.; Morel, J. Testing of the European Union exposure-response relationships and annoyance equivalents model for annoyance due to transportation noises: The need of revised exposure-response relationships and annoyance equivalents model. *Environ. Int.* **2016**, *94*, 83–94. [CrossRef]
5. Janssen, S.A.; Vos, H.; Eisses, A.R.; Pedersen, E. A comparison between exposure-response relationships for wind turbine annoyance and annoyance due to other noise sources. *J. Acoust. Soc. Am.* **2011**, *130*, 3743–3756. [CrossRef] [PubMed]
6. Takayuki, K.; Takashi, Y.; Sonoko, K.; Shinichi, S.; Hideki, T. Exposure-response relationship of wind turbine noise with self-reported symptoms of sleep and health problems: A nationwide socioacoustic survey in Japan. *Noise Health* **2016**, *18*, 53–61.
7. Fredianelli, L.; Gallo, P.; Licitra, G.; Carpita, S. Analytical assessment of wind turbine noise impact at receiver by means of residual noise determination without the wind farm shutdown. *Noise Control Eng. J.* **2017**, 417–433. [CrossRef]
8. Fredianelli, L.; Carpita, S.; Licitra, G. A procedure for deriving wind turbine noise limits by taking into account annoyance. *Sci. Total Environ.* **2019**, *648*, 728–736. [CrossRef] [PubMed]

9. Musial, W.; Butterfield, S.; Boone, A. Feasibility of Floating Platform Systems for Wind Turbines. In Proceedings of the 42nd AIAA Aerospace Sciences Meeting and Exhibit, Reno Nevada, NV, USA, 5–7 January 2004.
10. Karimirad, M.; Moan, T. Feasibility of the Application of a Spar-type Wind Turbine at a Moderate Water Depth. *Energy Procedia* **2012**, *24*, 340–350. [CrossRef]
11. Ahn, H.J.; Shin, H. Model test and numerical simulation of OC3 spar type floating offshore wind turbine. *Int. J. Naval Arch. Ocean Eng.* **2019**, *11*, 1–10. [CrossRef]
12. Tomasicchio, G.R.; D'Alessandro, F.; Avossa, A.M.; Riefolo, L.; Musci, E.; Ricciardelli, F.; Vicinanza, D. Experimental modelling of the dynamic behaviour of a spar buoy wind turbine. *Renew. Energy* **2018**, *127*, 412–432. [CrossRef]
13. Markus, L.; De-Prada-Gil, M.; Molins, C. The influence of different wind and wave conditions on the energy yield and downtime of a Spar-buoy floating wind turbine. *Renew. Energy* **2019**, *136*, 1–14.
14. Ma, Y.; Hu, Z.Q.; Xiao, L.F. Wind-wave induced dynamic response analysis for motions and mooring loads of a spar-type offshore floating wind turbine. *J. Hydrodyn.* **2015**, *26*, 865–874. [CrossRef]
15. Jeon, S.H.; Cho, Y.U.; Seo, M.W.; Cho, J.R.; Jeong, W.B. Dynamic response of floating substructure of spar-type offshore wind turbine with catenary mooring cables. *Ocean Eng.* **2013**, *72*, 356–364. [CrossRef]
16. Si, Y.L.; Hamid, R.K.; Gao, H.J. Modelling and optimization of a passive structural control design for a spar-type floating wind turbine. *Eng. Struct.* **2014**, *69*, 168–182. [CrossRef]
17. Ye, K.; Ji, J.C. Current, wave, wind and interaction induced dynamic response of a 5 MW spar-type offshore direct-drive wind turbine. *Eng. Struct.* **2019**, *178*, 395–409. [CrossRef]
18. Hong, S.; Lee, I.; Park, S.H.; Lee, C.; Chun, H.H.; Lim, H.C. An experimental study of the effect of mooring systems on the dynamics of a SPAR buoy-type floating offshore wind turbine. *Int. J. Naval Arch. Ocean Eng.* **2015**, *7*, 559–579. [CrossRef]
19. Subbulakshmi, A.; Sundaravadivelu, R. Heave damping of spar platform for offshore wind turbine with heave plate. *Ocean Eng.* **2016**, *121*, 24–36. [CrossRef]
20. Duan, F.; Hu, Z.Q.; Niedzwecki, J.M. Model test investigation of a spar floating wind turbine. *Mar. Struct.* **2016**, *49*, 76–96. [CrossRef]
21. American Petroleum Institute. *Recommended Practices for Planning, Designing, and Constructing Tension Leg Platforms*; American Petroleum Institute: Washington, DC, USA, 1997.
22. Vita, L.; Ramachandran, G.K.V.; Krieger, A.; Kvittem, M.I.; Merino, D.; Cross-Whiter, J.; Ackers, B.B. Comparison of Numerical Models and Verification Against Experimental Data, Using PelaStar TLP Concept. In Proceedings of the ASME 2015 34th International Conference on Ocean, Offshore and Arctic Engineering, St John's, NL, Canada, 31 May–5 June 2015.
23. Oguz, E.; Clelland, D.; Day, A.H.; Incecik, A.; López, J.A.; Sánchez, G.; Almeria, G.G. Experimental and numerical analysis of a TLP floating offshore wind turbine. *Ocean Eng.* **2018**, *147*, 591–605. [CrossRef]
24. Bulder, B.H.; Van, H.; Henderson, A.; Huijsmans, R.H.M.; Pierik, J.T.G.; Snijders, E.J.B.; Wijnants, G.H.; Wolf, M.J. *Studie Naar Haalbaarheid Van en Randvoorwaarden Voor Drijvende Offshore Windturbines*; Drijfwind Project Report; TNO: Delft, The Netherlands, 2002.
25. Robertson, A.N.; Jonkman, J.M. Loads analysis of several offshore floating wind turbine concepts. In Proceedings of the 21st International Offshore and Polar Engineering Conference, Maui, HI, USA, 19–24 June 2011; pp. 443–450.
26. Borg, M.; Wang, K.; Collu, M.; Moan, T. A comparison of two coupled model of dynamics for offshore floating vertical axis wind turbines. In Proceedings of the ASME 33rd International Conference on Ocean, Offshore and Arctic Engineering, San Francisco, CA, USA, 8–13 June 2014.
27. Beyer, F.; Choisnet, T.; Kretschmer, M.; Cheng, P.W. Coupled MBS-CFD Simulation of the Ideol Floating Offshore Wind Turbine Foundation Compared to Wave Tank Model Test Data. In Proceedings of the 25th International Ocean and Polar Engineering Conference, Kona, Big Island, HI, USA, 21–26 June 2015.
28. Robertson, A.; Jonkman, J.; Masciola, M.; Song, H.; Goupee, A.; Coulling, A.; Luan, C. *Definition of the Semisubmersible Floating System for Phase II of OC4 NREL/TP-5000–60601*; National Renewable Energy Laboratory (NREL): Golden, CO, USA, 2014.
29. Paquette, J.; Barone, M.; Paquette, J.; Barone, M. Innovative offshore vertical-axis wind turbine rotor project. In Proceedings of the EWEA 2012 Annual Event, Copenhagen, Denmark, 16–19 April 2012.
30. Jonkman, J.M. Influence of control on the pitch damping of a floating wind turbine. In Proceedings of the 2008 ASME Wind Energy Symposium, Reno, NV, USA, 7–10 January 2008.

31. Luan, C.; Gao, Z.; Moan, T. Modelling and analysis of a semi-submersible wind turbine with a central tower with emphasis on the brace system. In Proceedings of the 32nd International Conference on Ocean, Offshore and Arctic Engineering, Nantes, France, 9–14 June 2013.
32. Jonkman, J.M.; Buhl, M.L., Jr. Loads Analysis of a Floating Offshore Wind Turbine Using Fully Coupled Simulation. In Proceedings of the Wind Power 2007 Conference & Exhibition, Los Angeles, CA, USA, 3–6 June 2007.
33. Barooni, M.; Ali, N.A.; Ashuri, T. An open-source comprehensive numerical model for dynamic response and loads analysis of floating offshore wind turbines. *Energy* **2018**, *154*, 442–454. [CrossRef]
34. Viselli, A.M.; Goupee, A.J.; Dagher, H.J. Model test of a 1:8-scale floating wind turbine offshore in the gulf of maine. *J. Offsh. Mech. Arctic Eng.* **2015**, *137*, 041901. [CrossRef]
35. Jain, A.; Robertson, A.N.; Jonkman, J.M.; Goupee, A.J.; Swift, R.W. FAST code verification of scaling laws for DeepCwind floating wind system tests. In Proceedings of the 22nd International Offshore and Polar Engineering Conference, Rhodes, Greece, 17–22 June 2012; pp. 355–365.
36. Michailides, C.; Gao, Z.; Moan, T. Experimental study of the functionality of a semisubmersible wind turbine combined with flap-type Wave Energy Converters. *Renew. Energy* **2016**, *93*, 675–690. [CrossRef]
37. Li, L.; Gao, Y.; Hu, Z.Q.; Yuan, Z.M.; Sandy, D.; Li, H.R. Model test research of a semisubmersible floating wind turbine with an improved deficient thrust force correction approach. *Renew. Energy* **2018**, *119*, 95–105. [CrossRef]
38. Zambrano, T.; Cready, T.M.; Kiceniuk, T. Dynamic modeling of deepwater offshore wind turbine structures in Gulf of Mexico storm conditions. In Proceedings of the 25th International Conference on Offshore Mechanics and Arctic Engineering, Hamburg, Germany, 4–9 June 2006; pp. 1–6.
39. Shimada, K.; Miyakawa, M.; Ishihara, T. A Study on a Semi-Submersible Floating Offshore Wind Energy Conversion System. In Proceedings of the 16th International Offshore and Polar Engineering Conference, San Francisco, CA, USA, 28 May–2 June 2006; pp. 348–355.
40. Robertson, A.N.; Jonkman, J.M.; Masciola, M.D.; Molta, P.; Coulling, A.J.; Goupee, A.J. Summary of Conclusions and Recommendations Drawn from the DeepCWind Scaled Floating Offshore Wind System Test Campaign. In Proceedings of the ASME 2013 International Conference on Ocean, Offshore and Arctic Engineering, Nantes, France, 9–14 June 2013.
41. Coulling, A.J.; Goupee, A.J.; Robertson, A.N.; Jonkman, J.M.; Dagher, H.J. Validation of a FAST semi-submersible floating wind turbine numerical model with DeepCwind test data. *Renew. Sustain. Energy* **2013**, *3*. [CrossRef]
42. Hall, M.; Goupee, A. Validation of a lumped-mass mooring line model with DeepCwind semisubmersible model test data. *Ocean Eng.* **2015**, 590–603. [CrossRef]
43. Masciola, M.; Robertson, A.; Jonkman, J.; Coulling, A.; Goupee, A. Assessment of the Importance of Mooring Dynamics on the Global Response of the DeepCwind Floating Semisubmersible Offshore Wind Turbine. In Proceedings of the Twenty-third International Offshore and Polar Engineering Conference, Anchorage, AL, USA, 30 June–5 July 2013.
44. Roddier, D.; Cermelli, C.; Weinstein, A. WindFloat: A Floating Foundation for Offshore Wind Turbines-Part II: Hydrodynamic Analysis. In Proceedings of the 28th International Conference on Ocean, Offshore Mechanics and Arctic Engineering, Honolulu, HI, USA, 31 May–5 June 2009; pp. 1–9.
45. Huijs, F.; Bruijn, R.D.; Savenije, F. Concept design verification of a semi-submersible floating wind turbine using coupled simulations. *Energy Procedia* **2014**, *53*, 2–12. [CrossRef]
46. Zheng, X.Y.; Lei, Y. Stochastic Response Analysis for a Floating Offshore Wind Turbine Integrated with a Steel Fish Farming Cage. *Appl. Sci.* **2018**, *8*, 1229. [CrossRef]
47. Jonkman, J.M. *Dynamics Modeling and Loads Analysis of an Offshore Floating Wind Turbine*; Technical Report NREL/TP-500–41958; The National Renewable Energy Laboratory: Lakewood, CO, USA, 2007.
48. Ormberg, H.; Bachynski, E.E. Global analysis of floating wind turbines: Code development, model sensitivity and benchmark study. In Proceedings of the 22nd International Ocean and Polar Engineering Conference, Rhodes, Greece, 17–22 June 2012.
49. Karimirad, M. Stochastic Dynamic Response Analysis of Spar-Type Wind Turbines with Catenary or Taut Mooring Systems. Ph.D. Thesis, Norwegian University of Science and Technology, Trondheim, Norway, 2011.

50. Goupee, A.J.; Fowler, M.J.; Kimball, R.W.; Helder, J.; Ridder, E.J. Additional wind/wave basin testing of the DeepCwind semi-submersible with a performance-matched wind turbine. In Proceedings of the ASME 33rd International Conference on Ocean, Offshore and Arctic Engineering, San Francisco, CA, USA, 8–13 June 2014.
51. Skaare, B.; Hanson, T.D.; Nielsen, F.G.; Yttervik, R.; Hansen, A.M.; Thomsen, K.; Larsen, T.J. Integrated dynamic analysis of floating offshore wind turbines. In Proceedings of the European Wind Energy Conference, Milan, Italy, 7–10 May 2007; pp. 7–10.
52. Moriarty, P.J.; Hansen, C.A. *AeroDyn Theory Manual*; Technical Report; National Renewable Energy Laboratory: Golden, CO, USA, 2005.
53. Jonkman, J.M.; Hayman, G.J.; Jonlman, B.J.; Damiani, R.R.; Murray, R.E. *AeroDyn v15 User's Guide and Theory Manual*; National Renewable Energy Laboratory (NREL): Golden, CO, USA, 2016.
54. Masciola, M.; Jonkman, J.; Robertson, A. Implementation of a Multi-segmented, Quasi-Static cable model. In Proceedings of the 23rd International Ocean and Polar Engineering Conference, Anchorage, AL, USA, 30 June–5 July 2013.
55. Masciola, M.; Jonkman, J. *Instructional and Theory Guide to the Mooring Analysis Program Phase I: The Multi–Segmented Quasi–Static Model*; NREL Technical Report; National Renewable Energy Laboratory (NREL): Golden, CO, USA, 2013.
56. Jonkman, J.M.; Robertson, A.N.; Hayman, G.J. *HydroDyn User's Guide and Theory Manual*; National Renewable Energy Laboratory (NREL): Golden, CO, USA, 2015.
57. Wang, H.; Hu, Z.Q.; Meng, X.Y. Dynamic Performance Investigation of a Spar-Type Floating Wind Turbine Under Different Sea Conditions. *China Ocean Eng.* **2018**, *32*, 256–265. [CrossRef]
58. Butterfield, S.; Musial, W.; Scott, G. *Definition of a 5-MW Reference Wind Turbine for Offshore System Development*; Technical Report NREL/TP 500–38060; National Renewable Energy Laboratory (NREL): Golden, CO, USA, 2009.
59. Kim, H.C.; Kim, M.H. Comparison of simulated platform dynamics in steady/dynamic winds and irregular waves for OC4 semi-submersible 5MW wind-turbine against DeepCwind model-test results. *Ocean Syst. Eng.* **2016**, *1*, 1–21. [CrossRef]
60. Goupee, A.; Koo, B.; Lambrakos, K.; Kimball, R. Model tests for three floating wind turbine concepts. In Proceedings of the Offshore Technology Conference, Houston, TX, USA, 30 April–3 May 2012.
61. Palm, J.; Moura Paredes, G.; Eskilsson, C.; Taveira Pinto, F.; Bergdahl, L. Simulation of mooring cable dynamics using a discontinuous galerkin method. In Proceedings of the V International Conference on Computational Methods in Marine Engineering, Hamburg, Germany, 29–31 May 2013.
62. Liu, Y.Y.; Yoshida, S.; Yamamoto, H.; Toyofuku, A.; He, G.H.; Yang, S.H. Response Characteristics of the DeepCwind Floating Wind Turbine Moored by a Single-Point Mooring System. *Appl. Sci.* **2018**, *11*, 2306. [CrossRef]
63. Christian, K.; Frank, L.; Friedemann, B.; Denis, M.; Po, W.C. Validation of INNWIND.EU Scaled Model Tests of a Semisubmersible Floating Wind Turbine. In Proceedings of the International Ocean and Polar Engineering Conference, Rhodos, Greece, 26 June–1 July 2016.
64. Ramachandran, G.K.V.; Robertson, A.N.; Jonkman, J.M.; Masciola, M.D. Investigation of Response Amplitude Operators for Floating Offshore Wind Turbines. In Proceedings of the 23rd International Ocean, Offshore and Polar Engineering Conference, Anchorage, AL, USA, 30 June–5 July 2013.

energies

MDPI

Article

Reducing Tower Fatigue through Blade Back Twist and Active Pitch-to-Stall Control Strategy for a Semi-Submersible Floating Offshore Wind Turbine

Dawn Ward [1,*], Maurizio Collu [2] and Joy Sumner [3]

1 Department of Energy and Power, Cranfield University, Cranfield, Bedfordshire MK43 0AL, UK
2 Department of Naval Architecture, Ocean & Marine Engineering, University of Strathclyde, Henry Dyer Building, 100 Montrose Street, Glasgow G4 0LZ, UK; maurizio.collu@strath.ac.uk
3 Centre for Thermal Energy Systems and Materials, Cranfield University, Cranfield, Bedfordshire MK43 0AL, UK; j.sumner@cranfield.ac.uk
* Correspondence: Dawn.Ward@cranfield.ac.uk

Received: 11 April 2019; Accepted: 15 May 2019; Published: 18 May 2019

Abstract: The necessity of producing more electricity from renewable sources has been driven predominantly by the need to prevent irreversible climate chance. Currently, industry is looking towards floating offshore wind turbine solutions to form part of their future renewable portfolio. However, wind turbine loads are often increased when mounted on a floating rather than fixed platform. Negative damping must also be avoided to prevent tower oscillations. By presenting a turbine actively pitching-to-stall, the impact on the tower fore–aft bending moment of a blade with back twist towards feather as it approaches the tip was explored, utilizing the time domain FAST v8 simulation tool. The turbine was coupled to a floating semisubmersible platform, as this type of floater suffers from increased fore–aft oscillations of the tower, and therefore could benefit from this alternative control approach. Correlation between the responses of the blade's flapwise bending moment and the tower base's fore–aft moment was observed with this back-twisted pitch-to-stall blade. Negative damping was also avoided by utilizing a pitch-to-stall control strategy. At 13 and 18 m/s mean turbulent winds, a 20% and 5.8% increase in the tower axial fatigue life was achieved, respectively. Overall, it was shown that the proposed approach seems to be effective in diminishing detrimental oscillations of the power output and in enhancing the tower axial fatigue life.

Keywords: floating offshore wind turbine (FOWT); pitch-to-stall; blade back twist; tower fore–aft moments; negative damping; blade flapwise moment; tower axial fatigue life

1. Introduction

Worldwide, climate policies have encouraged the development of energy from renewable resources, such that, within Europe, 11.6% of energy demand was provided solely from wind energy in 2018 [1]. The installed capacity of offshore wind turbines has increased globally by 87%, between 2016 and 2017, rising to a total of nearly 19 GW [2], of which floating turbines currently play only a small part. Indeed, the world's first floating farm, with a capacity of 30 MW only came online in 2017. However, due to the limited availability of offshore sites with shallow beds and low seabed slope, the industry is looking towards floating solutions in the future, with predictions of 5 GW of installed energy from floating wind by 2030. This would amount to 5% of the offshore wind market [2].

Generally, for floating platforms, the horizontal axis wind turbine (HAWT) specifics used for prototypes, model experimentation, and simulation purposes are largely those that have been originally designed for onshore locations. Although they are further modified for the offshore environment, the turbines, including most components and the controllers, have generally been optimized for a foundation fixed into the seabed. This has implications for the applicability of these simulations to floating platforms. Turbine loads, for example, are often increased when the HAWT is mounted on a floating platform, although the exact load, magnitude, and origin will vary, depending upon the platform design and environmental factors. Prominent load increases are usually the induced tower base bending moments, with extreme loads increasing by a factor as high as 1.6 times that seen in a fixed, land-based turbine [3]. Thus, the increased tower base axial fatigue could cause a large reduction in the expected life of the turbine, and therefore, there is a need to have more costly tower and platform structures in order to align with the standard 25–30 yr design life of modern offshore wind farms.

Currently, all large (3 MW+), variable speed, variable pitch, employ HAWTs pitch-to-feather to control the power output above rated wind speeds. Whether these turbines are fixed on dry land, the seabed, or floating offshore, they can experience a phenomenon when wind speeds exceed the rated wind speed, which is generally referred to as "negative damping". This is due to a decrease in rotor thrust at constant rotor speed as the wind speed increases. This can then lead directly to tower fore–aft oscillation, hence, decreased tower axial fatigue life if not controlled, as illustrated in Figure 1 [4]. Fixed turbines can avoid the negative damping effect without a loss in performance as the natural frequency of the blade pitch control system is lower than the first resonance modes of the tower. However, for floating offshore wind turbines (FOWTs), the frequency of the motions (surge, sway, roll, and pitch) are one or more orders of magnitudes lower than for a fixed-base turbine, and therefore, the pitch controller frequency must be lower to avoid negative damping. This causes a reduction in the performance of the rotor speed when operating at above rated wind speeds, such that the ideal constant rotor speed fluctuates around the desired output, leading to more undesirable fluctuations in the generated power output [5].

Horizontal axis wind turbines with variable speed and variable pitch are now considered the industrial standard for large turbines. However, utilising active pitch-to-stall rather than pitch-to-feather control, where the blade is rotated so that the induced aerodynamic stall slows down the rotor to produce constant rotor speed, has largely been ignored. This omission is because the pitch-to-stall configuration can lead to increased blade deflections in larger, softer blades and reduced predictability in terms of induced lift and drag at a specific angle of attack. The issues related to the negative damping phenomenon can, however, be avoided by design when in pitch-to-stall operation, as the rotor thrust at constant rotor speed does not decrease as the wind speed increases. This increasing thrust allows the control frequency to be higher than typically expected on an FOWT, resulting in a much improved rotor speed performance and high-quality, regulated power. As such, utilising pitch-to-stall configurations could thus offer a solution to the performance reduction seen by current, industry-standard pitch-to-feather floating turbines.

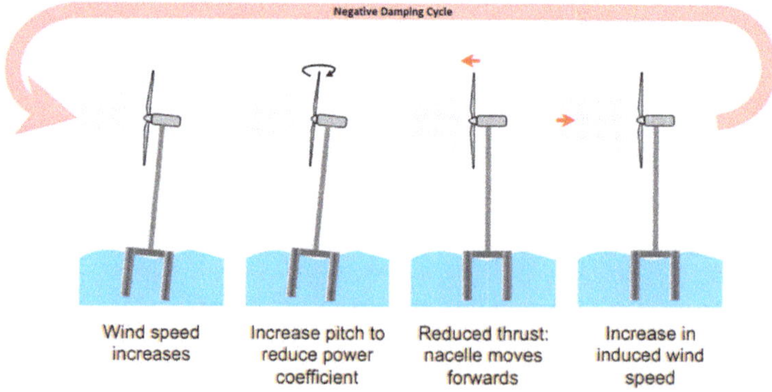

Figure 1. Tower oscillation propagation in a pitch-to-feather horizontal axis wind turbine (HAWT) due to the negative damping cycle. Adapted from Reference [4] Reprinted with permission [van der Veen]; 2012, American Control Conference.

Only two studies, by Larsen and Hanson [5] and Jonkman [6], have been conducted on large, variable speed, floating HAWTs that consider actively pitching the whole blade towards stall as a means to provide a near constant rotor speed at above rated wind speeds. This is partly due to the ambiguity that remains on the aerodynamic behavior of stalled blades [7]. Increased loads and deflection on stalling blades can also dictate that the blades are heavier [8], which would have a negative impact upon the rotor and turbine. The peak loads that can occur on a passive stall blade when power is not shed fast enough would, however, be more controlled when the stall blade is being actively pitched as the power control is more regulated at higher wind speed than in a pitch-to-feather control system. Therefore, although the thrust loads acting on a turbine are increased as the angle of attack increases, it also varies less with pitch angle changes, such that the thrust and the torque have increased stability when operating with pitch-to-stall blades [7]. This reduced variance in the thrust and torque responses will tend to have an overall benefit in terms of tower axial fatigue life. One of the studies simulated turbine, platform, and mooring responses of the Hywind spar concept for a 12 m/s mean turbulent stochastic wind, using a non-linear time domain simulation tool based on SIMO/RIFLEX, developed by Marintex, Trondheim, Norway and HAWC2, developed by DTU (Technical University of Denmark), Risø Campus, Denmark [5]. The simulation results for active pitch-to-stall control indicated not only improved regulation of rotor speed and power generation, but also saw more tower stability. These results were obtained even though the simulations used a blade designed for pitch-to-feather control. In fact, the authors, by accidentally making a pitch angle sign error in the pitch-to-feather control strategy, were using a pitch-to-feather wind turbine system in a pitch-to-stall mode. Hence, the design of the blades, controller, controller gains, and all components were from that original pitch-to-feather system. In the second study, Jonkman [6] looked at simulated responses for a floating barge in 18 m/s mean turbulent stochastic winds when actively pitching-to-stall as well as when actively pitching-to-feather. The results were generated with the aero-servo-elastic simulation package, FAST 7. In the simulations, rotor speed and power regulation again increased with stall control, as compared to feather control. However, this was at the expense of an increase in the tower motion. This difference in the impact of pitch-to-stall on tower stability between the two studies, such that for a spar response changes were beneficial, yet they were detrimental to the barge, indicates that the responses were very platform-type specific. The question of whether tower stability would be improved for a semi-submersible platform operating with a pitch-to-stall rotor has not been addressed to date. The impact of the platform-type is supported by the fact that the barge design also led to instability of the tower in pitch-to-feather control [6]. Although the blade used was also of a pitch-to-feather design (comparable to that used in the Larsen and Hanson [5] study), it had altered aerofoil lift coefficients, leading to a smoother change in lift coefficient being experienced over the range of angle-of-attack experienced along the blade length, during pitch-to-stall control. It was suggested by Jonkman [6] that a higher level of damping in pitch-to-stall could possibly be achieved if the aerofoil coefficients and rotor design were altered such that the rotor thrust continually increased from cut-in to cut-out wind speeds.

The optimisation of blade twist and its resultant effects have been extensively studied for pitch-to-feather and fixed-pitch stall configurations using a number of methods [9–11]. In such instances, the design criteria commonly focuses on maximising the power coefficient and reducing blade vibrations and fatigue. Stäblein et al. [10] achieved increased platform damping, and hence, reductions in tower fore–aft moment from blade twist modifications. For fixed-stall blades, a back twist was employed (the blade twists back towards feather at the tip for a proportion of the blade's length) which resulted in the air flow remaining attached at the blade tip for higher wind speeds, thus increasing the power production capabilities of these sections of the blade [9]. By contrast, the benefits of actively pitching blades are often obtained at the expense of an increased blade pitching rate, and hence, actuator action which can impact on component life and operational costs. However, damage to the pitch bearing was not found to be adversely proportional to pitch rate increase when the blade flapwise loading was also reduced [12]. Merz [9] performed in-depth research on fixed-pitch stall blades and designed a family of optimised blades to minimize the cost of a unit of energy using simulations in the linear frequency domain. Investigation of the effects of including a back twist of a few degrees, originating at different points along the blade's length were performed (starting from 5 to 22% of the blade length as measured from the tip). This resulted in a high level of aerodynamic damping at all operational wind speeds. Stäblein et al. [10] analyzed the effect of additional blade twist from simulations produced utilizing the non-linear time domain aero-servo-elastic simulation tool HAWCStab2, developed by DTU, Risø Campus, Denmark, specifically for wind turbines designed by DTU. This applied a twist to the blades of a land-based 10 MW DTU turbine under pitch-to-feather control. Models investigated changing the blade twist towards both feather and stall from as high up the blade as 70% of the blade radius from the tip. Reductions in both the blade root's flapwise moment and the tower base's fore–aft moment were seen, however, this study was not carried out for a floating turbine. Floating platforms need platform pitch frequencies to be below 0.05 Hz to avoid the wave excitation frequencies, thus the rotor speed would need to be actively controlled to dampen the platform's pitch motion [9].

The question of whether a variable pitch-to-stall, variable speed FOWT could reduce the tower base's fore–aft moment through blade design manipulation has not as yet been studied. Reductions in the blade root flapwise moments could decrease the tower base fore–aft moments, and therefore increase tower fatigue life. Improved rotor speed performance, and therefore higher quality energy regulation would also be anticipated due to the higher control frequency available to a variable pitch-to-stall controller. It is also worth noting that, whilst a dominant wind turbine floating platform type has not yet materialised on the market, it is desirable for research to have the potential to benefit more than one platform type.

To investigate the potential of active pitch-to-stall on a semi-submersible floating platform, this study investigates its possible benefits in terms of rotor speed performance, power generation, and tower base fore–aft moment-induced fatigue reduction. This study was conducted for a 5-MW floating turbine with a stall blade incorporating back twist (towards feather) nearer the tip. This research did not concern itself with the optimization of an actively pitching-to-stall blade or other systems loads, but rather focused on the effects of increasing back twist on tower fore–aft fatigue life, which can in turn be used to optimise the back twist for turbine life extension. This twist optimization was carried out for three specific wind and wave conditions. This is a method proposed by Stäblein et al. [13], as designing around an optimum tip speed ratio is not possible for a back-twisting blade. The 5 MW turbine was coupled to a floating semisubmersible platform, as this type of floater has not been previously investigated with an active pitch-to-stall control strategy and suffers from increased fore–aft oscillations of the tower, and therefore could benefit from this alternative control approach. Furthermore, the Carbon Trust's report for the Scottish government also found that semisubmersibles had a better levelized cost of energy (LCoE) overall, compared to other platform types [14]. Thus, in combination with other benefits (e.g., that a semisubmersible could possibly be towed back to shore for maintenance, repair, and commissioning), this floating platform type is an attractive option [15].

2. Methods

The properties of a semi-submersible floating platform coupled to a 5 MW HAWT are investigated by running simulations utilizing the non-linear time domain FAST v8 simulation tool. The FAST v8 allows aero-hydro-servo-elastic analysis of land, offshore fixed, and offshore floating HAWTs and was developed by NREL (National Renewable Energy Laboratory), Golden, CO, USA [16]. The FAST v8 solves equations of motion by numerical integration utilizing the rigid bodies of the system and from these, expressions for velocity and acceleration are derived. The aerodynamic models within FAST v8 require the user to specify a particular wind data profile, which FAST v8 uses to solve the aerodynamic loads on the blades, including dynamic stall and rotor and wake effects. For the analysis of offshore floating turbine analysis, the aerodynamic models not only generate the previously mentioned outputs, but are also linked to hydrodynamic models of the floating structure as well as dynamic structural and control models. This allows fully-coupled non linear aero-hydro-servo-elastic simulations to be performed within the time domain.

2.1. Model Parameters

For the purposes of this study, the semi-submersible floating platform was modelled with a three line catenary mooring system. This was based on a 1/50th scale experimental model tested in the MARIN (Maritime Research Institute Netherlands) wave basin, as part of the US DeepCWind project [17]. As a result of this, the platform model was used for the offshore simulation code comparison in the international OC4 Phase II study [18,19].

The wind turbine coupled to the semisubmersible was based on the NREL 5MW three-bladed HAWT [20], but with tower mode shapes that depict the expected flexibility of the tower when coupled to the semisubmersible floating platform [18]. The turbine had a hub height above sea water level (SWL) of 90 m, a rotor diameter of 126 m, and was rated at a wind speed of 11.4 m/s, which corresponded to a rated rotor speed of 12.1 rpm. It was designed to cut-in at 3 m/s (rotor speed of 6.9 rpm) and cut-out at 25 m/s. More details on the parameters of both models can be found in Reference [18]. The tower to platform coupling was represented in the turbine model as a cantilevered connection to capture the motions where the tower base joins to the platform. The environmental conditions were based on a site in the North Sea 100 miles northeast of Aberdeen with estimated water depths of 200 m. Such depths were, therefore, unsuitable for fixed base turbines [21]. Three main floating offshore wind turbine models were compared in this study:

1. A conventional "Feather Base Model" with a pitch-to-feather gain scheduling blade pitch controller. This model was used in the OC4 Phase II code comparison investigation [18]. The other models developed were compared to this baseline data.
2. A "Pitch-to-Stall Model" where the blade aerofoil lift coefficients were manipulated for stall operation in line with the blades used by Jonkman [6], such that a smoother change in lift coefficient occurred over the range of angle-of-attack experienced during pitch-to-stall control.
3. Three "Back Twist Pitch-to-Stall Models" further build on the blade properties of the "Pitch-to-Stall Model" by additionally imposing a back twist towards feather on the blade (of 3, 6, and 9°). This third model type operated below rated at a set pitch angle of −4°, as changing the setting of the constant pitch angle below rated wind speed had also been employed to reduce fatigue and optimize the energy output on passive stall turbines [22].

The collective blade pitch-to-stall control system that was designed for both the "Pitch-to-Stall Model" and the "Back Twist Pitch-to-Stall Model", was a relatively simple feedback proportional-integral (PI) controller with constant gains derived at 18 m/s steady wind conditions, as shown in Figure 2. Utilizing a gain scheduler that relies upon the pitch angle at any moment in time to predict the gain change required would be much more complex with an active pitch-to-stall control. This is due to the double-valued nature of the pitch angle within region 3, as highlighted within Section 3.1. This controller pitches the blades to achieve a constant rotor speed when winds are above 11.4 m/s and incorporates a pitch-rate limiter set to +/−8°. This pitch-to-stall controller was implemented in Simulink, which was coupled directly with the FAST v8 simulation code. It expanded upon a simple "constant gains feedback" PI pitch controller designed for pitch-to-feather control by Wright and Fingersh [22].

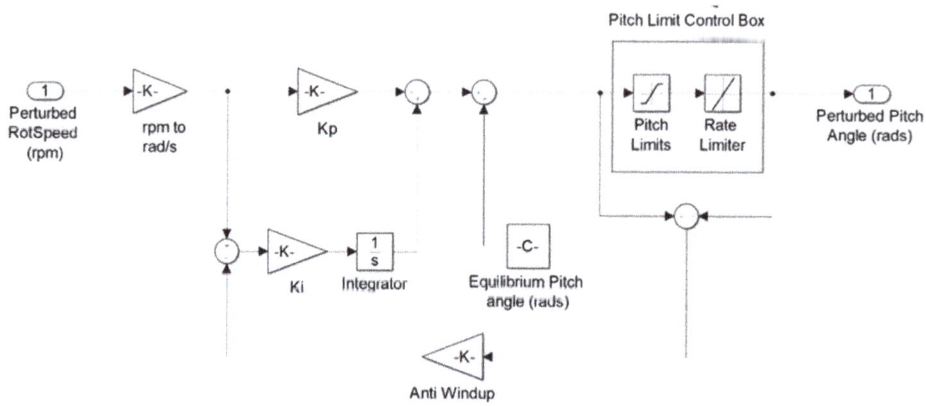

Figure 2. Pitch-to-Stall proportional-integral (PI) controller for the "Pitch-to-Stall Model" and all "Back Twist Pitch-to-Stall Models".

To derive the pitch-to-stall controller gains that will ensure that the performance of the turbine results in a higher quality regulated power output over the full range of operational wind speeds from rated to cut-out, a single degree of freedom (DOF) system (the angular shaft rotation speed) was considered as presented by Larsen and Hanson [5] and Jonkman [6]. From this, Equations (1)–(2) can be derived when the derivative gain, KD, neglected for simplicity, as proposed by Hanen et al. [23].

$$K_P = (2 \cdot I_{DRIVETRAIN} \cdot \Omega \cdot \xi_{\phi} \cdot w_{\phi n}) / N_{GEAR} \cdot (-(\text{d}P/\text{d}\emptyset)) \quad (1)$$

$$K_I = (I_{DRIVETRAIN} \cdot \Omega \cdot w_{\phi n}{}^2) / N_{GEAR} \cdot (-(\text{d}P/\text{d}\emptyset)) \quad (2)$$

where $I_{DRIVETRAIN}$ is the driveshaft inertia (kg/m^2) found from Equation (3); Ω is the rotational speed (rad/s); ξ_{ϕ} is the damping ratio related to the rotor speed error control system; $w_{\phi n}$ is the natural frequency of the control system (rad/s); N_{GEAR} is the gearbox ratio, and $\text{d}P/\text{d}\emptyset$ is the sensitivity of the aerodynamic power to the blade pitch angle (W/rad).

$$I_{DRIVETRAIN} = I_{ROTOR} + (N_{GEAR}{}^2 \cdot I_{GEN}) \quad (3)$$

where I_{ROTOR} is the rotor inertia (kg/m^2) and I_{GEN} is the generator inertia (kg/m^2).

The sensitivity of the aerodynamic power to the blade pitch angle for each model was derived in FAST v8.16.00 a-bjj through linearization of the turbine simulation model utilizing the pitch angles and rotor speed of each wind speed, as predicted in the periodic steady-state simulations detailed in Section 3.1. This followed the procedure presented for aerodynamic derivatives by Jonkman and Jonkman [24] such that all the structural DOF were disabled to eliminate convergence issues and frozen wake was applied. Values of 0.6 rad/s were used for the natural frequency of the control system and 0.7 for the damping ratio related to the rotor speed error control system as suggested in Hanen et al. [23]. This system frequency of the pitch-to-stall controller (0.6 rad/s) is much higher than the 0.2 rad/s used by the "Feather Base Model" pitch-to-feather controller, as negative damping is avoided with the pitch-to-stall control regime.

2.2. Blade Back Twist Configuration

The back twists employed in this study were chosen after investigative research of the altered dynamic coupling effects on the tower, due to the changes in the blade geometry, specifically in terms of tower base fore–aft moment. The initial selection process looked at back twist applied at initiation points from 5% to 35% in 10% intervals along the blade length (measured from the tip), with a "best guess" back twist angle of 3°, based on previous research involving back twist on passive stall blades [9] and blades that pitch-to-feather [10]. Responses during mean turbulent winds of 18 m/s for short 600 s simulations were examined. The effects on tower fore–aft bending moment were then analyzed when both increasing and decreasing the back-twist angle, which indicated a reduction in tower fore–aft moment range could be achieved with a higher back twist angle of 9°. The initiation point of the back twist was chosen to be at 25% along the blade length, from the tip. Figure 3 gives a visual reference to the three back twist angles applied to the "Back Twist Pitch-to-Stall Model", alongside the original twisted NREL 5 MW blade which is used for both the "Feather Base Model" and the "Pitch-to-Stall Model".

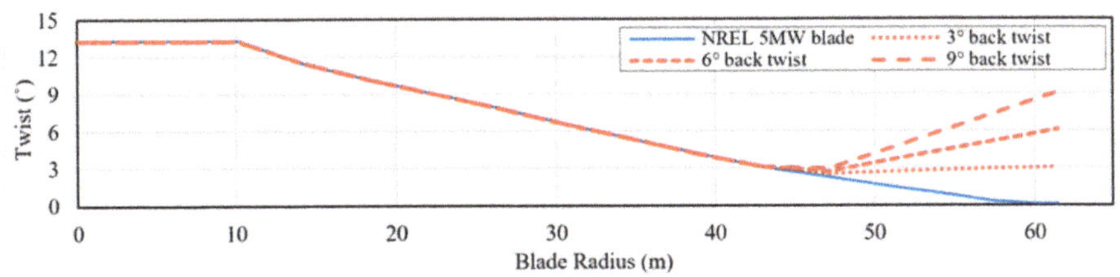

Figure 3. Twist profiles for pitch-to-stall blades in the tower axial fatigue life study.

A constant −4° pitch angle was also applied at all winds below rated to the three "Back Twist Pitch-to-Stall Models". This was introduced as both fatigue life and energy generation, which were increased when it was incorporated into the back-twist designs, as compared to a constant pitch angle of 0° or −2.5°. Model specifications and controller gains for each stall model are shown in Table 1.

Table 1. Model parameters in terms of back twist, pitch below rated setting, and blade pitch controller gains.

Model ID	Feather Base Model	Pitch-to-Stall (P-S) Model	3° Back Twist P-S Model	6° Back Twist P-S Model	9° Back Twist P-S Model
Constant pitch angle below rated (°)	0	0	−4°	−4°	−4°
Twist starting distance from blade root (%)	N/A	N/A	75	75	75
Back twist angle increase towards feather at the tip (°)	0	0	3	6	9
Proportional gain at 18mps, Kp (s)	Inbuilt gain scheduler *	−0.865	−0.917	−1.105	−1.414
Integral gain at 18mps, Ki		−0.371	−0.391	−0.474	−0.606

* As applicable to the OC4 (Offshore Code Comparison Collaboration Continuation) Phase II simulation model.

3. Results

3.1. Periodic Steady Wind Characteristics of a Blade Pitching-to-Stall

To compare the FOWT characteristics of the "Feather Base Model", represented by the blue dashed lines in Figure 4, with the "Pitch-to-Stall Model", represented by the solid red lines, periodic steady-state curves were derived from simulations. The periodic nature of the steady state was created as the variables did not have a constant value but a range of values that repeat over a given period. For all of the FOWT models presented here, the periodic influence was due to the tilt applied to the rotor shaft and the variables were repeated after each full rotation of the rotor. These simulations were run with steady uniform wind speeds from cut-in to cut-out, in 1 m/s increments, with all the platform DOF disabled (i.e., like a fixed land-based wind turbine) as suggested by Jonkman [6] to ensure that the semi-submersible floating platform is not excited by the applied wind and waves.

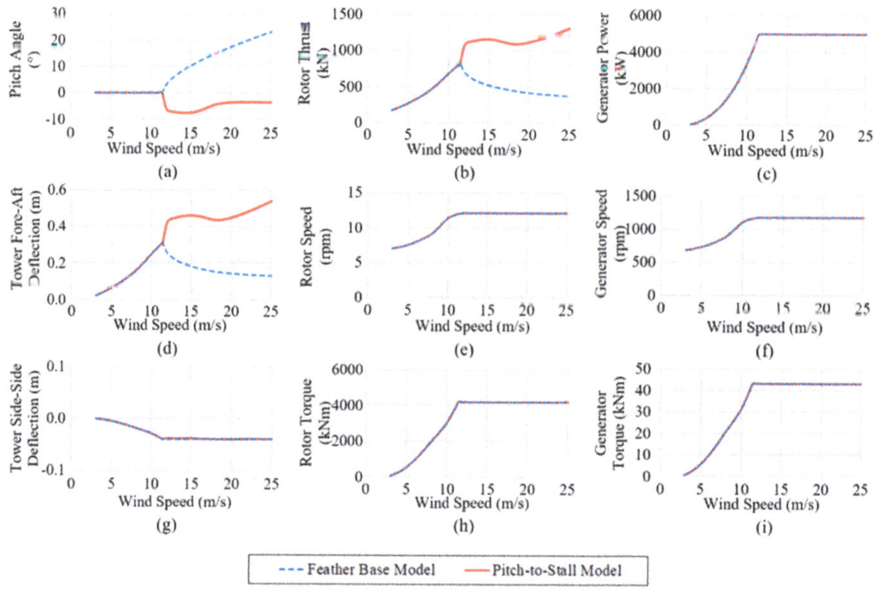

Figure 4. Simulated FOWT characteristics from periodic steady winds for two different blade models: (blue) pitch-to-feather and (red) pitch-to-stall. (**a**) pitch angle (°); (**b**) rotor thrust (kN); (**c**) generator power (kW); (**d**) tower fore-aft deflection (m); (**e**) rotor speed (rpm); (**f**) generator speed (rpm); (**g**) tower side-side deflection (m); (**h**) rotor torque (kNm); (**i**) generator torque (kNm).

Figure 4a highlights the difference in the pitch angles required for both control methods. For feather control, a continuous, smooth rise in the blade pitch angle required was seen from 0° at the rated wind speed of 11.4 m/s, to just exceeding 23° at the cut-out wind speed of 25 m/s. For stall control, the negative pitch angle requirement rose steeply at wind speeds just above rated, the area of operation often referred to as Region 2.5, and a maximum negative pitch angle (approaching −8°) was experienced at wind speeds as low as 15 m/s. The pitch angle then slowly decreased in magnitude again as wind speeds increase to over 20 m/s, after which the required blade pitch angle was maintained at a fairly constant level of approximately −3.5° before the FOWT cut-out at 25 m/s.

From the simulated rotor thrusts in Figure 4b, it can be seen that, at wind speeds above rated, the rotor thrust curves were the inverse of the required pitch angle curves for both the pitch-to-feather and pitch-to-stall control methods. The tower fore–aft deflections curves, shown in Figure 4d, followed the same pattern as the rotor thrusts experienced by the relative control method. The outputs of all other parameters observed in Figure 4 were identical and are presented to show the specific responses for both control methods.

3.2. Integrator Saturation Issue

Initial investigation into the response and load analysis of the "Pitch-to-Stall Model" in turbulent winds, when the floating wind turbine was operating near to the rated wind speed, highlighted a problem with the rotor speed increasing with simulation time to velocities beyond the design limit, as seen in Figure 5a. As rotor speed directly influences the generated power, this was also negatively affected, as seen in Figure 5b. This rotor speed increase occurred during sudden changes from wind speeds below rated to wind speeds above rated and was due to the pitch controller actuator reaching its upper limit. When this occurred, it caused the feedback loop to become an open loop, thus the saturated actuator remained at its upper limit even when the output changed. This error then continued to be integrated by the PI controller which caused the integrator to be "wound up", as observed by Wright and Fingersh [22].

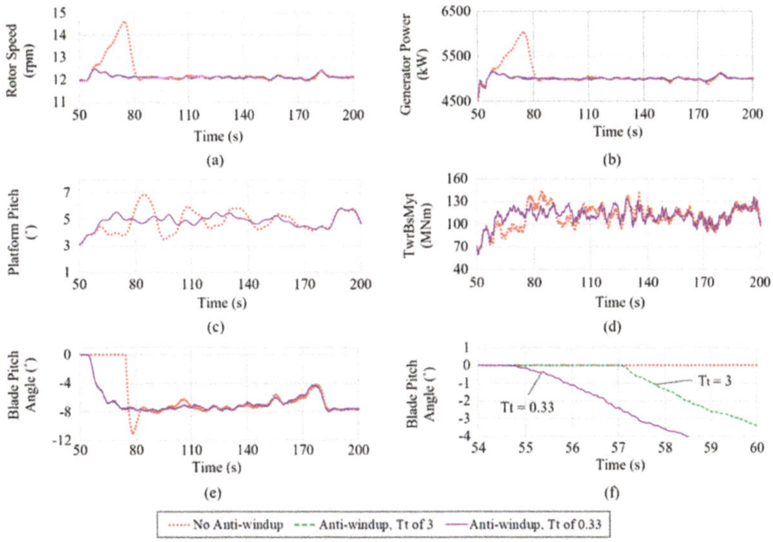

Figure 5. Simulated responses of the "Pitch-to-Stall Model" with and without anti-windup at 13 m/s mean turbulent winds. (**a**) rotor speed (rpm); (**b**) generator power (kW); (**c**) platform pitch (°); (**d**) tower base fore-aft bending moment (TwrBsMyt) (MNm); (**e**) blade pitch angle (°); (**f**) blade pitch angle (°).

As well as a large overshoot in the blade pitch angle, such that the angle increases from around −8° to nearly −12° (Figure 5e), oscillations were also seen as the system returned to a stable state. This can clearly be seen by the response shown in red in Figure 5c,d for the platform pitch motion and the tower fore–aft bending moment, respectively. These oscillations lasted for up to 100 s after each rotor over-speed occurrence. These undesirable responses, along with rotor speeds far exceeding the design limit, cannot be tolerated by the turbine, hence, an anti-windup filter was incorporated to feed back into the pitch-to-stall controller design (Figure 2). The anti-windup filter changes the gain value of the integral during saturation and prevents the rotor speed from erroneously over-shooting. The anti-windup filter is inactive except when the floating wind turbine is operating near to the rated wind speed, as the pitch angle is the same at the input and output of the pitch-limit control box under these conditions (Figure 2).

3.3. Normalized Outputs in Turbulent Wind Conditions

A selection of the most relevant load analysis results are presented as ratios against the "Feather Base Model", in Figure 6, obtained with all the DOF enabled except for the rotor-teeter DOF, which is not applicable to a three-bladed turbine. Each output was calculated from 3 × 3 hr simulations each generated with three different random wind and wave seeds, compiled from periodic winds files, such that the wind distribution was repeated every 10 min throughout the full simulation time. The 3 hr simulation length was chosen to improve the fatigue analysis, as the responses of FOWTs vary more with time than fixed turbines. The tower axial fatigue life was derived using a Rainflow cycle counting algorithm and Miner's Rule within MATLAB, developed by The MathWorks, Natick, Massachusetts, USA [25]. The stochastic turbulent mean winds were generated using the Kaimal wind spectrum. Results presented as a range represent the full range of motion experienced from the maximum to the minimum values for the combined simulations. Simulations are presented under mean winds of 8, 13, and 18 m/s, each with an IEC (International Electrotechnical Commission) turbulent intensity model B

(0.14) [20], a power law exponent of 0.14, and a surface roughness length of 0.03 m. All environmental conditions are the same for each model at each mean wind speed. The first 300 s of each simulation was discarded to avoid transient responses.

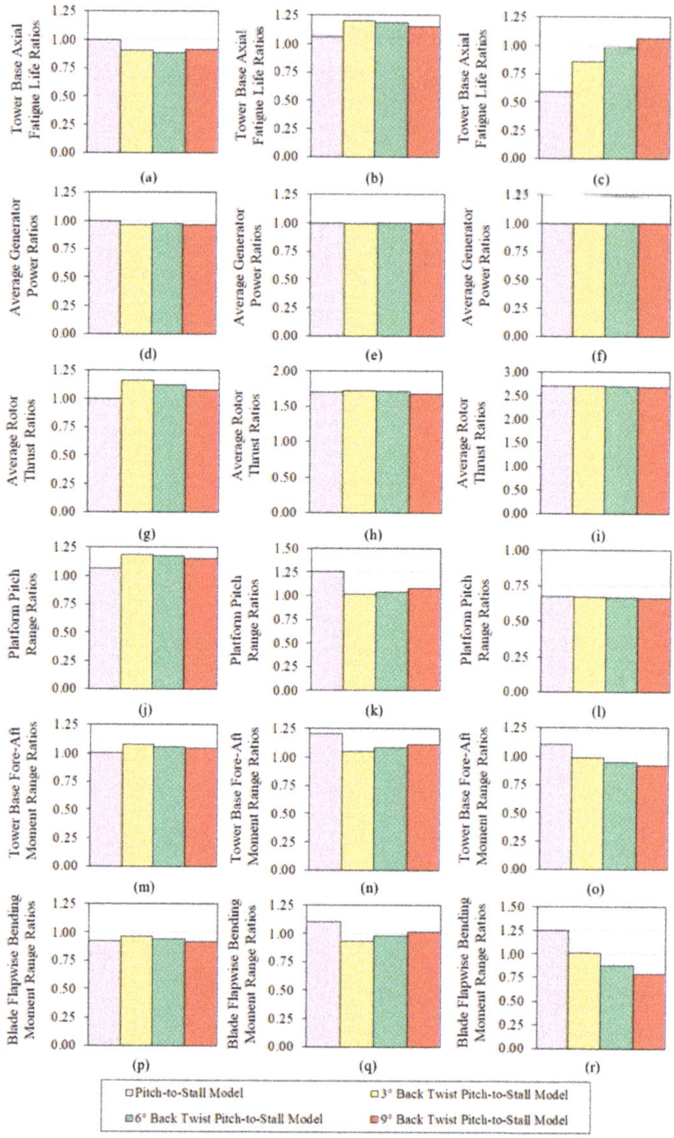

Figure 6. Simulation outputs for pitch-to-stall blade models with and without back twist. Outputs were normalized against the "Feather Base Model" and given for 8 (left), 13 (center), and 18 m/s (right) mean turbulent winds. (**a**) tower base axial fatigue life ratios (TBAFLR)(8mps); (**b**) (TBAFLR) (13mps); (**c**) (TBAFLR) (18mps); (**d**) average generator power ratios (AGPR) (8mps); (**e**) AGPR (13mps); (**f**) AGPR (18mps); (**g**) average rotor thrust ratios (ARTR) (8mps); (**h**) ARTR (13mps); (**i**) ARTR (18mps); (**j**) platform pitch range ratios (PPRR) (8mps); (**k**) PPRR (13mps); (**l**) PPRR (18mps); (**m**) tower base fore-aft moment range ratios (TBFARR) (8mps); (**n**) TBFARR (13mps); (**o**) TBFARR (18mps); (**p**) blade flapwise bending moment range ratios (BFBMRR) (8mps); (**q**) BFBMRR (13mps); (**r**) BFBMRR (18mps).

At 8 m/s mean wind speed, the tower base axial fatigue life of the untwisted Pitch-to-Stall Model, was seen to be identical to the "Feather Base Model" and higher than all "Back Twist Pitch-to Stall Models" (Figure 6a). For mean winds of 13 m/s, seen in Figure 6b, the tower axial fatigue life increased for all stall scenarios, with a maximum extension in life of over 20% when the back twist was 3°. However, at the 18 m/s mean wind speed, shown in Figure 6c, it was only with the more extreme back twist of 9° that the tower life expectancy exceeded that achieved with pitch-to-feather control. The increase in tower axial fatigue life for this high back twist was 5.8%; however, the average generated power had a slight decrease for the back twisted pitch-to-stall models at the below rated wind speed (Figure 6d).

The rotor thrust was higher for all the blade models with a back twist for mean winds of 8 m/s (Figure 6g); however, the thrust decreased as the twist angle increased. The variation in rotor thrust was higher when moving from pitch-to-feather to pitch-to-stall at both mean wind speeds above rated; an increase of around 1.7×. at mean winds of 13 m/s (Figure 6h), rising to 2.7× at 18 m/s (Figure 6i). The variation between pitch-to-stall models, however, was negligible. The platform pitch response at mean wind speeds of both 8 and 13 m/s, as seen in Figure 6j,k, was in accordance with the tower base fore–aft moment range (Figure 6m,n), and hence corresponded inversely to the tower base axial fatigue life (Figure 6a,b).

At the higher mean wind speed of 18 m/s, the platform pitch range, shown in Figure 6l, was much lower for all stall blades than for the "Feather Base Model" at a ratio of approximately 0.67 with only a 1.2% variation between stall blades' responses. Reductions in tower base fore–aft moment range were seen with back twist angle increases (Figure 6o). The blade flapwise bending moment was much reduced by the introduction of back twist when simulated in 18 m/s mean turbulent winds (Figure 6r) by over 20% for the most extreme back twist angle of 9°, when compared to the "Feather Base Model".

3.4. Performance Characteristics in Turbulent Winds

Relevant turbine responses for the "Feather Base Model" are shown alongside the original "Pitch-to-Stall Model" and the 9° "Back Twist Pitch-to-Stall Model" in Figure 7. These were derived with all DOF enabled (excluding rotor-teeter). The presented data are for one randomly selected wind and wave seed combination and highlight the performance of the relevant responses arising from mean wind speed, model, and controller variations. However, all numerical comparisons are the averages from the 3 × 3hr simulations, as detailed in Section 3.3. It should be noted that the results between the two stall models are within 1% of each other for the variables presented in Figure 7g–l, making no apparent distinction visible.

In Figure 7e, at mean turbulent winds of 13 m/s, the blade pitch angle range was reduced by over 2° with the 9° "Back Twist Pitch-to-Stall Model" compared to the "Pitch-to-Stall Model". At 13 m/s, the variation in the rotor speed response seen in Figure 7h was 4.7 rpm for the "Feather Base Model" and 1.4 rpm for all pitch-to-stall models. The rotor speed range at 18 m/s was approximately 0.2 rpm for the pitch-to stall models, whereas the "Feather Base Model" range was 4.8 rpm (Figure 7i). The generated power also had a smoother response with pitch-to-stall control in above rated winds, with an average variance for all pitch-to-stall controlled models of 1980 kW compared to 3718 kW for pitch-to-feather control at 13 m/s (Figure 7k). The variation in generated power at 18 m/s was approximately 101 kW for all pitch-to-stall models, compared to 1997 KW with pitch-to-feather control (Figure 7l). The rotor thrust response range shown in Figure 7n at 13 m/s was not lower overall for the pitch-to-stall models; however, the variation was reduced with the implementation of the 9° "Back Twist Pitch-to-Stall Model", which smoothed the response by 10% compared to the "Pitch-to-Stall Model". At 18 m/s mean winds shown in Figure 7o, the rotor thrust was more constant overall with pitch-to-stall control, with the 9° "Back Twist Pitch-to-Stall Model" again have a further 10% thrust reduction compared to the "Pitch-to-Stall Model" response.

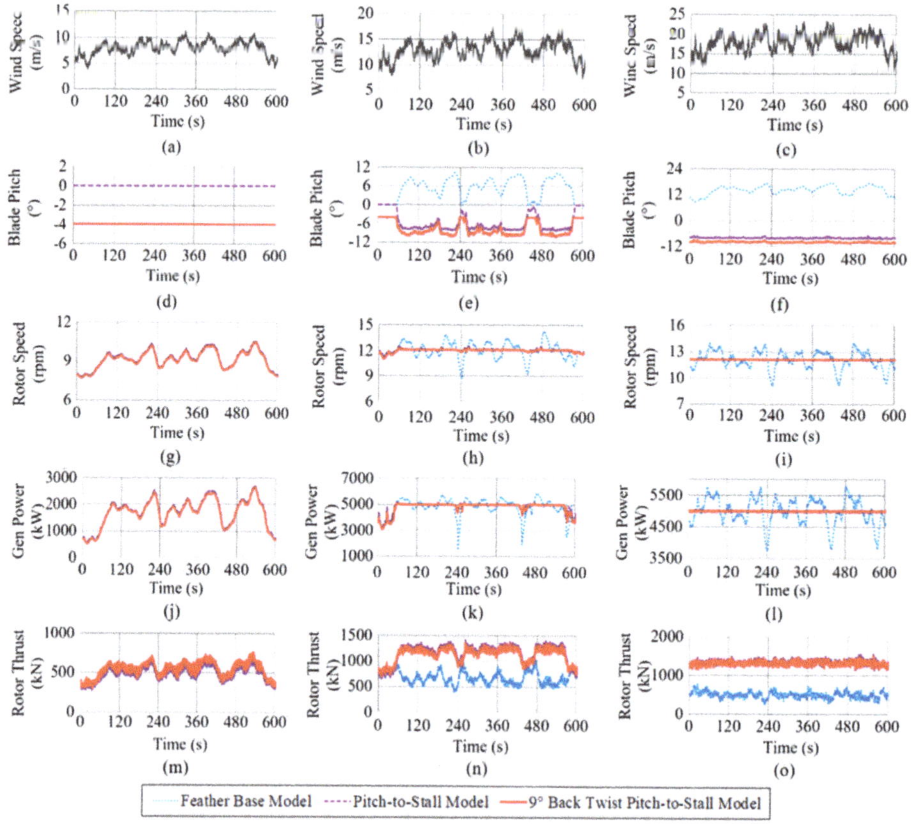

Figure 7. Simulation outputs for pitch-to-feather and pitch-to-stall blade models without and with 9° back twist at 8 (left), 13 (center), and 18m/s (right) mean turbulent winds. (**a**) wind speed, m/s (8mps); (**b**) wind speed, m/s (13mps); (**c**) wind speed, m/s (18mps); (**d**) blade pitch, ° (8mps); (**e**) blade pitch, ° (13mps); (**f**) blade pitch, ° (18mps); (**g**) rotor speed, rpm (8mps); (**h**) rotor speed, rpm (13mps); (**i**) rotor speed, rpm (18mps); (**j**) generator power, kW (8mps); (**k**) generator power, kW (13mps); (**l**) generator power, kW (18mps); (**m**) rotor thrust, kN (8mps); (**n**) rotor thrust, kN (13mps); (**o**) rotor thrust, kN (18mps).

4. Discussion

For stall operation, a big reduction in the blade pitch actuation range (defined as the pitch angle) was seen (Figure 4a), a requirement of approximately 8° compared to a 23° for feather pitch control. This does not, however, correlate into a reduction in blade pitch actuation rate, as this tends to increase whenever control frequencies are higher, as allowed in pitch-to-stall control.

For the "Feather Base Model", the continuous reduction in rotor thrust for above rated wind speeds (Figure 4b) would contribute to negative damping, and hence turbine tower instabilities and oscillations, if negative damping was not avoided by reducing the natural frequency of the pitch-to-feather control system. The "Pitch-to-Stall Model" also experiences a dip in the rotor thrust seen at around 17 to 18 m/s wind speeds. This negative derivative of the thrust versus the wind speed could introduce negative damping of the platform pitch mode, as the thrust does not increase within this section because excess power is discarded to keep the rotor speed constant. Ideally a pitch-to-stall model could be designed to produce a continuous rise in rotor thrust.

All turbine responses at 8 m/s mean turbulent winds, shown in the left hand column of Figure 6, are not influenced by the blade pitch control method as the pitch angle is constant below 11.4 m/s Therefore, the increase in the simulated platform pitch response in Figure 6j and decrease in blade flapwise bending moment range in Figure 6p experienced by the "Pitch-to-Stall Model" compared to the "Feather Base Model", must be solely due to the changes in blade aerofoil lift coefficients. The "Back Twist Pitch-to-Stall Models" are further influenced at 8 m/s mean wind speeds by the −4° constant blade pitch angle and the specific back twist angle for each model.

When operating in mean winds of 13 m/s, all the pitch-to-stall models experienced a beneficial increase in the tower axial fatigue life, shown in Figure 6b, when compared to the "Feather Base Model". This was further increased by the introduction of a back twist that extended the tower axial fatigue life by over 20% when the back twist was 3°. At this mean wind speed, a correlation between tower fore–aft moment's range reduction (Figure 6n) and platform pitch (Figure 6k) and blade flapwise bending moment (Figure 6q) reductions can be made.

At 18 m/s mean turbulent winds, the tower axial fatigue life of the "Pitch-to-Stall Model" was very low and the influence of applying and then increasing the back twist are very apparent. However, it is only with the more extreme 9° "Back Twist Pitch-to-Stall Model" that the tower axial life expectancy was exceeded, achieved by the "Feather Base Model" (Figure 6c). The increase in tower axial fatigue life is a very desirable 5.8%. A strong correlation between the tower fore–aft base moment (Figure 6o) and the blade flapwise bending moment (Figure 6r) was apparent at 18 m/s mean winds, such that as the flapwise bending range reduces so does the tower base fore–aft moment, and hence the fatigue life increases. However, this does not hold true when compared against the "Feather Base Model", as there are scenarios where the blade flapwise bending moment range is higher than that seen for feather control, but the fatigue life is still improved for the stall cases and vice versa. This could be due to the different profile shapes of the blade relative to the axial bending direction, when pitching-to-feather versus pitching-to stall. Further investigation would be necessary to explore this theory. The platform pitch range seen in Figure 6l does not influence the tower fore–aft moment, and hence the tower fatigue life at this mean wind speed.

An account of the site wind speed probability density function for the three wind profiles was taken, such that the tower axial fatigue life was weighted depending upon its frequency of occurrence for the three mean wind speeds examined. A larger increase in tower base axial fatigue life was achieved with this weighting; with the higher 9° "Back Twist Pitch-to-Stall Model", a 2.8% increase was achieved compared to the "Feather Base Model". This came at the expense of a decrease in the power generation of 1%.

Pitch-to-stall control of the HAWT coupled the semi-submersible floating platform provided a more regulated energy output than was achieved with the "Feather Base Model". The average variance for both pitch-to-stall models was 1980 kW compared to 3718 kW for the "Feather Base Model" at 13 m/s (Figure 7k), and 101 kW compared to 1997 KW at 18 m/s (Figure 7l). This is due to the higher pitch-to-stall control system frequency setting allowed, as negative damping is avoided by design.

5. Conclusions

In the present day, the requirement to increase the renewable portion of our energy supply worldwide is becoming more imperative each day. The offshore wind capacity has been increasing rapidly, by 87% between 2016 and 2017, in order to meet these needs [2] and due to the limited shallow bed and low seabed slope, the industry is looking more towards floating offshore wind solutions in the future. However, turbine tower loads are often increased when a turbine is mounted on a floating platform and unless these undesirable motions can be significantly reduced, then the strength of the tower must be increased.

This research is therefore concerned with reducing excessive motions and loads associated with FOWTs and their effects on the turbine tower, while improving energy generation by pitching the turbine to stall, instead of feather, to control rotor speed which avoids the negative damping issues by design, resulting in an improvement in the regulation of the power generated. The example FOWT presented is a semi-submersible floating platform, coupled to a 5 MW HAWT, investigated through simulation.

Overall, it has been shown that the proposed approach seems to be effective in diminishing detrimental oscillations of the power output and in enhancing the tower axial fatigue life. This is achieved by variable speed, variable pitch-to-stall control, reducing the negative damping phenomenon (more regular power output), and through manipulating the blade design by the introduction of a back twist towards feather (reductions in the blade root flapwise moment leading to reductions in the tower bending moments, and hence increasing the tower axial fatigue life). This research, therefore, highlights that active variable speed and variable pitch-to-stall control has potential benefits specific to FOWTs.

More specifically, tower base axial fatigue life was increased for all pitch-to-stall scenarios, when operating in 13 m/s mean turbulent winds. The introduction of a back twist added an additional desirable increase of over 20% when the back twist was 3°. This correlated with a corresponding decrease in both platform pitch and blade flapwise bending moment ranges of motion.

At 18 m/s mean winds, the more extreme 9° "Back Twist Pitch-to-Stall Model" provided a tower base life expectancy that exceeded that achieved by the "Feather Base Model" by 5.8%. No correlation between the platform pitch response and tower base fore–aft moment was apparent at this mean wind speed. However, a strong correlation between the ranges of motion from the blade flapwise bending and the tower base moment was seen, which strongly indicates that decreasing the blade flapwise bending range decreases the tower fore–aft moment range.

Overall, the 9° "Back Twist Pitch-to-Stall Model" achieved the best tower base axial fatigue life with a 2.8% increase on the "Feather Base Model" when accounting for the relevant frequencies of occurrence of the three mean turbulent wind speed profiles, for the specific site, at the expense of a decrease in the power generation of 1%. At wind speeds above rated, the performance of pitch-to-stall in terms of rotor speed regulation and generated power output far exceeds that achieved with pitch-to-feather.

Further investigation with more variations in terms of twist initiation point, back twist angle, and constant pitch below rated angles would be recommended. Additional research on a complete back-twisted stall blade design would also be required to enable an in-depth analysis of all the blade loads as well as loads on the nacelle, gearbox, drive train, and other system components. Hence, enabling an overall assessment of the pros and cons of a back twist blade for a FOWT with active pitch-to-stall control. A study on the benefits of this blade and pitch control system on other FOWT types would also be beneficial.

Author Contributions: Formal analysis, D.W.; investigation, D.W.; methodology, D.W.; supervision, M.C. and J.S.; validation, D.W.; writing—original draft, D.W.; writing—review and editing, M.C. and J.S.

Funding: This work was funded by the United Kingdom Engineering and Physical Sciences Research Council (EPSRC), grant EP/L016303/1, for Doctoral Training in Renewable Energy Marine Structures (REMS).

Conflicts of Interest: The authors declare no conflict of interest.

References

1. WindEurope. Available online: https://windeurope.org/about-wind/daily-wind/about (accessed on 24 January 2019).
2. Fried, L.; Qiao, L.; Sawyer, S. Global Wind Report: Annual Market Update 2017. Global Wind Energy Council. Available online: http://files.gwec.net/files/GWR2017.pdf (accessed on 10 April 2019).
3. Robertson, A.N.; Jonkman, J.M. Loads Analysis of Several Offshore Floating Wind Turbine Concepts. In Proceedings of the 21st International Society of Offshore and Polar Engineers Conference, Maui, HI, USA, 19–24 June 2011; pp. 443–450.
4. van der Veen, G.J.; Couchman, I.J.; Bowyer, R.O. Control of floating wind turbines. In Proceedings of the 2012 American Control Conference (ACC), Montreal, QC, Canada, 27–29 June 2012; pp. 3148–3153. [CrossRef]

5. Larsen, T.; Hanson, T.D. A Method to Avoid Negative Damped Low Frequent Tower Vibrations For A Floating, Pitch Controlled Wind Turbine. *J. Phys. Conf. Ser.* **2007**, *75*, 12073. [CrossRef]
6. Jonkman, J.M. *Dynamics Modeling and Loads Analysis of an Offshore Floating Wind Turbine*; Technical Report: NREL/TP-500-41958; National Renewable Energy Laboratory: Golden, CO, USA, November 2007.
7. Bossanyi, E.A. Wind Turbine Control for Load Reduction. *Wind Energy* **2003**, *6*, 229–244. [CrossRef]
8. Macquart, T.; Maheri, A. A Stall-Regulated Wind Turbine Design to Reduce Fatigue. *Renew. Energy* **2019**, *133*, 964–970. [CrossRef]
9. Merz, K.O. Conceptual Design of a Stall-Regulated Rotor for a Deepwater Offshore Wind Turbine. Ph.D. Thesis, Norwegian University of Science and Technology, Trondheim, Norway, June 2011.
10. Stäblein, A.R.; Hansen, M.H.; Verelst, D.R. Modal Properties and Stability of Bend–Twist Coupled Wind Turbine Blades. *Wind Energy Sci.* **2017**, *2*, 343–360. [CrossRef]
11. Méndez, J.; Greiner, D. Wind Blade Chord and Twist Angle Optimization Using Genetic Algorithms. In Proceedings of the Fifth International Conference on Engineering Computational Technology, Las Palmas de Gran Canaria, Spain, 12–15 September 2006. [CrossRef]
12. Barlas, A.; Bergami, L.; Hansen, M.H.; Pedersen, M.M.; Verelst, D.; Thomsen, K.; Aagaard, H. Load Alleviation Potential of Active Flaps and Individual Pitch Control in a Full Design Load Basis. In Proceedings of the EWEA Annual Event and Exhibition 2015, Paris, France, 17–20 November 2015; European Wind Energy Association (EWEA): Paris, France, 2015.
13. Stäblein, A.R. Analysis and Design of Bend-Twist Coupled Wind Turbine Blades. In *MARE-WINT*; Springer: Cham, Switzerland, 2016; pp. 67–80.
14. James, R.; Ros, M.C. *Floating Offshore Wind: Market and Technology Review*; Prepared for the Scottish Government June 2015; The Carbon Trust: London, UK, 2015.
15. Thiagarajan, K.P.; Dagher, H.J. A Review of Floating Platform Concepts for Offshore Wind Energy Generation. *J. Offshore Mech. Arct. Eng.* **2014**, *136*, 020903. [CrossRef]
16. Jonkman, J.M.; Buhl, M.L., Jr. *FAST User's Guide*; Technical Report: NREL/EL-500-38230; National Renewable Energy Laboratory: Golden, CO, USA, August 2005.
17. Goupee, A.J.; Fowler, M.J.; Kimball, R.W.; Helder, J.; de Ridder, E.-J. Additional Wind/Wave Basin Testing of the DeepCwind Semi-Submersible with a Performance-Matched Wind Turbine. In Proceedings of the ASME 2014 33rd International Conference on Ocean, Offshore and Arctic Engineering, San Francisco, CA, USA, 8–13 June 2014; Volume 9B: Ocean Renewable Energy. p. V09BT09A026. [CrossRef]
18. Robertson, A.; Jonkman, J.; Masciola, M.; Song, H.; Goupee, A.; Coulling, A.; Luan, C. *Definition of the Semisubmersible Floating System for Phase II of OC4*; Technical Report: NREL/TP-5000-60601; National Renewable Energy Laboratory: Golden, CO, USA, September 2014.
19. Robertson, A.; Jonkman, J.; Vorpahl, F.; Popko, W.; Qvist, J.; Frøyd, L.; Chen, X.; Azcona, J.; Uzunoglu, E.; Guedes Soares, C.; et al. Offshore Code Comparison Collaboration Continuum: Phase II Results Regarding a Floating Semisubmersible Wind System. In Proceedings of the 22nd International Society of Offshore and Polar Engineers Conference, Rhodes, Greeze, 17–22 June 2012. [CrossRef]
20. Jonkman, J.; Butterfield, S.; Musial, W.; Scott, G. *Definition of a 5-MW Reference Wind Turbine for Offshore System Development*; Technical Report: NREL/TP-500-38060; National Renewable Energy Laboratory: Golden, CO, USA, February 2009.
21. GEOS, F. *Wind and Wave Frequency Distributions for Sites Around the British Isles*; Offshore Technology Report 2001/030; HSE Books: Norwich, UK, 2001.
22. Wright, A.D.; Fingersh, L.J. *Advanced Control Design for Wind Turbines; Part I: Control Design, Implementation, and Initial Tests*; Technical Report: NREL/EL-500-42437; National Renewable Energy Laboratory: Golden, CO, USA, March 2008.
23. Hanen, M.H.; Hansen, A.; Larsen, T.J.; Øye, S.; Sørensen, P.; Fuglsang, P. *Control Design for a Pitch-Regulated, Variable Speed Wind Turbine*; Wind Energy Department Report: RisØ-R-1500(EN), Risø; National Laboratory: Roskilde, Denmark, January 2005.
24. Jonkman, J.; Jonkman, B. FAST Modularization Framework for Wind Turbine Simulation: Full-System Linearization. Presented at the 2016 Science of Making Torque from Wind Conference, Munich, Germany, 5–7 October 2016.

25. Kvittem, M.I.; Moan, T. Time Domain Analysis Procedures for Fatigue Assessment of a Semi-Submersible Wind Turbine. *Mar. Struct.* **2015**, *40*, 38–59. [CrossRef]
26. BS EN 61400-1. *Wind Turbines: Part 1: Design Requirements*; British Standards: London, UK, 2005.

energies

MDPI

Article

Assessment of Primary Energy Conversion of a Closed-Circuit OWC Wave Energy Converter

Pierre Benreguig [1,*], Vikram Pakrashi [1,2] and Jimmy Murphy [1]

1 MaREI Centre, Beaufort building, University College Cork, Haubowline Road, P43C573 Ringaskiddy, Co. Cork, Dublin 4, Ireland; vikram.pakrashi@ucd.ie (V.P.); jimmy.murphy@ucc.ie (J.M.)
2 Dynamical Systems and Risk Laboratory, School of Mechanical and Materials Engineering, University College Dublin, Dublin 4, Ireland
* Correspondence: pierre.benreguig@ucc.ie

Received: 12 March 2019; Accepted: 14 May 2019; Published: 22 May 2019

Abstract: Tupperwave is a wave energy device based on the Oscillating-Water-Column (OWC) concept. Unlike a conventional OWC, which creates a bidirectional air flow across the self-rectifying turbine, the Tupperwave device uses rectifying valves to create a smooth unidirectional air flow, which is harnessed by a unidirectional turbine. This paper deals with the development and validation of time-domain numerical models from wave to pneumatic power for the Tupperwave device and the conventional OWC device using the same floating spar buoy structure. The numerical models are built using coupled hydrodynamic and thermodynamic equations. The isentropic assumption is used to describe the thermodynamic processes. A tank testing campaign of the two devices at 1/24th scale is described, and the results are used to validate the numerical models. The capacity of the innovative Tupperwave OWC concept to convert wave energy into useful pneumatic energy to the turbine is assessed and compared to the corresponding conventional OWC.

Keywords: wave energy; oscillating water column; tank testing; valves; air compressibility

1. Introduction

Waves are an ocean energy resource, which has the potential to contribute in the offshore renewable energy mix. A multitude of different devices have been invented to convert the wave power into electrical power, but the challenges involved in the development of an economically-sustainable solution are huge. No device for large-scale energy production from the waves has yet reached the stage of commercialisation. The barriers include high maintenance costs and survivability in extreme sea states amongst others. Oscillating-Water-Columns (OWCs) are among the most promising wave energy devices because of their relative simplicity. The only moving part being the turbine connected in direct drive to the generator, low levels of maintenance are expected. Moreover, the air contained in the OWC chamber flows across the turbine at high speed, transforming the slow motion and high forces from the waves into a fast rotational speed and low torque at the turbine. This primary conversion of the wave power into pneumatic power acts as a non-mechanical gear-box and constitutes a huge advantage in terms of maintenance and survivability. This feature is not exclusive to OWC device and is found in other wave energy converters that use air as a conversion fluid and flexible membranes such as the sea clam [1] and the Bombora device [2], amongst others. OWC devices are among the most studied type of wave energy technologies, and their principle is used in various forms.

For the most common form of OWC devices, the air flow across the turbine is bidirectional, and the turbine is self-rectifying. Such a turbine can harness both flow directions, but is not as efficient as a unidirectional turbine. State-of-the-art self-rectifying air turbines reach a maximum efficiency in the order of 70–75% in constant flow conditions during scaled tests [3]. In real-sea conditions, the air flow across the turbine is highly fluctuating since it stops and changes direction every 3–5 s.

In such conditions, the average efficiency of self-rectifying turbines is 5–10% lower than their maximum efficiency [3]. Other forms of OWCs use non-return valves to rectify the flow in a single direction across a unidirectional turbine. Single-stage unidirectional turbines reach 85–90% efficiency in constant flow conditions [4,5]. The use of rectifying valves was successful for Masuda's commercial navigation buoy (1965) powered by wave energy [6], but the experience of the Kaimei device [7] in 1986 revealed that there were challenges associated with the use of valves in larger scale devices for power production where air flow rates are in the order of $100\,\mathrm{m}^3 \cdot \mathrm{s}^{-1}$. Such flow rates necessitate large valve dimensions, which are unsuited for the fast opening and closing response time required for the valves. The moderate success of the Kaimei was also due to the limited theoretical knowledge of wave energy absorption available at the time [8].

The idea of using a high-efficiency unidirectional turbine in an OWC device is however still appealing since it could potentially increase the device efficiency. The Tupperwave principle suggests another approach on the use of non-return valves for the generation of a closed-circuit air flow. The Tupperwave concept for OWCs consists of self-rectifying valves, two large air chambers that act as accumulators, and a unidirectional turbine. The Tupperwave principle is described in Figure 1 and can be applied to fixed or floating devices.

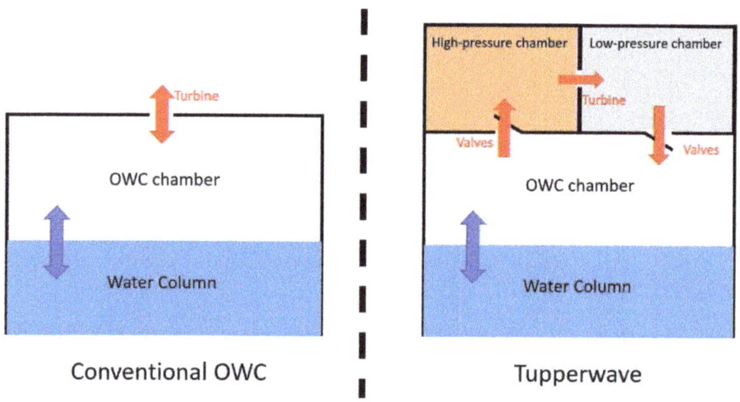

Figure 1. Schematic diagram of the conventional Oscillating-Water-Column (OWC) and Tupperwave device concepts.

Floating devices are however more suitable because of the available buoyancy volume, which can be divided in two parts and used as the High-Pressure (HP) chamber and Low-Pressure (LP) chambers. In this article, the Tupperwave principle is applied to a floating axisymmetric structure in the form of a spar buoy. In addition to a potential yield increase due to the high efficiency of the unidirectional turbine, this principle also aims at smoothing the flow across the turbine and hence maximising its efficiency and the power quality delivered by the device. An optimization study showed that the volume of these chambers should be maximized [9]. Thus, the whole buoyancy volume of the chosen spar buoy was used, and each chamber was $950\,\mathrm{m}^3$. Figure 2 displays the schematic for the Tupperwave device at full scale, as well as the corresponding conventional OWC using the same spar buoy geometry.

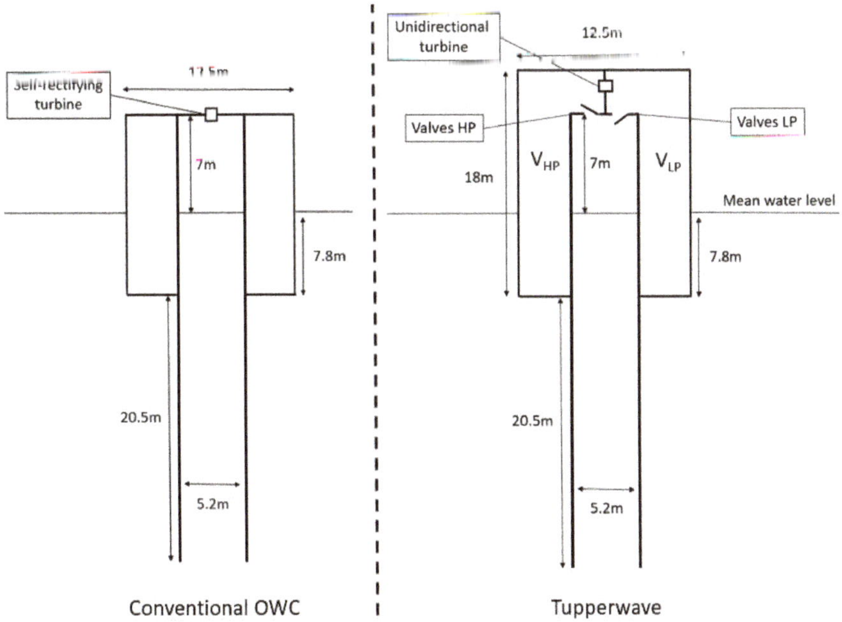

Figure 2. 2D schematic of the full-scale conventional OWC and Tupperwave devices. HP, High-Pressure; LP, Low-Pressure.

The power conversion chain of wave energy converters is achieved in various stages. In the case of OWC devices, the incoming wave power reaching the device is partly absorbed by the device. This absorbed power is then converted into useful pneumatic power available across the turbine. Finally, the power available to the turbine is converted into electrical power by the turbine and generator system. The development and validation of a model encompassing all power conversion stages is challenging due to complex interdependent physical phenomena happening at each stage and the cost of building a physical model with all the components [10]. Therefore, developers usually decide to study parts of the power conversion chain separately, as in [11–13].

In this paper, the first two conversion stages are considered: the Tupperwave device and the corresponding conventional OWC (Figure 2) are numerically modelled from wave to pneumatic power, and the results are validated against tank testing experiments. The objective is to assess, numerically and physically, the capacity of the Tupperwave device to convert wave power into smooth pneumatic power available to the turbine and compare it to the performance of the conventional OWC.

The hydrodynamic and thermodynamic equations forming the devices' numerical models from wave to pneumatic power are presented in Section 2. Section 3 describes the tank testing experiments carried out. Section 4 compares the physical model performances and the numerical and experimental results of both devices for the validation of the numerical models. Finally, the conclusions of the work are given in Section 5.

2. Numerical Models From Wave to Pneumatic Power

The time-domain numerical models from wave to pneumatic power for the two devices studied in this research consist of a number of coupled differential equations obtained via hydrodynamic and thermodynamic considerations.

2.1. Hydrodynamics

Both the Tupperwave device and the conventional OWC use the same spar buoy geometry. The hydrodynamic model is therefore the same for the two devices. The model is based on linear wave theory, using the assumptions that wave steepness and bodies motions are sufficiently small. To solve the linear hydrodynamic problems in the OWC device, several approaches have been developed. The two most popular approaches are the uniform surface pressure model [14] and the rigid piston model [15]. The former approach is exact because it makes no assumption on the generally warped Internal Water Surface (IWS) and assumes spatially-uniform pressure on the IWS. The latter approach approximates the IWS to be a thin rigid piston moving along the column of the device. This approximation is reasonable when the radius of the IWS is small compared to the wavelength. With this approach, the problem becomes a two-rigid-body problem (device floating structure-rigid piston) that can be solved using the Boundary Element Method (BEM) developed in the 1970s for the study of interaction between waves and floating bodies (ships) [16]. This approach is used in this research.

Spar buoys move essentially in heave due to the relatively large length of the submerged tail tube, and the heave motion is the main contribution for the power conversion of the axisymmetric device. Hence, to simplify the problem, only the heave motions of the two bodies are considered. This assumption is reasonable and commonly adopted in the literature [17–19]. The vertical displacement of the device structure and rigid piston are respectively denoted as (x_1, x_2), and the vertical axis is upward. The constant horizontal section of the IWS is denoted as S_2. The volume of the OWC chamber, denoted as V, varies with the relative vertical displacement of the bodies. A dot over a variable indicates the variable's derivative taken with respect to time. When the device is floating at rest on calm water ($x_1 = x_2 = 0$), the OWC chamber has a volume of V_0. We therefore have:

$$\begin{cases} V = V_0 + S_2(x_1 - x_2) \\ \dot{V} = S_2(\dot{x}_1 - \dot{x}_2) \end{cases} \tag{1}$$

In order to establish a time-domain model of the system's dynamic response, the hybrid frequency-time domain method described in [20] is used. Each body is subjected to the Cummins equation [21]. The reciprocating pressure force f_p due to the pressure building in the OWC chamber is added, along with the viscous drag force f_d. The coupled heave motions of the two-bodies can be written in time-domain as [17]:

$$\begin{cases} [m_1 + A_{11}(\infty)]\ddot{x}_1(t) + \int_0^t K_{11}(t-\tau)\dot{x}_1(t)d\tau + A_{12}(\infty)\ddot{x}_2(t) \\ \quad + \int_0^t K_{12}(t-\tau)\dot{x}_2(t)d\tau + c_1x_1(t) = f_1(t) + f_p(t) + f_{d1}(t) + f_{m1}(t) \qquad (a) \\ A_{21}(\infty)\ddot{x}_1(t) + \int_0^t K_{21}(t-\tau)\dot{x}_1(t)d\tau + [m_2 + A_{22}(\infty)]\ddot{x}_2(t) \\ \quad + \int_0^t K_{22}(t-\tau)\dot{x}_2(t)d\tau + c_2x_2(t) = f_2(t) - f_p(t) + f_{d2}(t) \qquad (b) \end{cases} \tag{2}$$

where m_i are the bodies masses; $A_{ij}(\infty)$ are the bodies heaving added masses at infinite frequency (including the proper and crossed modes); c_i are the hydrostatic stiffness terms and are calculated as $c_1 = \rho_w g S_1$ and $c_2 = \rho_w g S_2$, where ρ_w is the water density, g is the acceleration of gravity, S_1 is the constant horizontal cross-sectional area of the device structure at the undisturbed sea surface; K_{ij} are the impulse response functions for heave motions; f_i are the wave excitation forces; f_{m1} is the force applied by the mooring system on the device structure.

The impulse response functions can be obtained by the following formula:

$$K_{ij}(t) = \frac{2}{\pi}\int_0^\infty B_{ij}(\omega)cos(\omega t)d\omega \tag{3}$$

where B_{ij} is the radiation damping coefficient in the frequency domain.

The excitation forces are calculated as:

$$f_i(t) = \int_0^t K_{ex,i}(t-\tau)\eta(t)dt \tag{4}$$

where η is the external wave elevation and $K_{ex,i}$ is the excitation force impulse response function calculated as:

$$K_{ex,i}(t) = \frac{1}{\pi}\int_0^\infty Re\{\tilde{F}_{ex,i}(\omega)e^{-i\omega t}\}d\omega \tag{5}$$

where $\tilde{F}_{ex,i}(\omega)$ is the excitation force coefficient from the waves on body i. The frequency domain coefficients $A_{ij}(\infty)$, $B_{ij}(\omega)$, c_i, and $\tilde{F}_{ex,i}(\omega)$ are calculated using the commercial BEM solver WAMIT [22].

The reciprocating pressure force is calculated as $f_p = S_2 p_{owc}(t)$, and p_{owc} is the excess pressure relative to atmospheric pressure built in the OWC chamber. Because of the particular mooring system used in the experiments (see Section 3.2), the vertical component of the mooring force was neglected in the numerical model.

The viscous drag forces have an important role in wave energy converter dynamics as their misestimation would lead to numerical prediction of an unrealistic amplitude of motions, and thereby also energy absorption. The identification of drag coefficients for wave energy applications is particularly challenging because other sources of non-linearities may interfere with the isolation of the viscous drag force, causing uncertainties and inconsistency in the literature. Following the method suggested in [23], the drag force was calculated as $f_{di} = -C_{di}|\dot{x}_i(t)|\dot{x}_i(t)$ (based on Morison's equation [24]) where C_{di} is the equivalent drag coefficient and is estimated as the one that minimizes the error between the physical measurements and the numerical model. As a result, this equivalent drag coefficient incorporates the viscous drag effects, as well as all other non-linear effects such as the changes of the wetted surface subject to viscous effects, splashes, mooring line drag, etc. The values $C_{d1} = 150\,\mathrm{N}\cdot\mathrm{s}^2\cdot\mathrm{m}^{-2}$ and $C_{d2} = 40\,\mathrm{N}\cdot\mathrm{s}^2\cdot\mathrm{m}^{-2}$ were found to obtain the best fit between the vertical displacement of the bodies predicted numerically and the ones obtained physically.

According to Equation (2a,b), the hydrodynamic system can be described by the three main variables $\{x_1, x_2, p_{owc}\}$. A third differential equation verified by p_{owc} is necessary to solve the problem. This equation is established in the next section using thermodynamic equations.

2.2. Thermodynamics

In this section, the general thermodynamic equations ruling an open air chamber are first described. They are then applied to the modelling of the conventional OWC and Tupperwave device.

2.2.1. General Equations

We consider the following open thermodynamic system: an air chamber of variable volume V containing a mass m of air at the density ρ, at the temperature T and at pressure $p_{atm} + p$. The air mass flow rates w_{in} and w_{out} are the flows respectively in and out of the chamber and are functions of the air excess pressure p.

The mass balance equation in the system is:

$$\dot{m} = \rho\dot{V} + \dot{\rho}V = w_{in} - w_{out} \tag{6}$$

If the system is considered adiabatic and the transformations slow enough to be reversible, the isentropic density-pressure relation is applicable:

$$\rho = \rho_{atm}\left(1 + \frac{p}{p_{atm}}\right)^{\frac{1}{\gamma}} \tag{7}$$

where ρ_{atm} is the density of the air at atmospheric conditions and γ is the isentropic expansion factor.

Moreover, in the case where the excess pressure remains small compared to the atmospheric pressure, it is possible to linearise the isentropic relationship between density and pressure. Once linearised, Equation (7) leads to:

$$\rho = \rho_{atm}\left(1 + \frac{p}{\gamma p_{atm}}\right) \tag{8}$$

The derivation of Equation (8) with respect to time gives:

$$\dot{\rho} = \frac{\rho_{atm}}{\gamma p_{atm}}\dot{p} \tag{9}$$

Finally, the combination of Equations (6) and (9) leads to the differential Equation (10) governing the evolution of the pressure in an open air chamber during isentropic transformations:

$$\dot{p} = \frac{\gamma p_{atm}}{\rho_{atm} V}(w_{in} - w_{out} - \rho\dot{V}) \tag{10}$$

2.2.2. Conventional OWC Thermodynamics

The air chamber in a conventional OWC device is commonly modelled in the literature using the linearized isentropic assumption. The main justification is that the temperature oscillations in the air chamber are relatively small and their time scales (a few seconds) are too short for significant heat exchanges to occur across the chamber walls and across the air-water interface [6]. It was proven in [25] that the linearised isentropic assumption provides a satisfactory results for the modelling of conventional OWCs except possibly under very rough sea conditions. In this paper, the numerical model will be compared to model-scale experimental results where pressure and temperature oscillations are even smaller. The use of the linearised isentropic assumption is hence justified. Figure 3 displays a schematic of the OWC thermodynamic system.

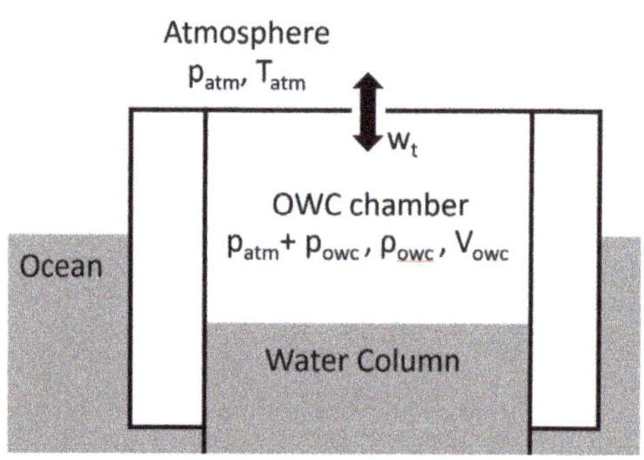

Figure 3. Conventional OWC schematic with thermodynamic variables.

At model scale, the turbine (of impulse type) connecting the inside of the chamber to the atmosphere is modelled using an orifice plate. In the conventional OWC, the flow is bidirectional through the orifice. Hence, the density of the air entering the turbine is either ρ_{atm} during the inhalation process ($p_{owc} < 0$) or ρ_{owc} during the exhalation process ($p_{owc} > 0$). Under the testing conditions, the maximum excess pressures observed in the OWC chamber were ± 1250 Pa, for which, according to Equation (8), ρ_{owc} reaches the maximum values of $\rho_{atm} \pm 0.8\%$. Hence, the air in the OWC chamber can be considered as quasi-incompressible, and the air flow rates are described using the volumetric flow rates ([26,27]):

$$\Delta p = k_t q_t^2 \quad \text{with} \quad k_t = \frac{\rho_{atm}}{2(\alpha A_t)^2} \tag{11}$$

where q_t is the volumetric flow rate across the orifice/turbine, k_t is the turbine damping coefficient, and αA_t is the effective area of the orifice representing the turbine.

Finally, according to Equations (10) and (11), the relationship between the hydrodynamic body motions (see Equation (1)) and the excess pressure in the OWC chamber is:

$$\dot{p}_{owc} = \begin{cases} \dfrac{\gamma p_{atm}}{\rho_{atm} V_{owc}} \left(-\rho_{owc}\sqrt{\dfrac{p_{owc}}{k_t}} - \rho_{owc}\dot{V}_{owc} \right) & \text{for} \quad p_{owc} > 0 \\ \dfrac{\gamma p_{atm}}{\rho_{atm} V_{owc}} \left(+\rho_{atm}\sqrt{\dfrac{p_{owc}}{k_t}} - \rho_{owc}\dot{V}_{owc} \right) & \text{for} \quad p_{owc} < 0 \end{cases} \tag{12}$$

2.2.3. Tupperwave Device Thermodynamics

In the Tupperwave device, air is exchanged between three different air chambers. Unlike for the conventional OWC where the OWC chamber is open to the atmosphere, the air in the Tupperwave device is flowing in a closed-circuit. This raises the issue of a possible air temperature increase in the device due to heat created by viscous effects, mainly happening in the turbine. It was shown in [28] that the heat created is dissipated through the device walls, and only a slight temperature increase (<1 °C) is observed in moderate sea states.

At model scale, the temperature increase is expected to be negligible. Moreover, the pressures reached in the device are expected to be in the same order as in the conventional OWC. Hence, the linearized isentropic assumption is assumed to be valid. Figure 4 displays a schematic of the three chambers, which constitute three interconnected thermodynamic systems.

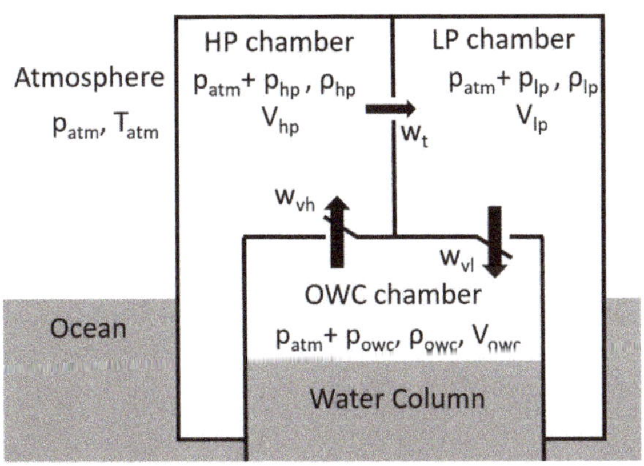

Figure 4. Tupperwave device schematic with thermodynamic variables.

Equation (10c) can be directly applied to the three chambers:

$$\begin{cases} \dot{p}_{owc} = \dfrac{\gamma p_{atm}}{\rho_{atm} V_{owc}} (w_{vl} - w_{vh} - \rho_{owc}\dot{V}_{owc}) & (a) \\ \dot{p}_{hp} = \dfrac{\gamma p_{atm}}{\rho_{atm} V_{hp}} (w_{vh} - w_t) & (b) \\ \dot{p}_{lp} = \dfrac{\gamma p_{atm}}{\rho_{atm} V_{lp}} (w_t - w_{vl}) & (c) \end{cases} \tag{13}$$

where w_t, w_{vh}, and w_{vl} are the mass air flow rates across the turbine, the HP valve, and the LP valve. The sign convention for the volumetric flow rates is given by the arrow directions in Figure 4.

As for the conventional OWC, the excess pressures reached in the different chambers at model scale do not exceed 1250 Pa. The variation of air density is therefore small, and the flow across the orifice is considered as incompressible:

$$w_t = \rho_{hp}\sqrt{\frac{p_{hp} - p_{lp}}{k_t}} \tag{14}$$

The valves are non-return valves that close under a certain opening pressure p_o. When opened, their model is similar to orifice plates of effective opening area αA_v and damping coefficient $k_v = \frac{\rho_{atm}}{2(\alpha A_v)^2}$:

$$w_{vh} = \begin{cases} 0 & \text{if} \quad p_{owc} - p_{hp} < p_o \\ \rho_{owc}\sqrt{\frac{p_{owc} - p_{hp} - p_o}{k_v}} & \text{if} \quad p_{owc} - p_{hp} > p_o \end{cases} \tag{15}$$

$$w_{vl} = \begin{cases} 0 & \text{if} \quad p_{lp} - p_{owc} < p_o \\ \rho_{lp}\sqrt{\frac{p_{lp} - p_{owc} - p_o}{k_v}} & \text{if} \quad p_{lp} - p_{owc} > p_o \end{cases} \tag{16}$$

The system of Equation (13a,b,c) is coupled with the hydrodynamic system of Equation (2a,b) and completes the model for the Tupperwave device.

2.3. Numerical Solution

The conventional OWC is governed by the coupled Equations (2a,b) and (12), while the Tupperwave device is governed by Equations (2a,b) and (13a,b,c).

The four convolution integrals $I_{ij}(t) = \int_0^t K_{ij}(t-\tau)\dot{x}_j(\tau)d\tau$ present in Equation (2a,b) are called memory effect integrals. Their values depend on the history of the system, which implies their recalculation at each time step and is not practical for the system resolution. By using the conventional Prony methods [29], it is possible to calculate each of these functions as the sum of N_p additional unknowns $\{I_{ij,k}, k = 1 : N_p\}$, which are the solutions of N_p additional first order equations that will be solved along with the system of Equation (2a,b). Details on the conventional Prony method are given in Appendix A. Following the recommendations made in [30], $N_p = 4$ was taken for the approximation of the impulse function of heave motion, which adds 16 first order equations to the system.

The solution of these equations was obtained numerically using the ordinary differential equation solver ode45 from the mathematical software MATLAB [31].

3. Physical Modelling in the Wave Tank

3.1. Physical Models Design and Fabrication

3.1.1. Scaling

The devices displayed in Figure 2 were built at model scale. For dynamic similarity in the water, all underwater dimensions were multiplied by the scaling factor $\epsilon = 0.0415 \simeq 1/24$ according to the Froude similarity law. However, if the volumes of the high- and low-pressure chambers were also scaled using Froude scaling (i.e., multiplied by ϵ^3), air compressibility effects occurring at full scale would not be reproduced at model scale [32]. Since the Tupperwave device working principle fully relies on air compressibility in the high- and low-pressure chambers, these effects had to be reproduced, and a different scaling method was implemented for the volumes of the chambers.

As shown in Equation (13b), the variation of mass in the HP chamber of volume V_{hp} is only due to air compressibility and directly related to the change in pressure:

$$\dot{p}_{hp} = \frac{\gamma p_{atm}}{\rho_{atm} V_{hp}} \dot{m}_{hp} \tag{17}$$

Equation (17) is valid both at full scale (*F*) and model scale (*M*). Hence:

$$\frac{\dot{p}_{hp,M}}{\dot{p}_{hp,F}} = \frac{p_{atm,M}\ \dot{m}_{hp,M}\ \rho_{atm,F}\ V_{hp,F}}{p_{atm,F}\ \dot{m}_{hp,F}\ \rho_{atm,M}\ V_{hp,M}} \tag{18}$$

For similitude to be achieved between full scale and model scale, Froude scaling laws need to be respected. This implies:

$$\begin{cases} \dfrac{\dot{p}_{hp,M}}{\dot{p}_{hp,F}} = \epsilon^{1/2} \\ \dfrac{\dot{m}_{hp,M}}{\dot{m}_{hp,F}} = \epsilon^{5/2} \end{cases} \tag{19}$$

Unless very specific infrastructures are used, the atmospheric conditions are the same at model scale as at full scale, hence:

$$\begin{cases} p_{atm,M} = p_{atm,F} \\ \rho_{atm,M} = \rho_{atm,F} \end{cases} \tag{20}$$

Combining Equation (18) with Equations (19) and (20) leads to the necessary condition on the chamber volume to satisfy similarity regarding compressibility effects between full scale and model scale:

$$\frac{V_{hp,M}}{V_{hp,F}} = \epsilon^2 \tag{21}$$

The same condition needs to be satisfied for the LP chamber. At full scale, the HP and LP chamber are 950 m^3. At model scale, the volumes of the chambers therefore are 1.64 m^3. Unlike full scale, it is impossible to fit both chambers on the device as their volume largely exceeds the overall volume of the device. The alternative at small scale is to locate the main volume of the HP and LP chambers outside of the device and connect them to two smaller chambers on the device with flexible pipes. Large reservoirs were used for the HP and LP chambers and located on the pedestrian bridge above the tank, see Figure 5. The flexible pipes were chosen as lightweight and flexible as possible to reduce their influence on the floating device motion. Part of the pipes' weight was supported by bungee ropes. A similar experimental setup was implemented in [33] to test a floating air bag wave energy converter at model scale.

This scaling method was suggested and applied in [32] to represent the spring-like effect of the air in the OWC chamber of a conventional OWC device properly. To implement this method, an additional flexible pipe connecting the OWC chamber to another reservoir outside of the device is required. It is however rarely implemented at model scale by conventional OWC developers since the air compressibility in the OWC chamber is not essential for the devices working principle, and it increases even more the testing difficulty, especially for a floating device. Therefore, in the present experiments, the volume of the OWC chamber was simply scaled down by the factor ϵ^3 using direct Froude scaling, and the spring-like effect of the air in the OWC chamber was not physically modelled in both devices. The air in the OWC chamber is therefore quasi-incompressible in the conditions of the tests, and it is acknowledged that the perfect similitude with the full-scale devices is not achieved. The power conversion performances of the full scale devices can therefore not be directly obtained from the results of the model tests, as this may result in unrealistic overestimations [34]. The present experiments are however still valuable to validate the Tupperwave concept and compare it to the conventional OWC.

Figure 5. Physical model of the Tupperwave device.

Since both devices use the same Spar structure, a single spar was built and used for both device. The schematics of the physical models are shown in Figure 6. The positions of the centre of gravity and centre of buoyancy, as well as the moments of inertia were assessed using the computer-aided design software Solidworks [35]. The position of the centre of gravity was also assessed experimentally by hanging the device on a vertical rope and finding the balance point. The values obtained experimentally after five repeated tests verified the centre of gravity to be within ±1 mm from the position indicated by Solidworks. The moment of inertia around the horizontal axis Y parallel to the wave front was also verified using the bifilar pendulum method, and the values obtained were within $\pm 0.1\ \mathrm{kg \cdot m^2}$ from the value indicated by Solidworks. The conventional OWC being initially lighter than the Tupperwave device, it was ballasted such that both devices had the exact same mass properties. The mass properties of the device are given in Table 1.

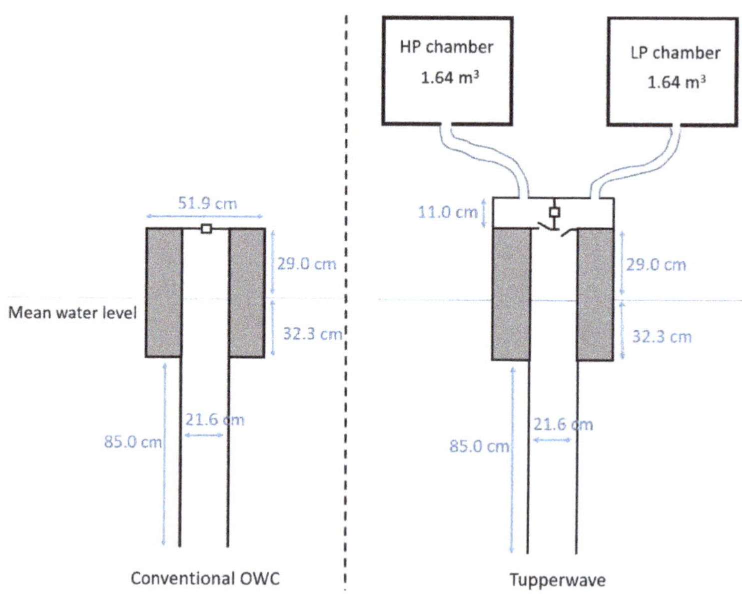

Figure 6. Schematic of the model scale conventional OWC and Tupperwave devices.

Table 1. Devices mass properties.

	Model Scale	Full Scale
Total mass (kg)	58.4	817×10^3
Distance device bottom, COG(m)	0.892	21.49
Distance device bottom, COB(m)	0.961	23.16
Ixx ($kg \cdot m^2$)	23	1.87×10^8
Iyy ($kg \cdot m^2$)	23.5	1.91×10^8
Izz ($kg \cdot m^2$)	2	1.62×10^7

3.1.2. Turbines

As mentioned in Section 2.2.2, the turbines were physically modelled using orifice plates. Preliminary numerical modelling showed that the optimal damping coefficients at full scale to maximise the pneumatic power output were close to $38\,Pa \cdot s^2 \cdot m^{-6}$ for the conventional OWC and $450\,Pa \cdot s^2 \cdot m^{-6}$ for the Tupperwave device [9]. Three orifice plates per device were built around those values scaled down using Froude scaling. Their exact damping coefficients and effective areas were then experimentally assessed prior to testing by forcing a known flow across the orifice and measuring the pressure drop [36]. Table 2 displays the orifice characteristics for the conventional OWC and the Tupperwave model scale devices. The damping coefficients were used during the tank testing to calculate the instantaneous volumetric flow across the orifice based on the measurement of the pressure drop according to Equation (14).

Table 2. Orifice characteristics for the conventional OWC and Tupperwave device.

	Model Scale			Full Scale	
Orifice	**Diameter (mm)**	**k_t ($Pa \cdot s^2 \cdot m^{-6}$)**	**αA (m^2)**	**k_t ($Pa \cdot s^2 \cdot m^{-6}$)**	**αA (m^2)**
OWC1	22.6	7.10×10^6	2.94×10^{-4}	21.1	1.71×10^{-1}
OWC2	20.6	10.4×10^6	2.42×10^{-4}	30.9	1.41×10^{-1}
OWC3	17.5	19.6×10^6	1.77×10^{-4}	58.3	1.03×10^{-1}
T1	11.5	0.70×10^8	9.33×10^{-5}	209	5.42×10^{-2}
T2	9.2	1.86×10^8	5.74×10^{-5}	552	3.33×10^{-2}
T3	7	4.85×10^8	3.55×10^{-5}	1439	2.06×10^{-2}

3.1.3. Valves

The Tupperwave working principle relies on the use of non-return valves. The valves are key components because they are likely to cause pneumatic power losses. They can either be passive or active. Passive valves mechanically open when a certain pressure difference is reached across the valves, while active valves are electrically activated. For the physical model, passive valves were chosen for their simplicity. The most appropriate valves found on the market were the Capricorn MiniHab HypAirBalance, see Figure 7.

They were passive normally closed air admittance valves from the plumbing market. A rubber membrane contained in the valve obstructs the opening of the valve due to gravity. When sufficient pressure is applied, the rubber membrane is lifted up, and the valve opens. Their opening pressure is 70 Pa (equivalent to 1686 Pa at full scale), and their light weight allowed their use in the small-scale Tupperwave physical model.

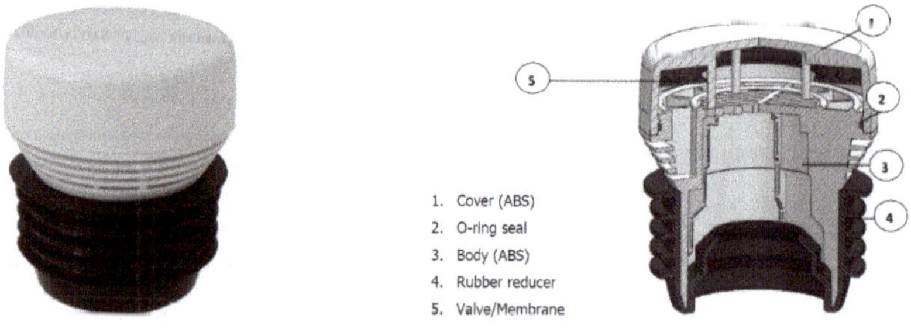

(**a**) Full device (**b**) Section

Figure 7. MiniHab HypAirBalance from Capricorn used in the Tupperwave small-scale model [37].

According to Equations (15) and (16), lower values of the valve damping coefficient k_v reduce the pressure drop across the valves and hence reduce the pneumatic power losses. The value of k_v is therefore fundamental for the device efficiency. Unlike for the orifices, the damping coefficient of the valves were experimentally assessed during each test by using Equations (15) and (16) and monitoring the pressure drop and air flow across the valves. The air flow rate across the valves was calculated from the measurement of the IWS elevation achieved with wave probes located inside the water column. Different valve damping values were obtained depending on the tests undertaken. Figure 8 displays the values of k_v for the HP valve obtained in regular waves.

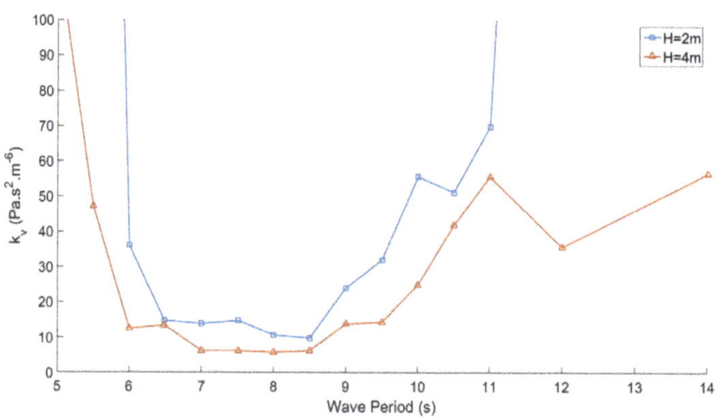

Figure 8. HP valve damping coefficient in regular waves (full-scale equivalent).

Wave height, wave period, and valve damping coefficients are given at full-scale equivalent. Variations of k_v with the wave height and wave period were observed. It was shown in [38] that the damping coefficient of those valves is highly dependant on the device excitation. Indeed, when the device is well excited (for $6.5\,\text{s} < T < 8.5\,\text{s}$), large pressure drops and flows across the valves are created. The valves therefore open fully, and their damping coefficient is small (large effective area). The lowest damping values reached by the valves were around $k_v = 7.5\,\text{Pa} \cdot \text{s}^2 \cdot \text{m}^{-6}$ at full-scale equivalent, which corresponds to an effective valve opening area of $0.286\,\text{m}^2$. This maximum opening area achieved was very small compared to the available $10.6\,\text{m}^2$ between the OWC chamber and the HP chamber. When the device is poorly excited (for $T < 6.5\,\text{s}$ and $T > 8.5\,\text{s}$), the valves do not open fully and create large damping (i.e., large losses). For $H = 4\,\text{m}$, the valves open fully on a larger range of wave periods than for $H = 2\,\text{m}$ since the device is more excited in bigger waves. The values of the valves' damping coefficient assessed for each regular wave physical test were fed into the numerical model for the corresponding numerical test.

In irregular waves, the excitation of the device varies with the incoming wave groups. Hence, the instantaneous damping of the valves fluctuates along the simulation. The k_v values in-putted in the numerical model for each irregular wave simulations were the average damping values obtained physically over the whole duration of the simulations.

3.2. Experimental Setup and Test Plan

The experiments took place in the Lir-National Ocean Test Facility (Lir-NOTF) of the MaREIcentre in Cork, Ireland. The devices were tested under regular and irregular sea states in the Deep Ocean Basin, which is 35 m long, 12 m wide, and has a movable floor with up to 3 m of depth. For the regular sea states and at full-scale equivalent, 2 wave heights (2 m and 4 m) were tested with periods ranging from 5–14 s. A set of 8 irregular sea states of various significant wave heights (2–5 m) and peak periods (5–14 s) was also tested. These sea states were chosen to represent a large variety of typical conditions in the Atlantic Ocean. The depth was set to 2.07 m, equivalent to 50 m at full scale.

The devices were moored using a 3-point mooring arrangement of the catenary type with 120 degrees between any two mooring lines. There were two bow mooring lines and one stern line. Each line was divided into two parts: a steel chain connects the anchor to a surface buoy that can support the chain weight. A neutrally-buoyant line then connects the surface buoy to the device at the mean water level. Figure 9 shows a schematic of one mooring line. With such a mooring configuration, the mooring forces applied on the device were principally horizontal, preventing the device from drifting in the waves, with a minimum impact on the heave motion from which the wave energy was absorbed. Therefore, the vertical component of the mooring force applied on the device was neglected in the numerical model presented in Section 2. The drag forces caused by the vertical motion of the neutrally buoyant lines with the device were taken into account in the numerical model by the equivalent drag coefficient C_d.

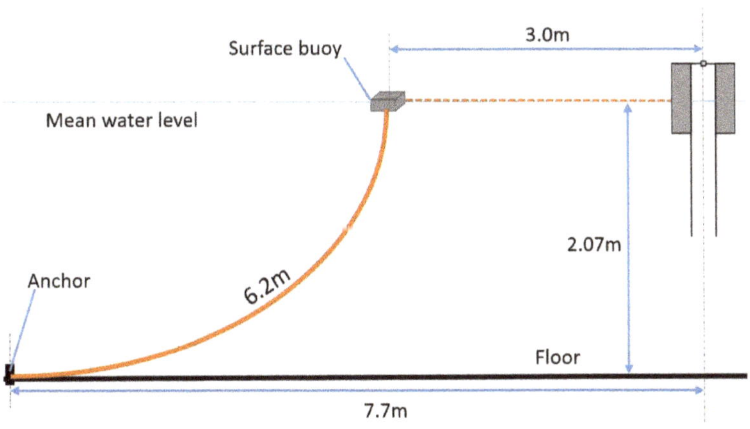

Figure 9. Schematic of a mooring line. The device was moored by 3 mooring lines with 120 degrees between any two mooring lines.

The devices were fully monitored using 3D cameras, pressure sensors, and wave probes, allowing the measurement of the devices' motions, IWS elevation, pressure drops, and volumetric flow rates across orifices and valves. Wave probes located outside of the device measured the incoming wave elevation. A wave probe located inside the column of the device measured the relative elevation of the water column relative to the buoy. Figure 10 displays a picture of the device, from above, equipped with pressure sensors and reflective markers for the 3D cameras.

The power absorbed from the waves by the devices is the power applied by internal water surface (IWS) on the air contained in the OWC chamber and was calculated, from those measurements, as:

$$P_{abs} = p_{owc}.\dot{V} = p_{owc} S_2 \frac{dx_{IWS}}{dt} \tag{22}$$

where $x_{IWS} = x_2 - x_1$ is the position of the IWS relative to the buoy. In testing conditions, the flows across the orifice or valves were incompressible. The pneumatic power available to the turbine P_t or to a valve P_v was therefore calculated as the product of the volumetric flow and the pressure drop:

$$P = \Delta p\, q \tag{23}$$

The regular wave tests were 125 s long, which is equivalent to 10 min at full scale. The irregular wave tests were 7 min long, which is equivalent to 35 min at full scale. This allows a full representation of the Bretschneider sea state.

For the analysis of regular wave tests, averaging of the key variables was made over several waves once a steady state was reached, practically between 90 and 115 s at model scale. The average values in irregular sea states were calculated over the full time of the simulation.

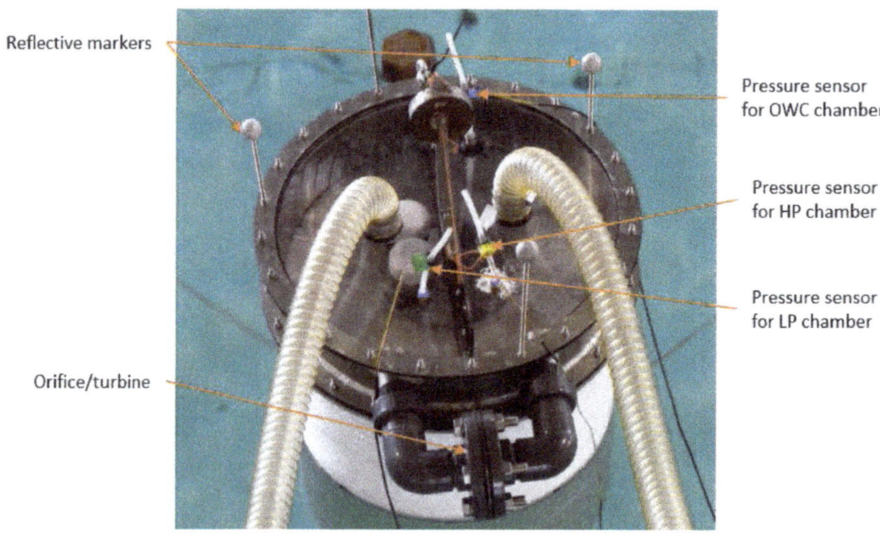

Figure 10. Top view of the Tupperwave device.

4. Results and Numerical Model Validation

Physical and numerical tests were carried out for the small-scale devices, but all results were given in full-scale equivalence to give the reader perspective.

4.1. Correction in the Tupperwave Numerical Model

During the tests, it was observed that the reservoirs used for the HP and LP chambers of the Tupperwave physical model did not have perfectly rigid walls and that the walls moved very slightly due to the pressure building inside the chambers. The HP chamber was observed to inflate with the build-up of a positive excess pressure inside, and the LP chamber was observed to deflate due to the negative excess pressure inside. Unfortunately, within the time frame and budget of the project, it was not possible to replace the chamber walls with stiffer material. As a result, the air compressibility in the HP and LP chambers was physically not modelled correctly, and the varying volumes of the chambers caused a dampening of the pressure variations. This required a correction in the initial numerical model described in Section 2.2.3 to take into account the slight volumetric changes of the HP and LP chambers.

A simple linear model of the chamber volumetric deformation as a function of the excess pressure was chosen:

$$\begin{cases} V_{hp} = V_0 + Cp_{hp} \\ V_{lp} = V_0 + Cp_{lp} \end{cases} \tag{24}$$

where C is the elastic stiffness of the chambers and $V_0 = 1.64\,\text{m}^3$ is their initial volume.

The same stiffness value was applied to the two chambers, and its value was calibrated using the experimental results. The value $C = 8.3 \times 10^{-5}\,\text{m}^3 \cdot \text{Pa}^{-1}$ was obtained by an iterative process to minimise the error with the experimental results in the different irregular sea states. Therefore, for an excess pressure of $p_{ref} = \pm 1000\,\text{Pa}$ (close to maximum pressure observed in the chambers), the volume variations of the chamber were $dV_{ref=} \pm 0.083\,\text{m}^3 << V_0$, which is equivalent to a ± 2 cm deformation of the edge lengths of the cubic chambers and corresponds to the approximate visual observations.

The system of Equation (13) describing the pressure evolution in the Tupperwave chambers was modified to take into account the HP and LP chambers deformations:

$$\begin{cases} \dot{p}_{owc} = \dfrac{\gamma p_{atm}}{\rho_{atm} V_{owc}} (w_{vl} - w_{vh} - \rho_{owc} \dot{V}_{owc}) & (a) \\ \dot{p}_{hp} = \dfrac{\gamma p_{atm}}{\rho_{atm} V_{hp}} (w_{vh} - w_t - \rho_{hp} \dot{V}_{hp}) & (b) \\ \dot{p}_{lp} = \dfrac{\gamma p_{atm}}{\rho_{atm} V_{lp}} (w_t - w_{vl} - \rho_{lp} \dot{V}_{lp}) & (c) \end{cases} \tag{25}$$

Figure 11 compares the time series of the pressure drop $\Delta P_t = p_{hp} - p_{lp}$ across the orifice of the Tupperwave device in regular waves of a 9-s period and heights of 2 m and 4 m obtained by the physical model, the initial numerical model, and the corrected model. In regular waves, the wave excitation on the device is steady, and the excess pressures in the HP and LP chambers increased until they stabilized to certain values and reached a steady state. In the initial numerical model where the HP and LP chambers were perfectly rigid, the pressure drop between the chambers built up quickly to reach the steady state. Small pressure drop oscillations with a period of half a wave period were also visible and were due to the oscillatory motion of the water column. In the physical model, the chambers deformed gradually by a corresponding small volume $\pm\Delta V$, much less than the initial volume V_0, and consequently delayed the pressure evolution. Hence, the slight deformation of the chambers did not influence the end results in regular waves, but simply delayed the system from reaching steady state. The smaller oscillations were also attenuated.

In irregular waves, however, the slight volumetric deformations of the chambers caused more visible effects. During a simulation, the pressures in the HP and LP chambers varied significantly when the device was excited by high or low energy wave groups. The chamber volumes were therefore changing between wave groups and dampened fast pressure variations. Figure 12 compares the time series ΔP_t in the irregular sea state $\{H_s = 3\text{ m};\ T_p = 8.5\text{ s}\}$ obtained physically against the time series obtained with the initial and corrected numerical model. The pressures in the rigid-wall chambers from the initial numerical model varied more rapidly with the wave groups than in the chambers used in the physical tests. As a result of the chambers' deformation, an extra and unrealistic smoothing effect of the pressure variations between the wave groups was observed in irregular waves. Accounting for the small volume variations of the HP and LP chambers in the numerical model very clearly enhanced the fidelity of the physical model.

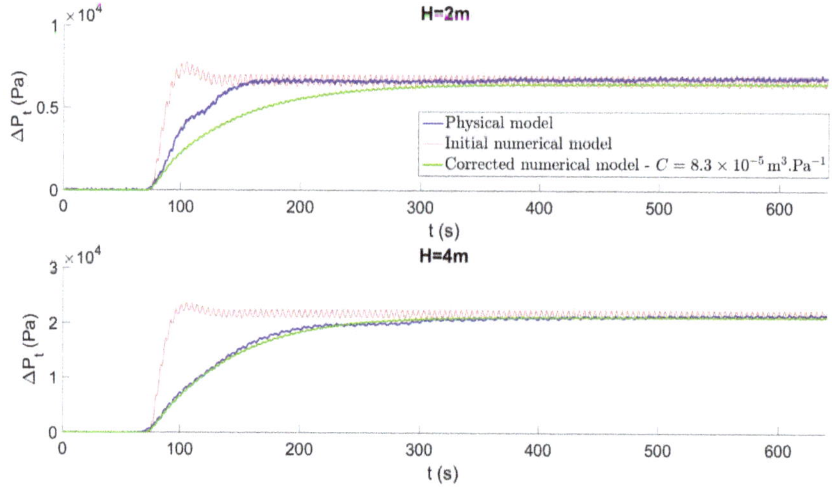

Figure 11. Pressure drop time series across Orifice 3 of the Tupperwave device obtained by the physical model, the initial numerical model, and the corrected model in the regular waves of a 9-s period and heights of 2 m and 4 m.

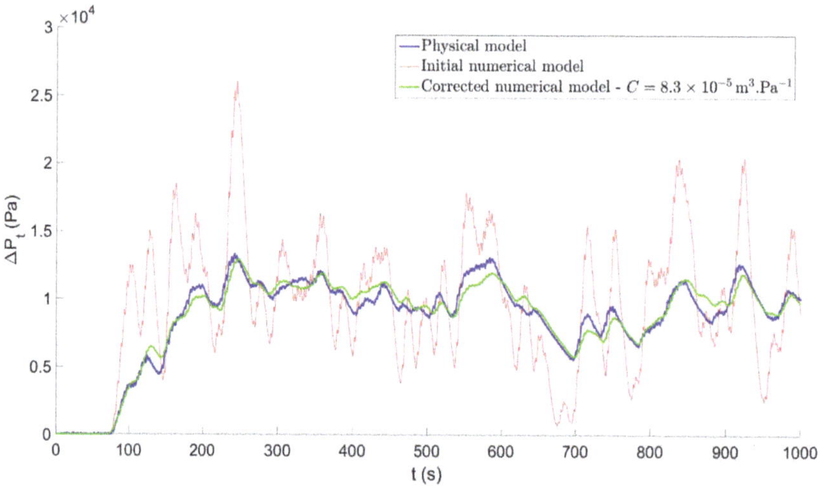

Figure 12. Pressure drop time series across Orifice 3 of the Tupperwave device obtained by the physical model, the initial numerical model, and the corrected model in the irregular sea state $\{H_s = 3$ m; $T_p = 8.5$ s$\}$.

Figure 13 displays the numerical results obtained with the following chamber stiffness values: $[4.2; 8.3; 12] \times 10^{-5}\,\mathrm{m}^3 \cdot \mathrm{Pa}^{-1}$, corresponding respectively to $\pm[1; 2; 3]$ cm deformation of the chambers' edge length for an excess pressure of $p_{ref} = \pm 1000$ Pa. This figure shows the sensitivity of the results on the chambers' stiffness value. With stiffer chambers ($C = 4.2 \times 10^{-5}\,\mathrm{m}^3 \cdot \mathrm{Pa}^{-1}$), the pressure variation was larger due to the smaller deformation of the walls. The pressure variations were more dampened with more flexible chambers ($C = 1.2 \times 10^{-4}\,\mathrm{m}^3 \cdot \mathrm{Pa}^{-1}$). Although the results were not extremely sensitive on the stiffness value, $C = 8.3 \times 10^{-5}\,\mathrm{m}^3 \cdot \mathrm{Pa}^{-1}$ obtained the closest results to the physical model for all sea states tested. Table 3 displays the Pearson correlation coefficients between the time series obtained physically and the time series obtained with the initial and corrected numerical models. Values closer to one indicate a better correlation between physical and numerical results.

Table 3. Pearson correlation coefficient between ΔP_t time series obtained physically and numerically for the various irregular sea states.

Sea State		Pearson Correlation Coefficient (-)	
H_s (m)	T_p (s)	Initial Model	Corrected Model $C = 8.3 \times 10^{-5}\,\mathrm{m^3 \cdot Pa^{-1}}$
2	5.7	0.72	0.85
3	7.1	0.68	0.93
3	8.5	0.69	0.93
5	8.5	0.65	0.94
3	10.6	0.70	0.94
5	10.6	0.65	0.93
5	12.7	0.62	0.93
3	14.1	0.70	0.86

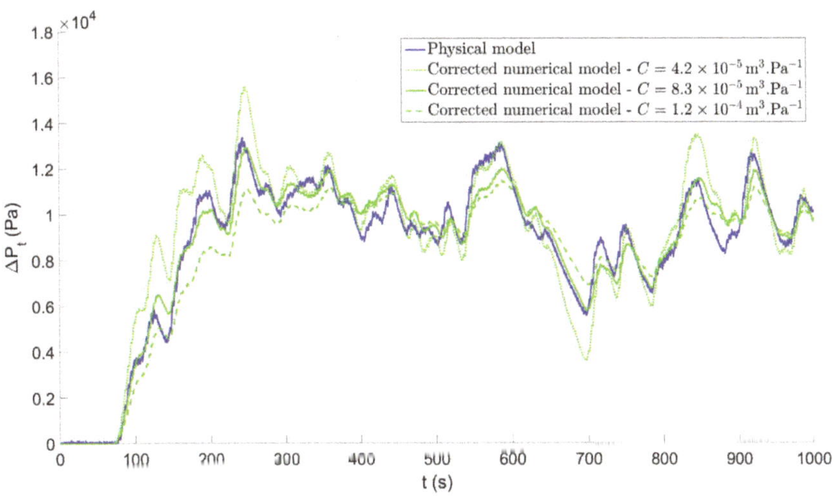

Figure 13. Pressure drop time series across Orifice 3 of the Tupperwave device obtained by the physical model and the corrected model in the irregular sea state $\{H_s = 3$ m; $T_p = 8.5$ s$\}$ with different chamber stiffness values.

From an energy point of view, the pneumatic energy stored in the HP and LP chamber was not only stored in the form of pressure, but also under the form of strain energy. This explains the lower pressure variations between wave groups, and the average value of the pressure drop across the orifice was not impacted; see Figure 12. In the next section, only the corrected numerical model of the model-scale device accounting for the chambers' deformation was used.

We note that the idea of storing pneumatic energy under the form of strain energy could be combined with the Tupperwave concept at full scale, by adding a spring and piston to the HP and LP chambers for example. This would enable further improving the pneumatic power smoothing capacity of the device or reducing the volumes of the reservoirs. Moreover, this would possibly introduce further control possibilities: changing the stiffness of the spring could enable the device to be tuned to the sea state to improve energy extraction (or de-tuned if desired). The concept variation of the Tupperwave device using variable volume HP and LP chambers will be investigated in further work and is not in the scope of this paper. With fixed volume chambers, the Tupperwave device was structurally and mechanically simple. History has shown that mechanical simplicity represents an advantage in the development of wave energy devices, particularly regarding the device cost, maintainability, and reliability. Developers commonly claim the simplicity of their wave energy device as an advantage [39–41].

4.2. Numerical Model Validation

For the validation process, the spar buoy and water column relative motions were first compared in regular waves. Figure 14 displays the Response Amplitude Operator (RAO) of the bodies' relative sinusoidal heave oscillations and their phase difference for the two devices. The numerical models agree generally well with the results obtained by the physical tests, and the influence of the different orifice damping tested is well predicted. The phase difference between the bodies in the Tupperwave device for short wave period was less accurately predicted by the numerical model than for the conventional OWC. It is interesting to note that, for the two devices, smaller orifices restricted both the relative motion amplitude and the phase difference between the bodies. In the conventional OWC, the larger damping of small orifices directly caused more resistance against the bodies' relative motions. In the Tupperwave device, the larger damping of small orifices created larger excess pressures in the HP and LP chamber, which increased the necessary OWC chamber pressure to open the valves and thus caused more resistance against the bodies' relative motions.

Furthermore, in comparison to the conventional OWC, the response of relative motion amplitude in the Tupperwave device was narrower, and the phase difference between the bodies was lower. This shows that the coupling between the structure and the water column was stiffer, and the bodies were more constrained to oscillate together in the Tupperwave device.

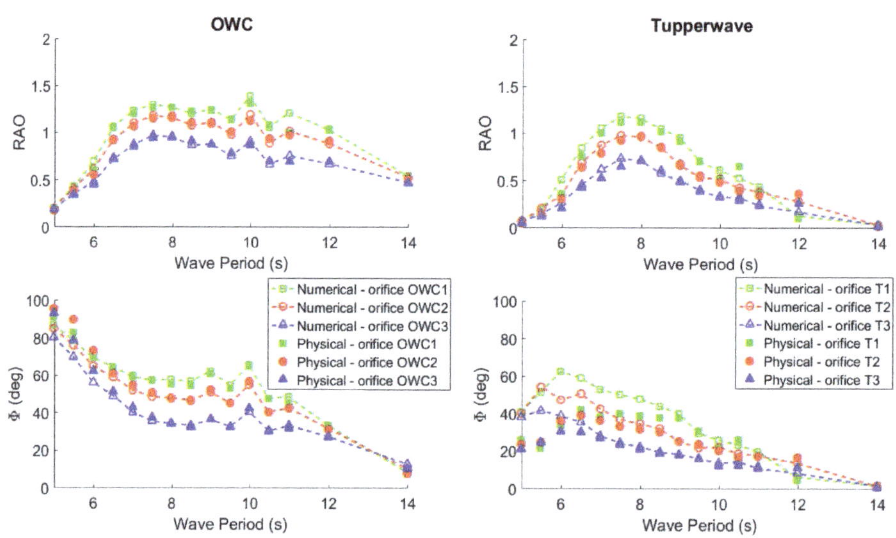

Figure 14. Response Amplitude Operator (RAO) of the buoy and water column relative motion and their phase difference for the conventional OWC and Tupperwave device in regular waves ($H = 2$ m).

The average pressure drop and volumetric flow across the orifices in regular waves are compared in Figure 15. Good agreement was obtained between numerical and physical results. In both devices, the pressure drop decreased with increasing orifice diameter, and the flow across the orifice increased. The Tupperwave device produced larger pressure drops and lower flow rates across the turbine than the conventional OWC.

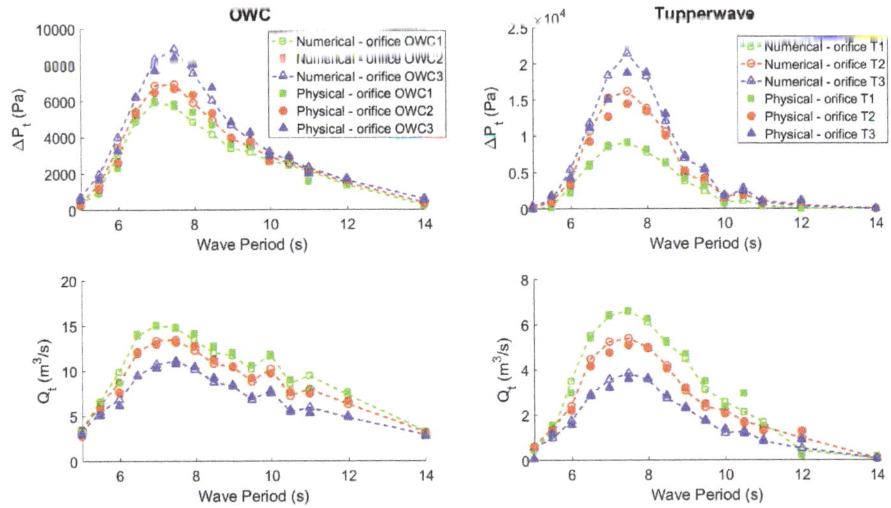

Figure 15. Average pressure drop and volumetric flow rate across the orifice for the conventional OWC and the Tupperwave device in two-meter high regular waves.

Figure 16 displays the average pneumatic power normalized by the significant wave height squared obtained numerically and physically by the two devices in the eight irregular sea states tested. Figure 17 displays the time series of the pneumatic power for the irregular sea state $\{H_s = 3\text{ m}; T_p = 7.1\text{ s}\}$. The two devices were equipped with their most efficient orifices (OWC2 and T2).

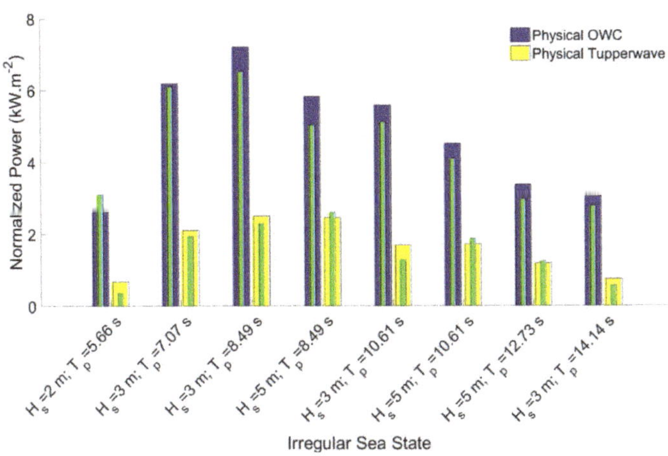

Figure 16. Average pneumatic power normalized by the significant wave height squared for the conventional OWC and the Tupperwave device in irregular sea states.

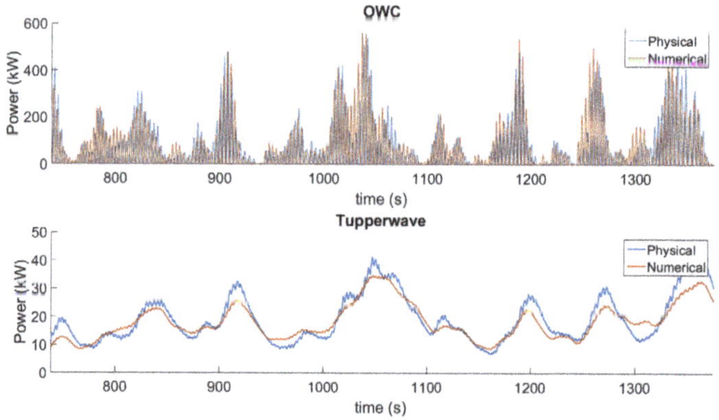

Figure 17. Pneumatic power time series for the conventional OWC and the Tupperwave device in the irregular sea state $\{H_s = 3\text{ m}; T_p = 7.1\text{ s}\}$.

Good overall agreement was obtained between physical and numerical results. The errors of the numerical models on average power prediction in irregular wave tests in Figure 16 were 10.1% for the conventional OWC and 16.5% for the Tupperwave device. The passive valves in the Tupperwave model are an additional component to model, and their fluctuating behaviour with the device excitation in irregular waves (see Section 3.1.3) is hence an additional source of error in numerical model fidelity.

Although moderate pitch and roll motions of the spar buoy were observed during the experiments, the satisfactory results of the numerical model prove, a posteriori, that considering heave only in the numerical model is a reasonable approach to assess the devices' power conversion.

4.3. Power Performance Comparison

Figure 18 displays the average power absorbed by the device from the waves $\overline{P_{abs}}$ and the average pneumatic power $\overline{P_{avail}}$ available to the turbine in both devices in 2 and 4 m-high regular waves. In the case of the conventional OWC, the absorbed power is, on average, entirely made available to the turbine $\overline{P_{abs}} = \overline{P_{avail}}$ [34]. The Tupperwave device absorbed 4–20% less power from the waves than the conventional OWC for wave periods between 6 and 9.5 s. This is probably due to the stiffer coupling between the structure and the water column in the Tupperwave device, which prevented optimal absorption. Moreover, unlike in the conventional OWC, only about 60% of the absorbed wave power was made available to the turbine, and the rest was dissipated in the valves [38]. This reveals the poor efficiency of the valves used during the tests. In the end, the Tupperwave device produced on average only about 40% of the available pneumatic power produced by the conventional OWC device for wave periods between 6 and 9.5 s. This went down to 33% on average in irregular waves; see Figure 16. The efficiency of the Tupperwave device to convert the absorbed wave power into available power to the turbine was studied in greater detail in [42].

The poor performance of the Tupperwave physical model relative to the conventional OWC was largely due to the pneumatic power losses occurring in the valves used in these tests. The valves, described in Section 3.1.3, were bought off-the-shelf from the plumbing market, and their poor performance was not representative of what could be obtained at full scale with purposely-designed valves. It is likely that purposely-designed valves would reach better performances. However, the available literature provides very little information on non-return valves for large OWC devices and no information on their achievable performances. This shortcoming of the available literature on OWC devices hinders definitive conclusions on the performance of flow-rectifying OWC devices like the Tupperwave device.

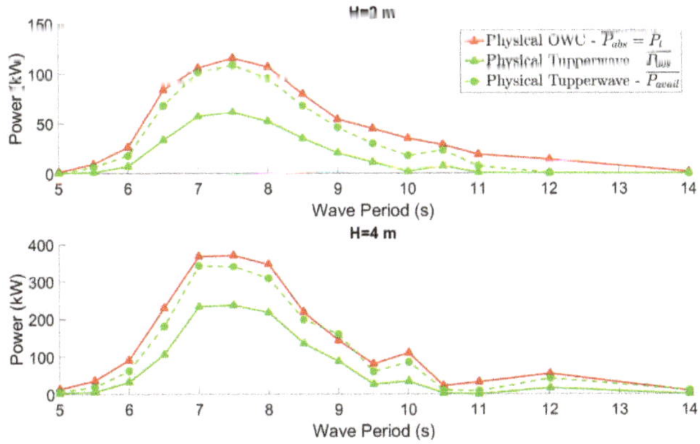

Figure 18. Average power absorbed from the waves P_{abs} and pneumatic power available to the turbine P_{avail} by the conventional OWC and the Tupperwave device in 2 and 4 m-high regular waves.

The valves used in these tests operated however sufficiently well to prove the capacity of the Tupperwave device to smoothen significantly the air flow across the turbine in comparison with the conventional OWC see Figure 17. In the latter, the air flow across the orifice stopped every half wave period to change direction, and the pneumatic power dropped to zero. Between two flow directional change, the pneumatic power reached a peak of 500 kW in about 3–4 s in a relative low-energetic sea state. At full scale, the efficiency of an air turbine to convert pneumatic power into mechanical power was largely dependant on the pneumatic power fluctuations. Large fluctuations of pneumatic power prevented the turbine from working at maximum efficiency, and the average efficiency of a self-rectifying turbine was lower in the real flow condition than in the constant flow condition, as shown in [3]. It is however to be kept in mind that the inertia of the turbines acted as short-term energy storage. Hence, the final electrical power output of the conventional OWC and of the Tupperwave device would be smoother than the pneumatic power flowing across their turbines presented in Figure 17.

The results show that the smoothing of the pneumatic power achieved by the Tupperwave device was significant, but made at the expense of important pneumatic power losses in the valves, resulting in lower power performance of the Tupperwave device. The smoothing of the pneumatic power would however play a role in the rest of the power conversion chain, as it will simplify the pneumatic to mechanical power conversion by the turbine and may pay back in terms of overall electrical power performance and power quality.

5. Conclusions

In this work, time-domain numerical models of a conventional OWC and of the Tupperwave device were developed. The models considered the conversion of wave power to pneumatic power and were built using hydrodynamic and thermodynamic equations. A tank testing campaign was carried out for the two devices at 1/24th scale to prove the Tupperwave working principle and provide data for the numerical models' validation.

Air compressibility is essential in the Tupperwave device working principle, and it was attempted to model it physically by scaling down the HP and LP chambers by ϵ^2. This was however not done correctly since the walls of the additional chambers used in the Tupperwave device were not perfectly rigid, which invalidated the exact reproduction of the air compressibility in the HP and LP chambers. The chambers being very slightly deformable, it added a phenomenon of pneumatic power storing in the chambers under the form of strain energy and caused an additional unrealistic smoothing of pressure variations in the chambers. Nevertheless, the working principle of the Tupperwave device was validated, and the device operated reasonably well. The numerical model of the Tupperwave device was corrected to take into account the chamber deformation by adding a linear model of the chamber volumetric deformation as a function of the excess pressure.

With this correction, the numerical models predicted correctly the influence of the orifice damping, wave period, and wave height on the body motions, pressure, and pneumatic power. In irregular waves, average pneumatic power productions were predicted by the numerical models with 10.1% and 16.5% error relative to the physical results for the conventional OWC and the Tupperwave device, respectively. The numerical model of the Tupperwave device showed less accuracy due to the fluctuating behaviour of the passive valves, which was revealed to be highly dependant on the device excitation.

Since the two devices were using the same Spar buoy, direct comparison in terms of pneumatic power performance was also possible. The physical model of the Tupperwave device produced on average only about one-third of the available pneumatic power produced by the conventional OWC device in irregular waves. The main reason for that is the large pneumatic power losses caused by the rectifying valves used in the tests, which were revealed to be of poor efficiency, as only about 40% of the absorbed wave power was effectively made available to the turbine. The poor performance of those valves is not representative of what could be achieved with purposely-designed valves at full scale, but reveals the importance of the valve design in the Tupperwave device performance. However, the pneumatic power available to the turbine of the Tupperwave device is much smoother than in the conventional OWC. The pneumatic power smoothing of the Tupperwave device demonstrated in this paper is likely to have a positive influence on the turbine efficiency and on the overall electrical power production and quality.

The benefits of the pneumatic power smoothing on the Tupperwave device operation will be studied and quantified in future works, where a turbine and a generator will be added to the numerical model to build a complete wave-to-wire model.

Author Contributions: Conceptualization, P.B.; data curation, P.B.; formal analysis, P.B.; funding acquisition, J.M.; investigation, P.B.; methodology, P.B.; resources, J.M.; supervision, J.M. and V.P.; validation, V.P.; writing, original draft, P.B.; writing, review and editing, P.B., V.P., and J.M.

Funding: The authors would like to acknowledge funding received through the OCEANERA-NETEuropean Network (OCN/00028).

Acknowledgments: The authors would also like to thanks M. Tahar, T. Walsh, and F. Thiebaut for their help in the physical tank testing campaign.

Conflicts of Interest: The authors declare no conflict of interest. The funders had no role in the design of the study; in the collection, analyses, or interpretation of data; in the writing of the manuscript; nor in the decision to publish the results.

Appendix A. Prony's Method

Prony's method allows the estimation of the impulse response function K as the sum of N_p damped complex exponentials:

$$K(t) \simeq \sum_{k=1}^{N_p} \alpha_k e^{\beta_k t} \tag{A1}$$

where α_k and β_k are complex coefficients and N_p is the order of the Prony function.

The memory effect integral I can then therefore be calculated as the sum of N_p functions I_k:

$$I(t) = \int_0^t K(t-\tau)\dot{x}(\tau)d\tau \simeq \sum_{k=1}^{N_p} I_k(t) \tag{A2}$$

where:

$$I_k(t) = \alpha_k e^{\beta_k t} \int_0^t e^{-\beta_k \tau}\dot{x}(\tau)d\tau \tag{A3}$$

Differentiating Equation (A3) leads to the differential equation:

$$\dot{I}_k(t) = \beta_k I_k(t) + \alpha_k \dot{x}(t) \quad \text{(A4)}$$

The memory effect integral is therefore calculated as the sum of the N_p additional function, $\{I_k, k = 1 : N_p\}$, which are the solutions of the N_p additional first order equations.

References

1. Bellamy, N. The circular sea clam wave energy converter. In *Hydrodynamics of Ocean Wave-Energy Utilization*; Springer: Berlin/Heidelberg, Germany, 1986; pp. 69–79.
2. Ryan, S.; Algie, C.; Macfarlane, G.J.; Fleming, A.N.; Penesis, I.; King, A. The Bombora wave energy converter: A novel multi-purpose device for electricity, coastal protection and surf breaks. In Proceedings of the Australasian Coasts & Ports Conference 2015: 22nd Australasian Coastal and Ocean Engineering Conference and the 15th Australasian Port And Harbour Conference, Auckland, New Zealand, 15–18 September 2015; Engineers Australia and IPENZ: Auckland, New Zealand, 2015; p. 541.
3. Falcão, A.F.O.; Henriques, J.C.C.; Gato, L.M.C. Self-rectifying air turbines for wave energy conversion: A comparative analysis. *Renew. Sustain. Energy Rev.* **2018**, *91*, 1231–1241. [CrossRef]
4. Lopes, B. Construction and Testing of a Double Rotor Self-Rectifying Air Turbine Model for Wave Energy Recovery Systems. Master's Thesis, Tecnico Lisboa, Lisboa, Portugal, 2017. (In Portuguese)
5. Borges, J. Three-Dimensional Design of Turbomachinery. Ph.D. Thesis, University of Cambridge, Cambridge, UK, 1986.
6. Falcão, A.F.O.; Henriques, J.C.C. Oscillating-water-column wave energy converters and air turbines: A review. *Renew. Energy* **2016**, *85*, 1391–1424. [CrossRef]
7. Masuda, Y.; McCormick, M.E. Experiences in pneumatic wave energy conversion in Japan. In *Utilization of Ocean Waves—Wave to Energy Conversion*; ASCE: Reston, VA, USA, 1986; pp. 1–33.
8. Falcão, A.F.O. Wave energy utilization: A review of the technologies. *Renew. Sustain. Energy Rev.* **2010**, *14*, 899–918. [CrossRef]
9. Vicente, M.; Benreguig, P.; Crowley, S.; Murphy, J. Tupperwave-preliminary numerical modelling of a floating OWC equipped with a unidirectional turbine. In Proceedings of the 12th European Wave and Tidal Energy Conference (EWTEC), Cork, Ireland, 27 August–1 September 2017.
10. Penalba, M.; Ringwood, J. A review of wave-to-wire models for wave energy converters. *Energies* **2016**, *9*, 506. [CrossRef]
11. Penalba, M.; Sell, N.; Hillis, A.; Ringwood, J. Validating a wave-to-wire model for a wave energy converter—Part I: The Hydraulic Transmission System. *Energies* **2017**, *10*, 977. [CrossRef]
12. Penalba, M.; Cortajarena, J.A.; Ringwood, J. Validating a wave-to-wire model for a wave energy converter—Part II: The electrical system. *Energies* **2017**, *10*, 1002. [CrossRef]
13. Kelly, J.F.; Wright, W.M.; Sheng, W.; O'Sullivan, K. Implementation and verification of a wave-to-wire model of an oscillating water column with impulse turbine. *IEEE Trans. Sustain. Energy* **2016**, *7*, 546–553. [CrossRef]
14. Evans, D. Wave-power absorption by systems of oscillating surface pressure distributions. *J. Fluid Mech.* **1982**, *114*, 481–499. [CrossRef]
15. Evans, D. The oscillating water column wave-energy device. *IMA J. Appl. Math.* **1978**, *22*, 423–433. [CrossRef]
16. Cheng, A.H.D.; Cheng, D.T. Heritage and early history of the boundary element method. *Eng. Anal. Bound. Elem.* **2005**, *29*, 268–302. [CrossRef]
17. Sheng, W.; Alcorn, R.; Lewis, A. Assessment of primary energy conversions of oscillating water columns. I. Hydrodynamic analysis. *J. Renew. Sustain. Energy* **2014**, *6*, 053113. [CrossRef]
18. Falcão, A.F.; Henriques, J.C.; Cândido, J.J. Dynamics and optimization of the OWC spar buoy wave energy converter. *Renew. Energy* **2012**, *48*, 369–381. [CrossRef]
19. Henriques, J.; Falcao, A.; Gomes, R.; Gato, L. Air turbine and primary converter matching in spar-buoy oscillating water column wave energy device. In Proceedings of the ASME 2013 32nd International Conference on Ocean, Offshore and Arctic Engineering, Nantes, France, 9–14 June 2013; p. V008T09A077.
20. Taghipour, R.; Perez, T.; Moan, T. Hybrid frequency–time domain models for dynamic response analysis of marine structures. *Ocean Eng.* **2008**, *35*, 685–705. [CrossRef]

21. Cummins, W. *The Impulse Response Function and Ship Motions*; Technical Report; David Taylor Model Basin: Washington, DC, USA, 1962.
22. Lee, C.H. *WAMIT Theory Manual*; Massachusetts Institute of Technology, Department of Ocean Engineering: Cambridge, MA, USA, 1995.
23. Giorgi, G.; Ringwood, J.V. Consistency of viscous drag identification tests for wave energy applications. In Proceedings of the 12th European Wave and Tidal Energy Conference (EWTEC), Cork, Ireland, 27 August–1 September 2017.
24. Morison, J.; Johnson, J.; Schaaf, S. The force exerted by surface waves on piles. *J. Pet. Technol.* **1950**, *2*, 149–154. [CrossRef]
25. Falcao, A.F.O.; Justino, P.A.P. OWC wave energy devices with air flow control. *Ocean Eng.* **1999**, *26*, 1275–1295. [CrossRef]
26. Sheng, W.; Alcorn, R.; Lewis, A. On thermodynamics in the primary power conversion of oscillating water column wave energy converters. *J. Renew. Sustain. Energy* **2013**, *5*, 023105. [CrossRef]
27. López, I.; Pereiras, B.; Castro, F.; Iglesias, G. Optimisation of turbine-induced damping for an OWC wave energy converter using a RANS–VOF numerical model. *Appl. Energy* **2014**, *127*, 105–114. [CrossRef]
28. Benreguig, P.; Murphy, J.; Vicente, M.; Crowley, S. Wave-to-Wire model of the Tupperwave device and performance comparison with conventional OWC. In Proceedings of the RENEW 2018 3rd International Conference on Renewable Energies Offshore, Lisbon, Portugal, 8–10 October 2018.
29. Duclos, G.; Clément, A.H.; Chatry, G. Absorption of outgoing waves in a numerical wave tank using a self-adaptive boundary condition. *Int. J. Offshore Polar Eng.* **2001**, *11*, 168–175.
30. Sheng, W.; Alcorn, R.; Lewis, A. A new method for radiation forces for floating platforms in waves. *Ocean Eng.* **2015**, *105*, 43–53. [CrossRef]
31. MATLAB. *Version 7.10.0 (R2010a)*; The MathWorks Inc.: Natick, MA, USA, 2010.
32. Falcão, A.F.O.; Henriques, J.C.C. Model-prototype similarity of oscillating-water-column wave energy converters. *Int. J. Mar. Energy* **2014**, *6*, 18–34. [CrossRef]
33. Kurniawan, A.; Chaplin, J.; Greaves, D.; Hann, M. Wave energy absorption by a floating air bag. *J. Fluid Mech.* **2017**, *812*, 294–320. [CrossRef]
34. Falcão, A.F.O.; Henriques, J.C.C. The Spring-Like Air Compressibility Effect in OWC Wave Energy Converters: Hydro-, Thermo-and Aerodynamic Analyses. In Proceedings of the ASME 2018 37th International Conference on Ocean, Offshore and Arctic Engineering, Madrid, Spain, 17–22 June 2018; American Society of Mechanical Engineers: New York, NY, USA, 2018.
35. SOLIDWORKS. Version 2017; Dassault Systèmes SE. Available online: https://www.solidworks.com/ (accessed on 21 May 2019).
36. Benreguig, P.; Thiebaut, F.; Murphy, J. Pneumatic orifice calibration, investigation into the influence of test rig characteritics on calibration results. In Proceedings of the CORE Conference, Glasgow, UK, 12–14 September 2016.
37. Capricorn, HypAir Balance, Product Technical Data Sheet (ver. 001/08.2013). 2013. Available online: http://www.capricorn.pl/upload/files/20150904/napowietrzacz-hipair-balance-karta-techniczna-en.pdf (accessed on 21 May 2019).
38. Benreguig, P.; Murphy, J.; Sheng, W. Model scale testing of the Tupperwave device with comparison to a conventional OWC. In Proceedings of the ASME 2018 37th International Conference on Ocean, Offshore and Arctic Engineering OMAE2018, Madrid, Spain, 17–22 June 2018; American Society of Mechanical Engineers: New York, NY, USA, 2018.
39. Hodgins, N.; Keysan, O.; McDonald, A.S.; Mueller, M.A. Design and testing of a linear generator for wave-energy applications. *IEEE Trans. Ind. Electron.* **2012**, *59*, 2094–2103. [CrossRef]
40. Prudell, J.; Stoddard, M.; Amon, E.; Brekken, T.K.; Von Jouanne, A. A permanent-magnet tubular linear generator for ocean wave energy conversion. *IEEE Trans. Ind. Appl.* **2010**, *46*, 2392–2400. [CrossRef]
41. Baker, N.; Mueller, M.A. Direct drive wave energy converters. *Rev. Energy Renew. Power Eng.* **2001**, *1*, 1–7.
42. Benreguig, P.; Vicente, M.; Dunne, A.; Murphy, J. Modelling Approaches of a Closed-Circuit OWC Wave Energy Converter. *J. Mar. Sci. Eng.* **2019**, *7*, 23. [CrossRef]

energies

MDPI

Article

Wave-to-Wire Model Development and Validation for Two OWC Type Wave Energy Converters

Pierre Benreguig [1,*], James Kelly [1], Vikram Pakrashi [2,3,4] and Jimmy Murphy [1]

1. MaREI Centre, Beaufort Building, University College Cork, Haubowline Road, P43C573 Ringaskiddy, Co. Cork, Ireland; James.Kelly@ucc.ie (J.K.); Jimmy.Murphy@ucc.ie (J.M.)
2. Dynamical Systems and Risk Laboratory, School of Mechanical and Materials Engineering, University College Dublin, D04V1W8 Dublin, Ireland; Vikram.Pakrashi@ucd.ie
3. SFI MaREI Centre, University College Dublin, D04V1W8 Dublin, Ireland
4. The Energy Institute, University College Dublin, D04V1W8 Dublin, Ireland

* Correspondence: pierre.benreguig@ucc.ie

Received: 28 August 2019; Accepted: 14 October 2019; Published: 18 October 2019

Abstract: The Tupperwave device is a closed-circuit oscillating water column (OWC) wave energy converter that uses non-return valves and two large fixed-volume accumulator chambers to create a smooth unidirectional air flow, harnessed by a unidirectional turbine. In this paper, the relevance of the Tupperwave concept against the conventional OWC concept, that uses a self-rectifying turbine, is investigated. For this purpose, wave-to-wire numerical models of the Tupperwave device and a corresponding conventional OWC device are developed and validated against experimental tests. Both devices have the same floating spar buoy structure and a similar turbine technology. The models include wave-structure hydrodynamic interaction, air turbines and generators, along with their control laws in order to encompass all power conversion stages from wave to electrical power. Hardware-in-the-loop is used to physically emulate the last power conversion stage from mechanic to electrical power and hence validate the control law and the generator numerical model. The dimensioning methodology for turbines and generators for power optimisation is explained. Eventually, the validated wave-to-wire numerical models of the conventional OWC and the Tupperwave device are used to assess and compare the performances of these two OWC type wave energy device concepts in the same wave climate. The benefits of pneumatic power smoothing by the Tupperwave device are discussed and the required efficiency of the non-return valves is investigated.

Keywords: wave energy; oscillating water column; air turbine; valves; wave-to-wire model

1. Introduction

Harnessing wave energy to produce electrical energy in an economically sustainable way requires the development of efficient and reliable wave energy converters. Despite significant research and development, the concepts for converting wave energy into electricity still have not converged to any favoured solution [1]. The oscillating water column (OWC) concept is among the most promising types of devices due to its simplicity and robustness and is therefore the most extensively studied [2]. An OWC consists of a partially submerged fixed or floating hollow structure, open to the sea below the water surface, that traps air between the inner free-surface and the top of the structure; under the wave excitation, the internal water column oscillates in the structure and alternately compresses and decompresses the trapped air which is forced in and out of the structure through a turbine coupled to a generator.

In the most common form of OWC, the compression and decompression of the air in the chamber directly creates a bidirectional flow across a self-rectifying turbine opened to the atmosphere. Such a turbine can harness both in-coming and out-coming flows. Their rotational direction remains unchanged regardless of the direction of the air flow. Several types of such special turbines have been developed: Wells and impulse turbines are the two main types of self-rectifying turbines [2]. Their maximum total-to-static efficiency in constant flow condition varies between 30% and 72% depending on their level of complexity and cost [3]. To date, the biradial and twin-rotor turbines are the best performing self-rectifying turbines and reach, respectively, about 79% and 74% efficiency in steady flow conditions [4,5]. In real ocean conditions, the flow across the turbine is, however, highly fluctuant and stops at every half-wave period to change direction. In these conditions, the average efficiency of self-rectifying turbines drops by 5 to 10% [3].

Other forms of OWC devices use non-return valves to rectify the flow across a unidirectional turbine. The incentive is that unidirectional turbines are more efficient than self-rectifying turbines with efficiencies higher than 85% [6,7]. Various OWC devices using a unidirectional turbine have been studied and different methods for rectifying the air flow have been considered: The Masuda's navigation buoy [2], the Kaimei [8], the Leancon [9] and the vented OWC from wave swell energy [10] all use a different air flow rectification method. Rectifying valves, however, induce pneumatic power losses [11] and can also be unpractical at full scale [8].

The Tupperwave concept, described in [12], is equipped with two non-return valves, two large air chambers that act as accumulators and a unidirectional turbine. The vertical motion of the internal water surface (IWS) alternatively compresses the air into the high-pressure chamber (HP chamber) and decompresses the air in the low-pressure chamber (LP chamber). This creates a differential of pressure between the HP and LP chambers which are connected via the unidirectional turbine. The air flows in a closed-circuit in the device.

In order to study the relevance of the Tupperwave principle against the conventional OWC principle, two devices using each principle and the same floating spar buoy structure are compared in this article. The chosen floating structure is an axisymmetric spar buoy which suits both working principles. The volume of the Tupperwave HP and LP chambers is maximized according to an optimization study [13] and hence the whole buoyancy volume is used. Each chamber is 950 m^3. The electrical power performance of each device in terms of electrical energy production and power quality are to be assessed and compared. For this purpose, complete wave-to-wire numerical models of the two devices presented in Figure 1 are required.

The power conversion chain of an OWC type device is split in four main stages. Part of the wave power P_w reaching the device is absorbed by the floating device. The absorbed power P_{abs} is the power applied by IWS on the air contained in the OWC chamber. It is then converted into pneumatic power P_{avail} available across the turbine. The pneumatic power is converted into mechanical power P_m by the turbine and further converted into electrical power P_e by the generator. Accurate wave-to-wire modelling requires the modelling of each power conversion stage and a validation against physical results. Such modelling and validation has already been achieved for OWC devices. This was done, for example, in [14], where the model is verified using experimental data from open sea device deployment. The model can then be used to test control strategies in specific conditions. However, if the device is at an earlier development stage, as is the case in this paper, no large scale prototype data are yet available. The wave-to-wire model validation therefore has to be done step-by-step across the power conversion stages using scaled lab tests. The power performances of the device can then be predicted using the validated model before the building and deployment of a large scale prototype.

In [15], numerical models from wave power to pneumatic power of the Tupperwave device and corresponding conventional OWC were developed and the results were validated against physical tank testing at 1/24th scale. The results also demonstrated the capacity of the Tupperwave concept to produce a smoother pneumatic power made available to the turbine. In this paper, the models from [15] are extended to build complete wave-to-wire models which are validated step-by-step as illustrated in Figure 2. The models are then used to compare the power performance of both devices and conclude on the relevance of the innovative Tupperwave concept.

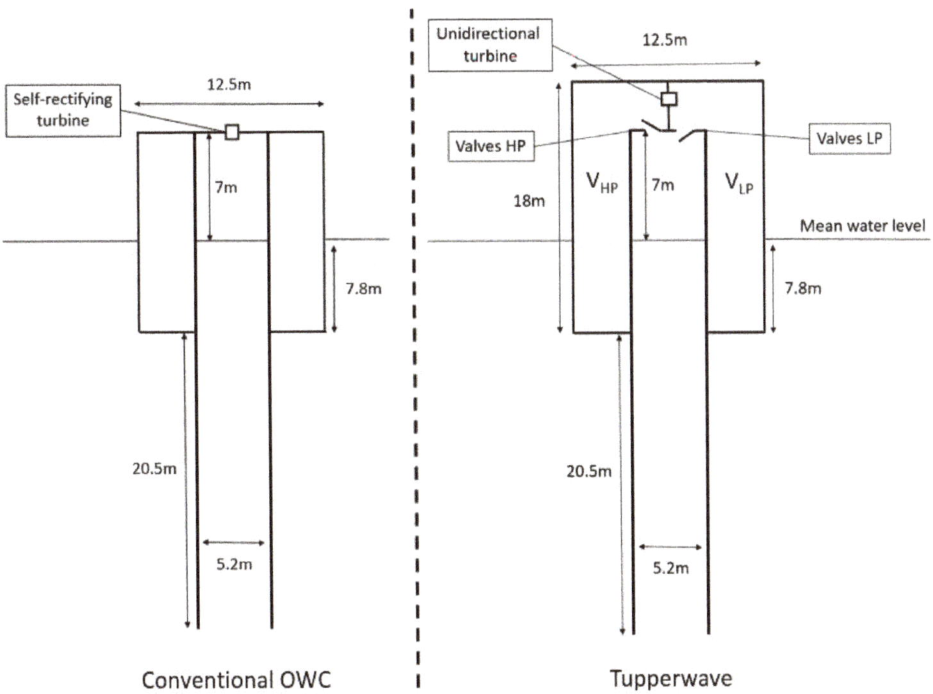

Figure 1. Two-dimensional schematic of the full scale conventional oscillating water column (OWC) and Tupperwave devices.

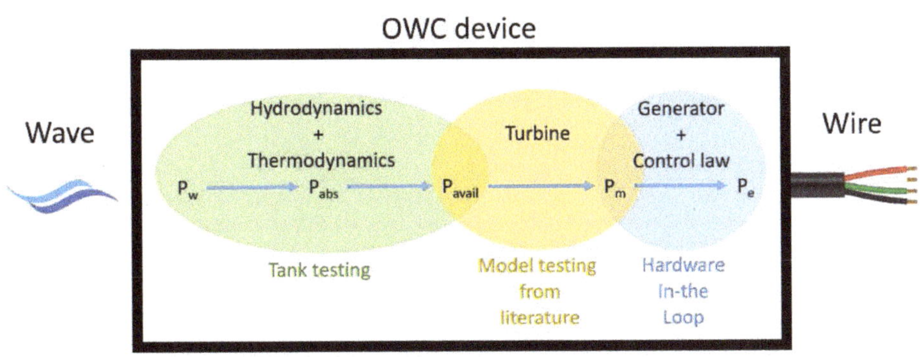

Figure 2. Schematic of power conversion chain of OWC device from wave to wire and step-by-step validation methods achieved in [6,15] and the present paper.

In Section 2 of this paper, the rest of the power conversion chain is considered and added to the existing models by the addition of a turbine and generator model to obtain the complete wave-to-wire models of the two devices. Section 3 explains the dimensioning methodology used on the turbine-generator systems for the devices' power performance optimisation in the wave climate of the EMEC test site located in Orkney, Scotland. Hardware-in-the-loop experiments were carried out to provide physical validation for the last power conversion stage and are described in Section 4. Eventually, the complete wave-to-wire models are used in Section 5 to assess and compare the power performances of the Tupperwave device and conventional OWC equipped with state of the art turbines in the EMEC wave climate. The impacts of the pneumatic power smoothing capacity of the Tupperwave working principle are identified.

2. Wave-to-Wire Models

The wave-to-wire models are based on several considerations. Hydrodynamic and thermodynamic equations were used in [15] for the development of wave-to-pneumatic power models and are first summarized for the two full scale devices. The rest of the power conversion chain is then described. State-of-the-art turbines are chosen and modelled using physical testing results from the literature. Finally the generator model and its control are presented.

2.1. Hydrodynamics

The Tupperwave device and the conventional OWC use the same floating spar structure. Therefore, the hydrodynamic wave-structure interactions for both devices can be described by the same set of equations within the linear wave theory. The structure and the water column are considered as two rigid bodies moving in heave only in the waves relatively to each other. The coupled heave motions of the two bodies (noted with indices 1 and 2) can be written in time-domain as [15,16]:

$$\begin{cases} [m_1 + A_{11}(\infty)]\ddot{x}_1(t) + \int_0^t K_{11}(t-\tau)\dot{x}_1(t)d\tau + A_{12}(\infty)\ddot{x}_2(t) \\ \quad + \int_0^t K_{12}(t-\tau)\dot{x}_2(t)d\tau + c_1 x_1(t) = f_1(t) + f_p(t) + f_{d1}(t) \quad (a) \\ A_{21}(\infty)\ddot{x}_1(t) + \int_0^t K_{21}(t-\tau)\dot{x}_1(t)d\tau + [m_2 + A_{22}(\infty)]\ddot{x}_2(t) \\ \quad + \int_0^t K_{22}(t-\tau)\dot{x}_2(t)d\tau + c_2 x_2(t) = f_2(t) - f_p(t) + f_{d2}(t) \quad (b) \end{cases} \tag{1}$$

where m_i are the bodies' masses; $A_{ij}(\infty)$ are the bodies' heave motion added masses at infinite frequency; c_i are the restoring force coefficients; K_{ij} are the impulse response functions for heave motions and their interactions; f_i are the wave excitation forces. f_p is the reciprocating pressure force acting on both bodies and is calculated as: $f_p = Sp_{owc}(t)$ where S is the internal water free surface in the water column and p_{owc} is the excess pressure relatively to atmospheric pressure built in the OWC chamber. The viscous drag forces f_{d1} and f_{d2} are calculated as $f_{di} = -C_{di}|\dot{x}_i(t)|\dot{x}_i(t)$ where C_{di} is the equivalent drag coefficient. This force includes the viscous drag effects and all non-linear viscous effects [17]. The equivalent drag coefficients at full scale are unknown. The values of $C_{d1} = 150$ N·s^2·m^{-2} and $C_{d2} = 40$ N·s^2·m^{-2} were established using the experimental results from the tank testing at 1/24th scale [15]. It is, however, to be noted that possible differences on these coefficients between model scale and full scale may arise from different flow characteristics between the two scales. The scaled-up coefficients provide reference values that were used for the numerical models at full scale.

2.2. Thermodynamics

The general thermodynamic differential equation relating the excess pressure p and volume V of air considered as a perfect gas in a chamber during isentropic transformations was derived in [15]:

$$\dot{p} = \frac{\gamma p_{atm}}{\rho_{atm} V}(w_{in} - w_{out} - \rho\dot{V}) \tag{2}$$

where p_{atm} and ρ_{atm} are the pressure and air density in atmospheric conditions; w_{in} and w_{out} are the mass flow rates of air flowing, respectively, in and out of the system.

If the system is considered adiabatic and the transformations slow enough to be reversible, the isentropic density-pressure relation is applicable:

$$\rho = \rho_{atm} \left(1 + \frac{p}{p_{atm}}\right)^{\frac{1}{\gamma}} \tag{3}$$

where γ is the isentropic expansion factor.

If the excess pressure p remains small compared to the atmospheric pressure p_{atm}, Equation (3) can be linearised:

$$\rho = \rho_{atm} \left(1 + \frac{p}{\gamma p_{atm}}\right) \tag{4}$$

In [18], the linearised isentropic assumption was shown to provide a satisfactory results for the modelling of conventional OWCs except possibly under very rough sea conditions. Hence, the transformations happening in a conventional OWC are commonly modelled in the literature using Equation (4). In the present work, the maximum pressures reached in the extreme sea states tested were in the order of 3×10^4 Pa. In those conditions, the error introduced by the linearisation of the isentropic relationship between density and pressure does not exceed 0.7%. Thus, for simplicity, Equation (4) is also used in this work.

At full scale, the turbine and the generator need to be protected from possible overloading happening in high energy sea-states. Bypass valves are commonly used as security system in OWC devices for this purpose [18]. The valve is normally closed and located in parallel to the turbine. When it opens, the flow splits between the turbine and the bypass valve, reducing the flow through the turbine and alleviate the load on this latter. The control of the bypass valve is described in Section 2.5.

Figure 3 displays a schematic of the OWC thermodynamic system. The value of mass flow rate w_t and w_{bypass} crossing the turbine and the bypass valves are considered positive whatever the flow direction. Equation (2) applied on the OWC chamber becomes:

$$\dot{p}_{owc} = \begin{cases} \frac{\gamma p_{atm}}{\rho_{atm} V_{owc}} \left(-(w_t + w_{bypass}) - \rho_{owc} \dot{V}_{owc} \right) & \text{for} \quad p_{owc} > 0 \\ \frac{\gamma p_{atm}}{\rho_{atm} V_{owc}} \left(+(w_t + w_{bypass}) - \rho_{owc} \dot{V}_{owc} \right) & \text{for} \quad p_{owc} < 0 \end{cases} \tag{5}$$

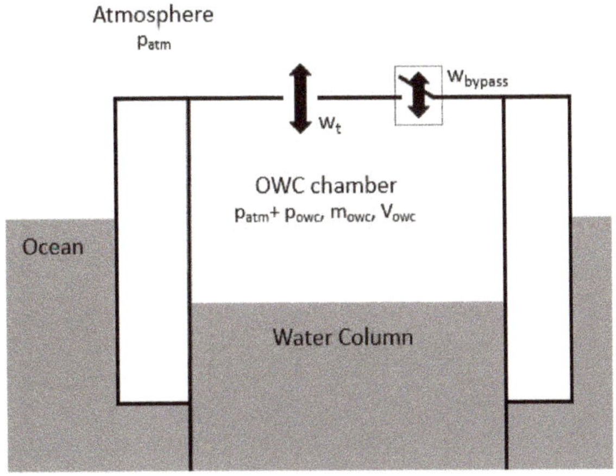

Figure 3. Conventional OWC schematic with thermodynamic variables.

Figure 1 displays a schematic of the three chambers of the Tupperwave device which constitute three interconnected thermodynamic systems. Unlike in the conventional OWC, in which the air is partly renovated during the inhalation part of the cycle, the air in the Tupperwave device is exchanged between the chambers in closed-circuit with no exchange with the atmosphere. This brings into question the possible air temperature increase in the device due to energy dissipation by viscous effects across the turbine and the valves. The study of such irreversible processes is not in the scope of this research and it is assumed that the heat transfer through the device walls into the environment is sufficient to prevent the temperature in the device to rise significantly above atmospheric conditions. Moreover, the pressure conditions in the Tupperwave device being of similar order to the ones in the conventional OWC, the linearised isentropic assumption is also adopted for the Tupperwave device and Equation (2) is directly applied to the three chambers:

$$\begin{cases} \dot{p}_{owc} = \dfrac{\gamma p_{atm}}{\rho_{atm} V_{owc}}\left(w_{vl} - w_{vh} - \rho_{owc}\dot{V}_{owc}\right) & (a) \\ \dot{p}_{hp} = \dfrac{\gamma p_{atm}}{\rho_{atm} V_{hp}}\left(w_{vh} - (w_t + w_{bypass})\right) & (b) \\ \dot{p}_{lp} = \dfrac{\gamma p_{atm}}{\rho_{atm} V_{lp}}\left((w_t + w_{bypass}) - w_{vl}\right) & (c) \end{cases} \tag{6}$$

where w_t, w_{bypass}, w_{vh} and w_{vl} are the air mass flow rates across the turbine, the HP valve and the LP valve. The expression of these flows as functions of the pressure are established using the turbine and valve models.

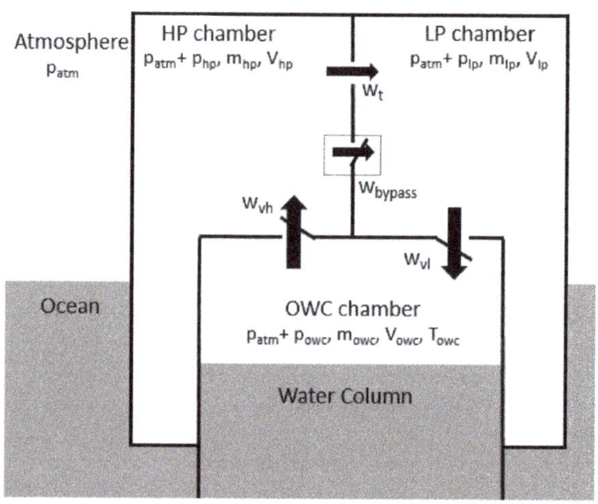

Figure 4. Tupperwave device schematic with thermodynamic variables.

2.3. Tupperwave Non-Return Valves

The valves are a key components in the Tupperwave device efficiency because they cause pneumatic power losses. The effective area of the valves and the opening pressure are the key parameters for the device efficiency [11]. Passive valves do not require to be electrically activated and are therefore very easy to implement. The use of passive valves in the full-scale Tupperwave device is justified to reduce the device mechanical and electrical complexity and hence to reduce its price. In this paper, the same passive valve model as in [15] is used. The valves only allow air flow in one direction when the pressure head $\Delta p_v = p_{in} - p_{out}$ is larger than their opening pressure p_o. In high energy sea states, the Mach number of the flow across the valves rises above 0.3. Hence, the flow is considered as compressible and a steady-state subsonic compressible flow model is adopted to calculate the air mass flow rates across the valves [19]:

$$w_{v,in\to out} = \begin{cases} 0 & \text{if} \quad p_{in} - p_{out} < p_0 \\ \alpha A_v \sqrt{\dfrac{2\gamma}{\gamma - 1} \rho_{in} (p_{in} - p_0) \left(r^{\frac{2}{\gamma}} - r^{\frac{\gamma+1}{\gamma}} \right)} & \text{if} \quad p_{in} - p_{out} > p_0 \end{cases} \tag{7}$$

where $r = \dfrac{p_{out}}{p_{in} - p_0}$ is the pressure ratio over the valve when open; αA_v is the effective opening area of the valves when fully opened [18]. To maximise the efficiency of the valves, the effective area should be maximised while the opening pressure should be minimised. No valves were designed and studied for the full scale device. Moreover, hardly any information on non-return valves for large OWC devices is available in the literature. Therefore, it is hard to know what effective area and opening pressure are practically achievable at full scale. During physical model testing carried out at model scale in [11,15], the valves caused very important pneumatic power losses, dissipating up to 60% of the absorbed wave power. They operated at best with a effective area of 0.286 m^2 at full scale equivalent. The area available for the valve connecting the OWC chamber to the HP chamber (resp. LP chamber) is about 10 m^2. With such a large available area, it is very likely that properly designed full scale valves could reach larger effective opening area than 0.286 m^2 = 53 × 53 cm. Nevertheless, this value will be used in the wave-to-wire model as a reference value. Regarding the opening pressure, a pressure differential of 150 Pa between both sides of the valve is equivalent to a force of 43 N = 4.4 kgf acting on the valve. Considering the small dimensions of the valve, such force should be sufficient to open it. The value of $p_0 = 150$ Pa was therefore chosen as a reasonable pressure value to activate the opening of the valves. The influence of the values of αA_v and p_0 are investigated in Section 5.

2.4. Turbine Model

The conventional OWC and the Tupperwave device are equipped with two different turbines, which will be described in Section 3.1. However, the numerical modelling remains the same for both turbines. The mechanical power P_{mech} extracted by a turbine from an air flow is a function of its diameter D, its rotational speed Ω, its pressure head $\Delta p_t = p_{in} - p_{out}$ between inlet and outlet, and of the volumetric air flow rate q_t flowing across a turbine. These relationships are usually presented in dimensionless form, assuming incompressible flow and using the dimensionless flow coefficient Φ, dimensionless pressure head Ψ and dimensionless turbine power output Π which are defined as [2]:

$$\Phi = \frac{q_t}{\Omega D^3}; \quad \Psi = \frac{\Delta p_t}{\rho_{in} \Omega^2 D^2}; \quad \Pi = \frac{P_{mech}}{\rho_{in} \Omega^3 D^5} \tag{8}$$

The dimensionless coefficients can be related by the polynomial functions f_Ψ and f_Π, established at model scale during laboratory tests:

$$\Psi = f_\Psi(\Phi)\,; \quad \Pi = f_\Pi(\Phi) \tag{9}$$

Neglecting the influence of the Reynolds and Mach numbers, the dimensionless turbine representation enables to describe any geometrically similar scaled-up version of the turbine.

From Equations (8) and (9), the mass flow rate crossing the turbine can be expressed as a function of pressure head and its expression can be used in Equations (5) and (6):

$$w_t = \rho_{in} q_t = \rho_{in} \Omega D^3 f_\Psi^{-1} \left(\frac{p_{in} - p_{out}}{\rho_{in} \Omega^2 D^2} \right) \tag{10}$$

The total-to-static turbine efficiency η_t is obtained by the ratio of the mechanical power of the turbine P_m and the pneumatic power available to the turbine P_{avail}:

$$\eta_t = \frac{P_m}{P_{avail}} = \frac{P_m}{q_t \times \Delta p_t} = \frac{\Pi}{\Phi \times \Psi} = f_\eta(\Phi) \quad (11)$$

The turbine reaches a maximum of efficiency for an optimal dimensionless flow coefficient Φ_{opt}.

2.5. Generator Model, Bypass Valve and Control Law

The mechanical power from the turbine is finally converted into electrical power by the generator. Newton's law applied on the generator rotor gives:

$$I\dot{\Omega} = T_{turb} - T_{gen} - T_{wind} \quad (12)$$

where I is the inertia of the turbine-generator system; T_{gen} is the electromagnetic braking torque applied by the generator; T_{wind} is the aerodynamic friction torque due to windage losses of the twin-rotor turbine (only relevant for the conventional OWC, see Section 3.1); T_{turb} is the mechanical torque applied by the turbine obtained from Equations (8) and (9) by:

$$T_{turb} = \frac{P_{mech}}{\Omega} = \rho_{hp}\Omega^2 D^5 f_\Pi(\Phi) \quad (13)$$

The control of the turbine-generator system is achieved via the generator torque. In this paper, the control strategy implemented is a maximum power point tracking (MPPT) based on optimal torque control. It is sought to optimise the instantaneous turbine efficiency by matching the generator braking torque to the torque expected to be produced by the turbine at maximum efficiency. According to Equation (8), the value of T_{gen} is obtained by:

$$T_{gen} = T_{turb}(\Phi_{opt}) = \frac{P_m(\Phi_{opt})}{\Omega} = \rho_{in}\Omega^2 D^5 f_\Pi(\Phi_{opt}) \quad (14)$$

We note that with this control law, the generator torque only depends on the rotational speed such that $T_{gen} = a\Omega^2$ with $a = \rho_{in} D^5 f_\Pi(\Phi_{opt})$. It is a fast, robust, simple and well-established control law to implement since it does not require any additional sensor [20]. Moreover, no power is required from the grid and energy only flows in one direction. For example, a nearly identical strategy to the MPPT control was used as a base case control strategy for comparing more complex control algorithms during sea trials of a OWC device in [21].

Care must, however, be taken so that the generator rotational speed remains within its operational range $[\Omega_{min}; \Omega_{max}]$. In low energy sea states, the generator braking torque is dropped to zero for $\Omega < \Omega_{min}$ and no electrical power is produced. In high energy sea states, the generator is at risk to be over spun. The generator and the power electronics are limited by their rating power, the maximum braking torque $T_{gen,max} = P_{rated}/\Omega_{max}$ is reached when $\Omega > \Omega_{max}$. The normally closed bypass valve, located in parallel to the turbine, is electrically activated and opens fully when the generator is overloaded to reduce the flow across the turbine until the condition $\Omega < \Omega_{max}$ is satisfied again. This security system dissipates the excess energy and prevents the over spinning of the generator. The same valve model as in Equation (7) is used to calculate the mass flow rate $w_{v,bypass}$ across the bypass valve. For quick relief of the turbine pressure head, the opening area of the bypass valve $\alpha A_{v,bypass}$ was chosen of $0.15\,\mathrm{m}^2$.

Then, the electromagnetic power $P_{em} = \Omega T_{gen}$ is converted into electricity by the generator with the efficiency η_{gen}. The efficiency of a generator is largely dependant on its load $\Lambda = \frac{P_{em}}{P_{rated}}$ where P_{rated} is the generator rated power. The efficiency drops very sharply for partial loads. η_{gen} was taken from [22] and is displayed in Figure 5 as a function of the load. The dimensioning of the generator is discussed in Section 3.

Finally, the electrical power produced by the generator is conditioned by power electronics before being delivered to the grid. The influence of the conditioning stage on the device efficiency is neglected here. As shown in [23], this assumption is reasonable in applications such as control parameters optimisation and power production assessment. The electrical power produced by the generator is therefore simply calculated as:

$$P_{elec} = \eta_{gen} P_{em} \tag{15}$$

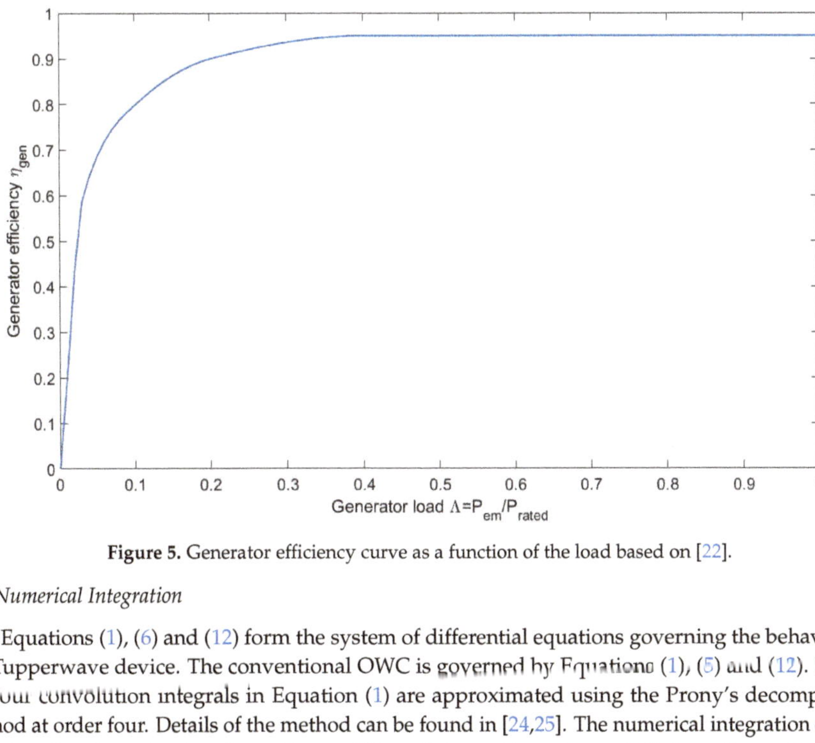

Figure 5. Generator efficiency curve as a function of the load based on [22].

2.6. Numerical Integration

Equations (1), (6) and (12) form the system of differential equations governing the behaviour of the Tupperwave device. The conventional OWC is governed by Equations (1), (5) and (12). Each of the four convolution integrals in Equation (1) are approximated using the Prony's decomposition method at order four. Details of the method can be found in [24,25]. The numerical integration of these equations is achieved using the ordinary differential equation solver ode45 from the mathematical software MATLAB [26].

3. Turbine-Generator Systems Dimensioning

3.1. Turbines

The objective of the paper being to assess and compare the power performances of the Tupperwave device against a conventional OWC, both devices are equipped with the state-of-the-art turbines.

The twin-rotor turbine, displayed in Figure 6a, is among the most efficient self-rectifying turbines [3]. It uses the idea that a bidirectional flow can be harnessed by two conventional air turbines in parallel: It consists on a pair of conventional radial-inflow rotors mounted on the same shaft in opposite direction, complemented by the corresponding guide vane rows and by a two-position cylindrical valve which, according to the flow direction, orientates the flow through one rotor or the other [27]. At all times, one rotor is driven by the air flow, while the other spins in no flow. The latter creates windage losses due to aerodynamic drag. Due to its symmetry, half of the turbine was built at model scale in [5] in order to assess its performance experimentally (Figure 6b). The resulting unidirectional single-rotor radial inflow unidirectional turbine was tested and the polynomial functions f_{Γ} and f_{Π} relating the dimensionless coefficients were established. The experimental assessment of the windage losses allowed the establishment of the twin-rotor turbine dimensionless coefficients.

In this paper, the conventional OWC and the Tupperwave device are, respectively, equipped with the twin-rotor turbine and the corresponding single-rotor unidirectional turbine. Both turbines are therefore very similar and based on the same aerodynamic design. The unidirectional turbine is, however, less mechanically complex with no need for the fast-acting and electrically activated two-position valve. Moreover, due to the windage losses, the twin-rotor turbine is less efficient than the unidirectional turbine: It reaches 72.7% while the unidirectional turbine reaches 83.9%. Their total-to-static efficiencies in constant flow condition are compared in Figure 7.

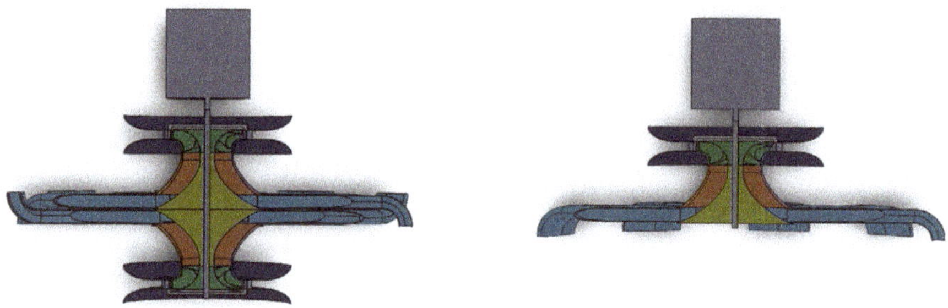

(a) (b)

Figure 6. Schematics of the radial-inflow twin-rotor turbine modelled in the conventional OWC device and corresponding unidirectional radial-inflow turbine modelled in the Tupperwave device [5]. (**a**) Twin-rotor turbine. (**b**) Single rotor turbine.

In [15], the devices absorbed most power out of the waves for wave periods from 6 to 9 s. They are therefore well adapted for the EMEC wave energy test site located off the coast of Scotland where such wave periods prevail. The scatter diagram of the EMEC wave climate is given in Figure 8.

For a fair comparison between the two devices, the turbines were optimised using the same methodology described below. The method described has no pretension of the being the best optimisation method. The turbine parameters to optimise are the diameter and the rotational speed. Their optimisation aims at maximising the device power output in the wave climate in which they are tested.

The optimisation of the turbines to maximise electrical power production in the EMEC wave climate was made in four steps:

1. Identification of the sea states for which the devices are the most productive over the year.
2. Assessment of optimal damping coefficients for the most productive sea states.
3. Assessment of turbine diameter and rotational speed to achieve optimal damping.
4. Verification that the damping achieved by the chosen turbine is close to optimal.

Basic turbine models were first used. A quadratic relationship was assumed between flow rate and pressure drop across the turbine such that:

$$\Delta P_t = k_t q_t^2 \tag{16}$$

where k_t is the damping coefficient of the turbine. The damping of the turbine is fundamental for the efficiency an OWC device in absorbing the wave energy. As a starting point, the turbine damping coefficients obtained by parametric optimisation in [13] carried out in regular waves were first tried over the whole EMEC wave climate and showed that the sea states of significant wave height H_s = 2–3 m and peak period T_p = 8–9 s are the most productive sea states over the year. Due to the large discrepancy of the sea states energy density, the most productive sea states are not the one that occur most often.

The turbine characteristics were then optimised for the most productive sea states. The damping coefficients maximising the devices' available power in the turbine P_{avail} for those sea states were investigated in detail. Figures 9 and 10 show the evolution of the average pneumatic power flowing across the turbine P_{avail} over the turbine damping coefficient in sea state $\{H_s = 2\,\text{m};\ T_p = 8\,\text{s}\}$ and $(H_s = 3\,\text{m};\ T_p = 9\,\text{s})$ for the two devices, respectively.

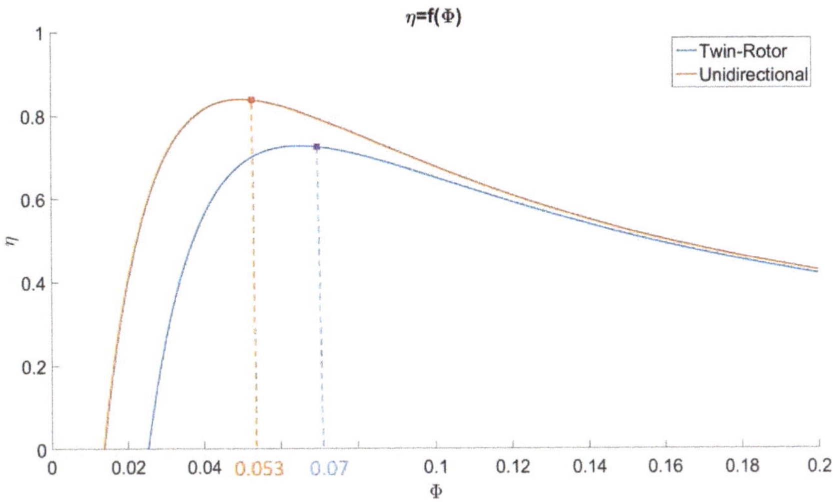

Figure 7. Total-to-static efficiencies for the twin-rotor and unidirectional turbines as functions of the flow coefficient.

It is observed that the ranges of damping coefficients maximising the power absorption from the waves are $k_t = 35\text{–}75\,\text{Pa}\cdot\text{s}^2\cdot\text{m}^{-6}$ for the conventional OWC and $k_t = 1000\text{–}2500\,\text{Pa}\cdot\text{s}^2\cdot\text{m}^{-6}$ for the Tupperwave device. The average flow rates achieved across the turbines in those conditions are denoted as $\overline{q_t}$.

It is fundamental that the turbines work close to their design flow coefficients Φ_{opt} (highlighted in Figure 7) which maximise their efficiencies. Therefore, for each turbine, a design rotational speed Ω_d is first chosen arbitrarily and its diameter is calculated from Equation (8) by:

$$D = \sqrt[3]{\frac{\overline{q_t}}{\Omega_d \Phi_{opt}}} \tag{17}$$

The obtained turbine characteristics are given in Table 1.

Table 1. Chosen turbine characteristics.

	Unidirectional	**Twin-Rotor**
Φ_{opt}	0.053	0.07
Ω_d (rpm)	4000	1000
$\overline{q_t}$ ($\text{m}^3\cdot\text{s}^{-1}$)	2.8	9.8
D (m)	0.5	1.10

It is then necessary to verify if the diameter and rotational speed selected induce a damping close to optimal for different turbine speeds. Figures 11 and 12 display the relationships between flow rates and pressure drops of the turbine models with the chosen parameters for different rotational speeds at which the turbines are likely to spin. They are compared with the simplified laws of optimised constant damping k_t. The damping achieved by the turbines for the different rotational speeds falls right in the range of optimum damping maximising wave absorption for sea states between $\{H_s = 2\,\text{m};\ T_p = 8\,\text{s}\}$ and $(H_s = 3\,\text{m};\ T_p = 9\,\text{s})$. This confirms that the diameters and rotational speeds of the turbine are well established. If it was not the case, the rotational speed of the turbines should be reconsidered until good agreement is obtained.

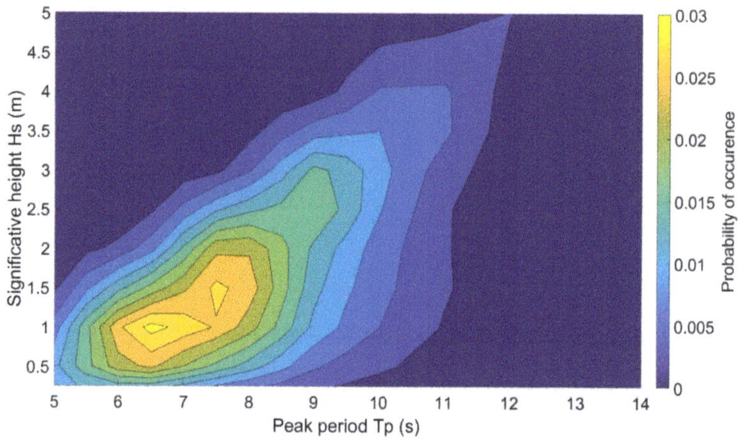

Figure 8. Scatter diagram of wave data statistic for the EMEC wave energy test site.

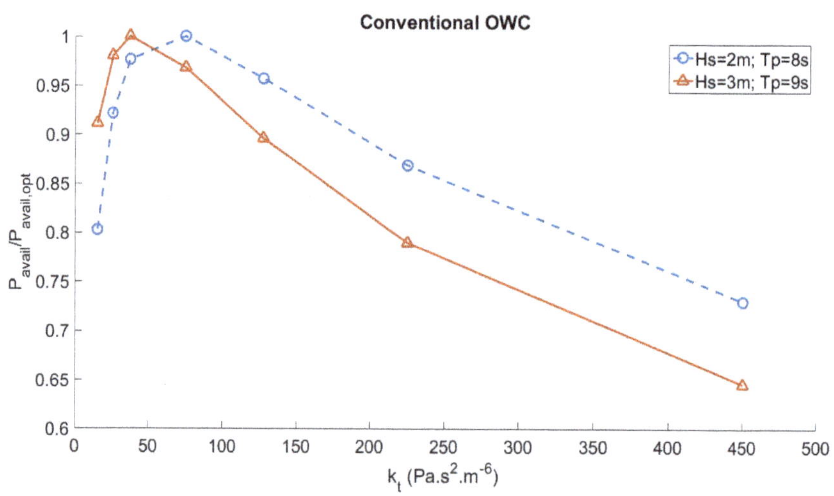

Figure 9. Evolution of pneumatic power available to the turbine with turbine damping coefficient for the conventional OWC.

It is observed in Figures 11 and 12 that the damping of the turbines increases slightly with the rotational speed. It is, however, interesting to note that, according to Figures 9 and 10, the optimal turbine damping becomes smaller for larger sea states where the turbine is likely to be spun faster.

This shows that speed control of the turbines depending on the wave condition would theoretically enable the turbine damping to be matched with the optimal damping and hence maximise the absorbed power and the real delivered power production. In the case of the conventional OWC, the flow across the turbine very quickly varies and stops every 3–5 s to change direction. In this flow condition, a fixed-speed generator would experience severe shock loads on the generator shaft, whereas if the speed were allowed to increase, the inertia of the system would absorb some of the extra power input. This consideration led to the adoption of the more simple variable speed control law described in Section 2.5. With such a control law, the turbines gain to be designed so that their damping stays relatively constant with various rotational speeds.

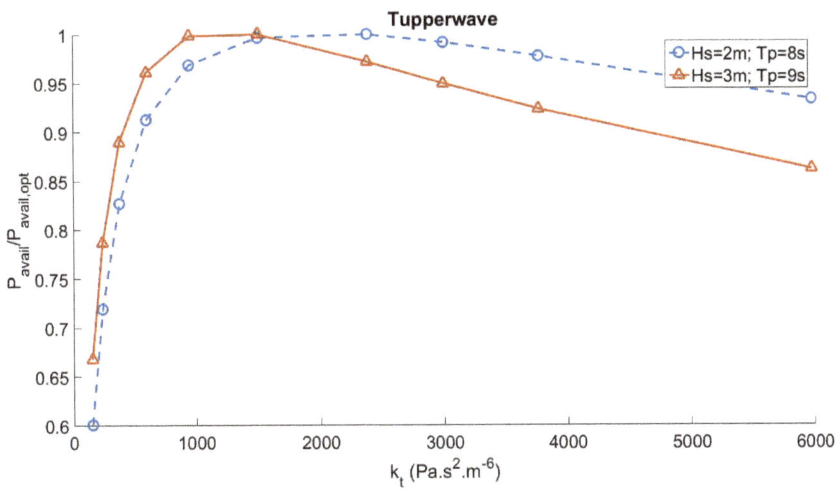

Figure 10. Evolution of pneumatic power available to the turbine with turbine damping coefficient for the Tupperwave device.

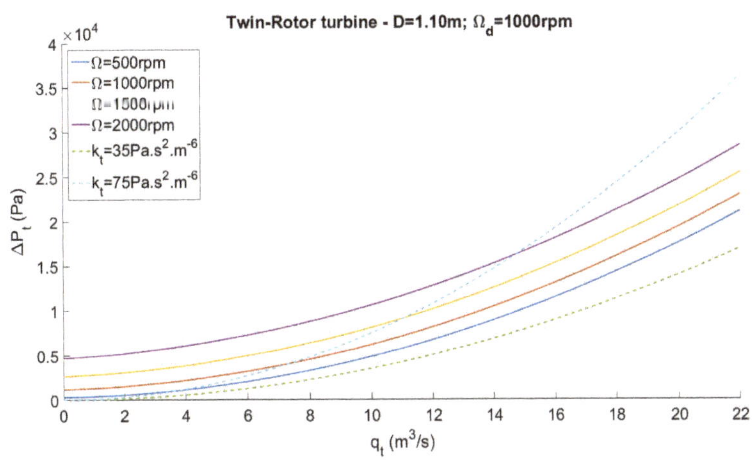

Figure 11. Relationship between pressure head ΔP_t and volumetric flow rate q_t for the twin-rotor turbine at various rotational speeds Ω compared to the optimal fixed damping relationships.

The adaptation of the unidirectional turbine tested in [6] for the pneumatic damping needs of the Tupperwave turbine and the conventional OWC led to very different results on what concerns the size (diameter D) and the rotational speed. This is due to the different working conditions of the turbines in the two devices. The turbine in the Tupperwave device has a small diameter of 0.5 m but a very high design rotational speed of 4000 rpm. In addition of being less mechanically complex because of the absence of valve and of a single rotor, the chosen Tupperwave turbine is half the size of the conventional OWC turbine, it is therefore likely to be much cheaper. Unfortunately, a gearbox is necessary to bring down its high design speed to a commonly used generator design speed (1000 rpm) and the investment cost and maintenance issues associated with a gearbox are undesirable.

The Cordier diagram, see [28], is an empirical diagram commonly used as a tool by turbine designers. It indicates that if the optimum operating conditions are kept constant and if the design speed of the turbine is lowered, then the diameter of the turbine increases. This shows that the use of a gearbox could be avoided if a different turbine design with larger diameter was adopted. The Cordier diagram does not, however, provide any information on the blade shape (i.e., angles and blade width) and such designing exercise is out of the scope of the present research.

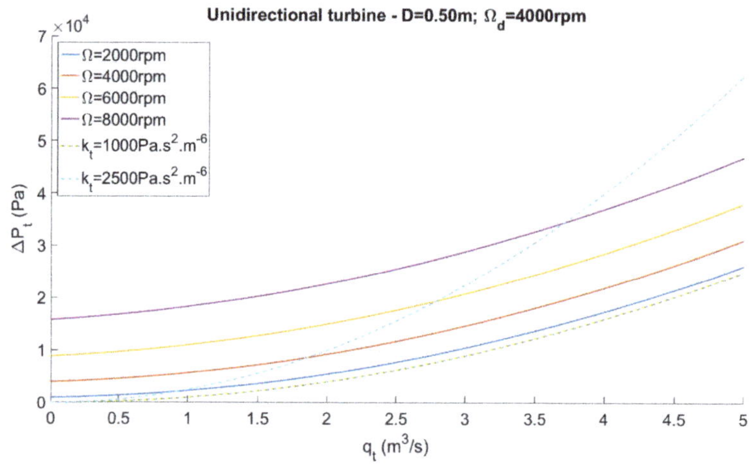

Figure 12. Relationship between pressure head ΔP_t and volumetric flow rate q_t for the unidirectional turbine at various rotational speeds Ω compared to the optimal fixed damping relationships.

3.2. Generator

As described in Section 2.5, the efficiency of conventional electrical generators decays markedly for loads less than one-third of the rated power. During operation the electrical equipment, especially the power electronics, cannot be subjected to overloads. It is the purpose of the bypass valve to protect the electrical equipment from overloads. This security system should be used the least possible to reduce the risk of failure and increase the reliability of the device. For these reasons, the generator rated power needs to be carefully chosen. Similarly to the turbine, the generator should be sized according to the sea states of the site's wave climate for which the device extract the most wave power over the year. In those sea states, the generator should work at maximum efficiency but its rated power should be such that it is able to withstand the power peaks when high energy is available.

The most productive sea states of the EMEC wave climate for the devices studied in this paper are sea states of significant wave height $H_s = 2$–$3\,\text{m}$ and peak period $T_p = 8$–$9\,\text{s}$. According to Figure 5, the generator works at maximum efficiency for loads from $\Lambda = 0.35$ onwards. The rated power of the generator is therefore chosen so that the average load of the generator in these sea states is close to 0.35. It will be seen in Section 5 that the conventional OWC and the Tupperwave device produce similar amount of electrical power and a generator of 100 kW rated power was chosen for both devices. Its inertia was estimated from an exiting generator of similar rated power.

Table 2 displays the optimised turbine and generator characteristics for the two devices studied. The turbines were assumed to be built in aluminium and their inertia were approximated based on their diameter.

Table 2. Turbine and generator parameters used in the wave-to-wire models.

		Tupperwave	Conventional OWC
Turbine	Type	Unidirectional radial inflow turbine	Self-rectifying radial inflow twin-rotor turbine
	Diameter (m)	0.50	1.10
	Inertia (kg·m^2)	1.7	38
	Max. efficiency (%)	86.6	73.9
Gearbox	Gearing Ratio	4	1
Generator	Rated power (kW)	100	
	Inertia (kg · m^2)	3.6	
	Design speed (rpm)	1000	
	Max. speed (rpm)	2000	
	Min. speed (rpm)	400	

4. Numerical Model Validation

The wave-to-wire models are validated in different power conversion steps. In [15], the models from wave to useful pneumatic power available to the turbine were validated. The turbine models added in this paper were established by the experimental tests carried out in [6] providing accurate modelling of the pneumatic to mechanical power conversion. The last power conversion step from mechanical to electrical power achieved by the generator is validated experimentally in this paper.

4.1. Objective and Method

The objective is to emulate the mechanical to electrical power conversion happening at full scale in the two devices studied and compare the results with the numerical models. This is achieved using Hardware-in-the-Loop (HIL) to simulate physically and in real time the turbine action on a real generator. The hardware is the rotary test rig of the Lir National Ocean Test Facilities (Lir-NOTF), MaREI Center, Ireland. The rig is basically composed of two 22 kW coupled electrical machines. One is used as a motor and acts as the prime mover (turbine). The other acts as the electrical generator and is connected to the local electrical grid using an off-the-shelf back-to-back converter. A picture of the rig is shown in Figure 13. Detailed descriptions of the rig, its capabilities and its limitations are described in [29,30]. The generator of the rotary test rig in the Lir-NOTF includes several different configurations. For the HIL tests performed for validation, the selected generator configuration was the squirrel cage induction generator (SCIG) as suggested in [31].

The generator control law presented in Section 2.5 is directly applied on the real generator. This type of simulation is generally undertaken in the development of full scale devices. The controlled environment enables to simulate the equipment and control system in offshore-like conditions at a significantly lower cost prior to offshore tests.

The HIL system includes a Matlab–Simulink numerical model of the wave energy device from wave to mechanical power. The torque T_{turb} created by the turbine on the shaft is solved based on the measured rotational speed of the rig. This torque is then applied in real time by the motor on the shaft. The connected generator converts the mechanical power into electrical power which is sent onto the grid. Since the numerical model depends on the turbine state, the measured rotational speed of the system is fed back into the numerical model at each time step.

The numerical model from wave to mechanical power is run at full scale while the hardware is run at model scale. The model scaling is determined by the hardware installation limits in power and speed. The scaling is executed through a two-step process. Since surface waves are gravity driven, the Froude similarity law of coefficient λ is first used to scale down the turbine output power to fit on the rig installation which is limited to 22 kW. For that, the Froude similarity law is applied on the turbine output torque T_{turb} and on the turbine speed Ω. As a result, the torque is reduced, but the speed is increased. To adapt the rotational speed of the model in line with the limitations of the rig (2200 rpm), the second step of the scaling process is the implementation of a virtual gearbox to the system with a virtual gearbox coefficient G, where a decrease in speed is exchanged for an increase in torque. The flow chart presenting the HIL system and the scaling method is displayed in Figure 14. As explained in Section 1, all parts of the device model (hydrodynamic, thermodynamic, turbine, generator) are strongly coupled and this is represented by the bidirectional arrows.

Figure 13. Rotary test rig of the Lir National Ocean Test Facilities (Lir-NOTF), used to emulate the turbine-generator systems of the devices.

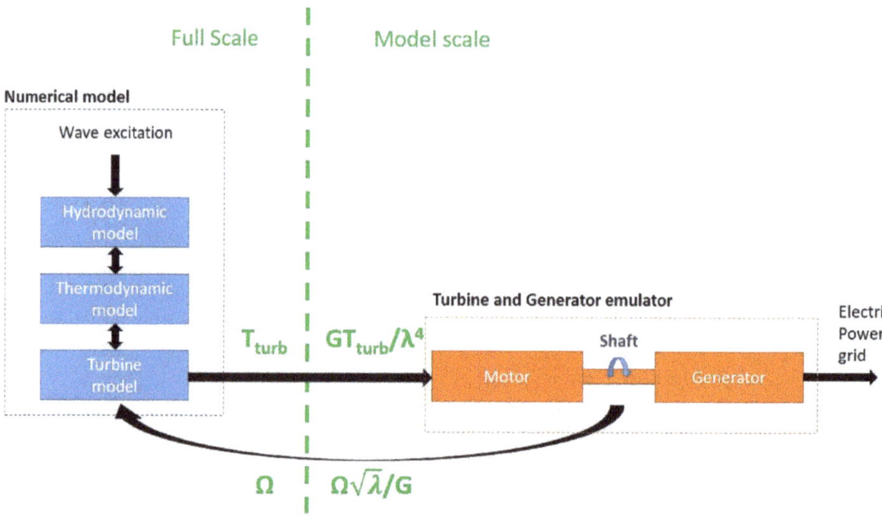

Figure 14. Hardware-in-the-Loop flow chart with scaling laws applied to the turbine speed Ω and output torque T_{turb}.

4.2. Hardware-in-the-Loop Results

HIL tests were carried out for three common sea states of the EMEC wave climate and compared to the fully numerical results. The Pearson correlation coefficient for the electrical power P_e and generator rotational speed Ω_{gen} were calculated and are displayed in Table 3. The coefficients are close to 1 and indicate good agreement between the fully numerical model and the tests with HIL.

Table 3. Pearson correlation coefficients between Hardware-in-the-Loop (HIL) and fully numerical results for the time series of electrical power P_e and generator rotational speed Ω_{gen} in three common sea states of the EMEC wave climate.

	Tupperwave		Conventional OWC	
	P_e	Ω_{gen}	P_e	Ω_{gen}
H_s = 1.5 m; T_p = 7.5 s	0.954	0.956	0.953	0.990
H_s = 2 m; T_p = 8 s	0.978	0.958	0.959	0.990
H_s = 3 m; T_p = 9 s	0.974	0.961	0.944	0.985

For a more illustrative comparison, time series are displayed in Figures 15 and 16. The results in Table 3 and very good visual agreement are obtained; Figures 15 and 16 validated the last power conversion step of the fully numerical wave-to-wire models. It is noticeable for the conventional OWC that the high power peaks are reduced by the HIL due to limitations in the power output of the prime-mover motor of the rotary rig and the low power are increased by the HIL due to a baseline power output that the Regen drive needs for output to the grid. These effects are caused by intrinsic limitations of the rotary rig and are more noticeable for the conventional OWC than for the Tupperwave device because of the larger power fluctuations.

For the two devices, the generator control law gives satisfactory results since it allows the generator to spin close its design speed of 1000 rpm.

This validates the last power conversion stage and concludes the validation by experimental testing of the entire power conversion chain, as illustrated in Figure 2.

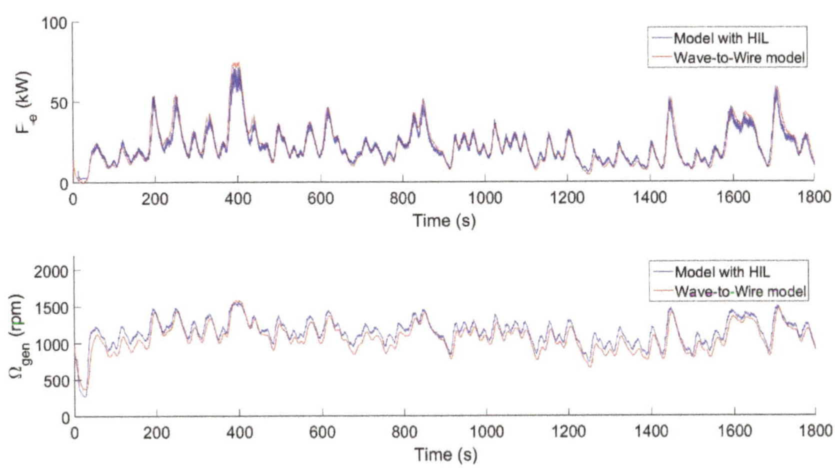

Figure 15. Time series of electrical power P_e and generator rotational speed Ω_{gen} obtained for the Tupperwave device with HIL and fully numerical model in sea state $\{H_s = 2\,\mathrm{m};\ T_p = 8\,\mathrm{s}\}$.

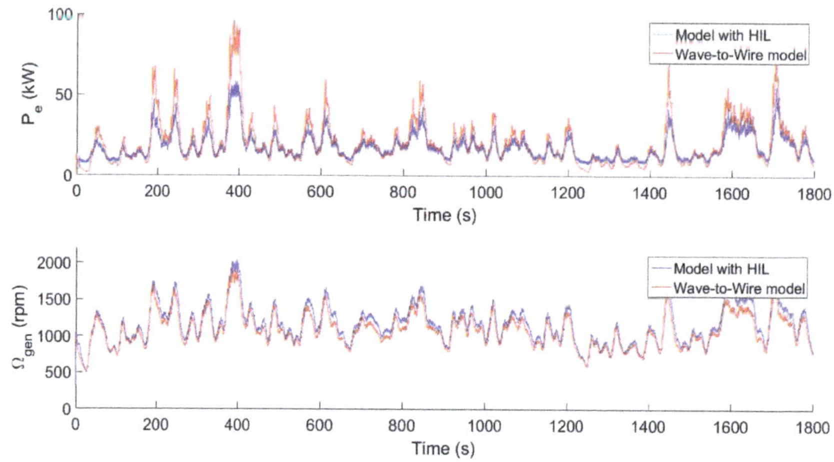

Figure 16. Time series of electrical power P_e and generator rotational speed Ω_{gen} obtained for the conventional OWC with HIL and fully numerical model in sea state $\{H_s = 2\,\text{m}; T_p = 8\,\text{s}\}$.

5. Wave-to-Wire Models Results

Once, the Tupperwave and conventional OWC wave-to-wire models were validated against physical testing for all power conversion stages, their results are used to compare the device performances. The different power conversion processes are first compared for a single sea state. The devices' yearly performance in the EMEC test site are then compared both in terms of average electrical power output and power quality. Since both devices have the same floating structure, the same turbine aerodynamic geometry, the same generator and the same control law, this comparison aims at comparing the two devices' working principles.

The performances of the Tupperwave device are largely dependant on the non-return valves characteristics. Since the available literature provides no information on the achievable performances of non-return valves, three cases of valve characteristics were considered and detailed in Table 4.

Table 4. Non-return valve characteristics considered in the Tupperwave device.

Tupperwave Valves	p_0 (Pa)	αA_v (m^2)
Case 1	1700	0.286
Case 2	150	0.286
Case 3	150	1.3

The experience on non-return valves gained from the tank testing campaign of the Tupperwave device carried out in [15] is valuable. Even though their performances were poor, the characteristics of the valves used in [15] can be used as a starting point. In case 1, the valves opening pressure p_0 and effective opening area αA_v corresponds to the ones obtained, at best, during the tank testing campaign. The values obtained at model scale were simply scaled up with Froude similarity law. In case 2, the opening area is kept unchanged and the opening pressure is set to 150 Pa. This pressure applied on the area of the valve correspond to an opening force of 5 kgf which seems reasonable for such a small valve. This case was used for the turbine-generator dimensioning described in Section 3. In the third case, the effective opening area of the valves is increased to 1.3 m^2. We recall that the wall surface available between the OWC chamber and the HP (resp. LP) chamber is 10.6 m^2. An effective opening area of 1.3 m^2 therefore seems easily achievable.

5.1. Along the Power Conversion Chain

It will be seen in Section 5.2 that both devices produce annually the most in the sea state $\{H_s = 3\,\text{m}; T_p = 9\,\text{s}\}$. This sea state is considered in this section and the powers along the conversion chain are observed.

The amount of incident wave power reaching the devices is the same and only about 11% is effectively absorbed from the motion of the IWS in the devices. This absorbed power P_{abs} is the product of the volumetric flow rate q_{IWS} displaced by the motion of the IWS (counted positive for upward motion) by the pressure in the OWC chamber and is the rate of work applied by the IWS on the air in the OWC chamber:

$$P_{abs} = p_{owc} q_{IWS} = -p_{owc} \dot{V}_{owc} \tag{18}$$

Figure 17 displays the absorbed power by the two devices. The power mainly flows from the IWS to the air in the chamber but it is alternatively positive and negative, meaning that energy is exchanged back and forth between the IWS and the air in the OWC chamber. At each stroke of IWS, a slight part of the power is stored by the air under the form internal energy. This power is then released back in the IWS at the next stroke in the opposite direction. This is caused by the spring-like air compressibility effect in the OWC chamber. This phenomenon is clearly described in [32]. Larger compressibility effect is observed in the OWC chamber of the Tupperwave device. The average power absorbed by the Tupperwave device is 5% lower than the power absorbed by the conventional OWC.

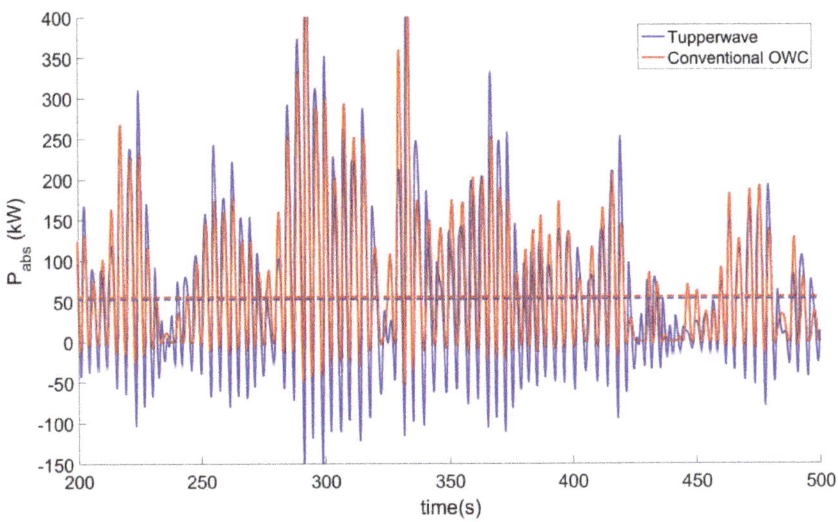

Figure 17. Power applied by the internal water surface (IWS) on the air in the OWC chamber (or absorbed wave power) in a conventional OWC and Tupperwave device (case 2) in sea state ($H_s = 3\,\text{m}$; $T_p = 9\,\text{s}$). Solid lines: Time series; dash lines: Average values.

The absorbed power is then converted into useful pneumatic power or power available at the turbine P_{avail} which, in compressible flow, is the rate of enthalpy drop of the air across the turbine:

$$P_{avail} = w_t(h_{in} - h_{out,s}) \tag{19}$$

where h_{in} is the specific enthalpy of the air at the entrance of the turbine, and $h_{out,s}$ is the specific enthalpy at the exit of the turbine for an isentropic expansion. Since the process is assumed adiabatic and isentropic without any dissipative losses, the absorbed power is, on average, entirely converted into power available at the turbine in the case of the conventional OWC: $\overline{P_{avail}} = \overline{P_{abs}}$. For the Tupperwave device, the air flow is rectified by the non-return valves to build a pressure differential between two accumulator chambers. This provides a smoothing of the pneumatic power across the turbine at the expense of pneumatic power losses across the valves. We define the valves efficiency as:

$$\eta_v = \frac{\overline{P_{avail}}}{\overline{P_{abs}}} \tag{20}$$

In the case presented in Figure 17, for the Tupperwave device, 80.5% of the absorbed power is actually converted into power available at the turbine and the rest is dissipated in the valves. The valves are thus 80.5% efficient. Figure 18 displays the resulting pneumatic powers available to the turbines in the Tupperwave device (case 2) and in the conventional OWC in sea state H_s = 3 m and T_p = 9 s.

The power available at the turbines is then converted into mechanical power by the turbine. The smooth pneumatic power available at the turbine of the Tupperwave device is easier to harness and the turbine works very close to maximum efficiency while the turbine of the conventional OWC works on average 10% away from its maximum efficiency. Figure 19 illustrates this fact by showing the density of the working points along the turbines efficiency curves during the sea state. The unidirectional turbine of the Tupperwave device works 84% of the time within 1% of its maximum efficiency while the twin-rotor turbine only works in this state for 17% of the time.

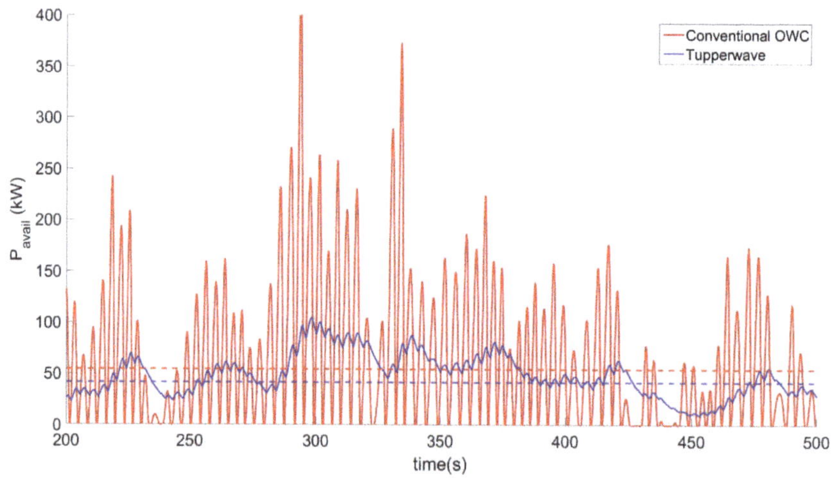

Figure 18. Pneumatic power available to the turbines in a conventional OWC and Tupperwave device (case 2) in sea state (H_s = 3 m; T_p = 9 s). Solid lines: Time series; dash lines: Average values.

Finally the smoothed mechanical power available to the generator is converted into electrical power with an efficiency close to 94% for both devices. The time-series of the devices' electrical power production are displayed in Figure 20.

The significant electrical power fluctuations, delivered by wave energy converters in general, may have a negative impact on the power quality of the local grid to which the wave farms is connected [33]. They are an issue of concern for grid operators as they can introduce undesirable effects on the grid such as voltage and frequency variations and flickers [34]. Regarding Figure 20, the power fluctuations can be split in two groups: The wave-by-wave fluctuations or the wave-group by wave-group. Wave-by-wave power fluctuations cause the small peaks at every half wave-period visible for the conventional OWC and are due the high pneumatic power fluctuation between waves. These peaks are smoothed for the Tupperwave device. The Tupperwave accumulator system also smoothen the power between wave groups: Part of the energy of a wave group is stored under the form of pressure differential between the accumulators and is slowly released through the turbine until the next wave group. This also introduces a phase shift between the two curves. As a result the electrical power delivered by the Tupperwave device is smoother than for the conventional OWC. For the smoothing of the power, the conventional OWC relies only on the inertia of its turbine. The larger inertia of the twin-rotor turbine due to its larger diameter is an asset and the turbine is able to absorb temporarily the pneumatic power under the form of kinetic energy and alleviate the load variations on the generator. The inertia of the unidirectional turbine being smaller, the Tupperwave device relies almost only on the pneumatic power storing capacity to smoothen the turbine mechanical output power. The pneumatic power smoothing achieved by the Tupperwave device leads to smoother electrical power than the mechanical power smoothing achieved by the conventional OWC's turbine inertia.

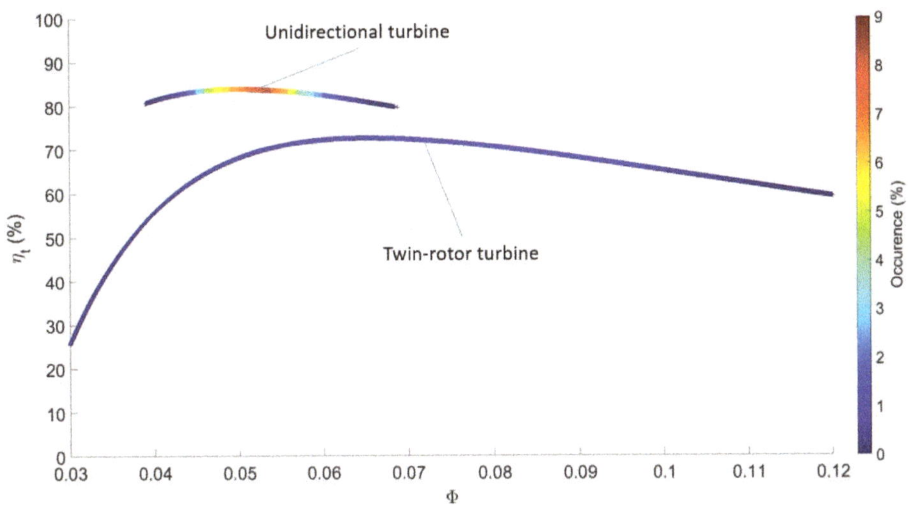

Figure 19. Percentage of occurrence of turbine working points in sea state ($H_s = 3$ m; $T_p = 9$ s).

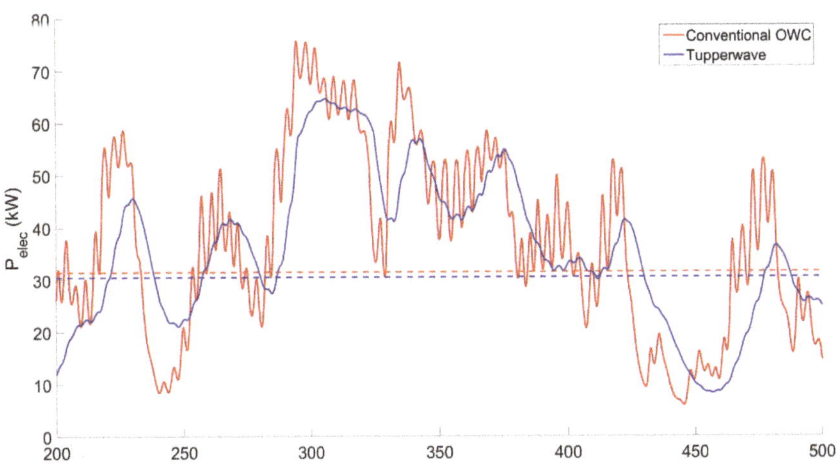

Figure 20. Electrical power produced by the conventional OWC and Tupperwave device (case 2) in sea state ($H_s = 3$ m; $T_p = 9$ s). Solid lines: Time series; dash lines: Average values.

By adding a flywheel of inertia 150 N · m^2 to the twin-rotor turbine, it is possible to obtain the same level of smoothness with the conventional OWC. Such inertia can be obtained with a 1.2 m diameter and 0.1 m thick iron wheel which weighs approximately 820 kg. This naturally adds to the initial cost of the turbine and the added mass could lead to additional maintenance costs due to larger stresses on the bearings. It also causes a 3% reduction in power production.

A smaller flywheel of inertia 36.3 N·m^2 can be added to the Tupperwave power-take off (PTO) system so that the resulting PTO inertia is the same as in the conventional OWC (see Table 2). This represents a 0.85 m diameter and 0.1 m thick iron wheel of about 400 kg. The time-series of the devices' electrical power production are displayed in Figure 21. The pneumatic power smoothing of the accumulator chambers is combined with the mechanical power smoothing of the inertial PTO and the resulting electrical power is remarkably smooth. The addition of the flywheel only causes a 0.3% reduction in average electrical power production for this sea state.

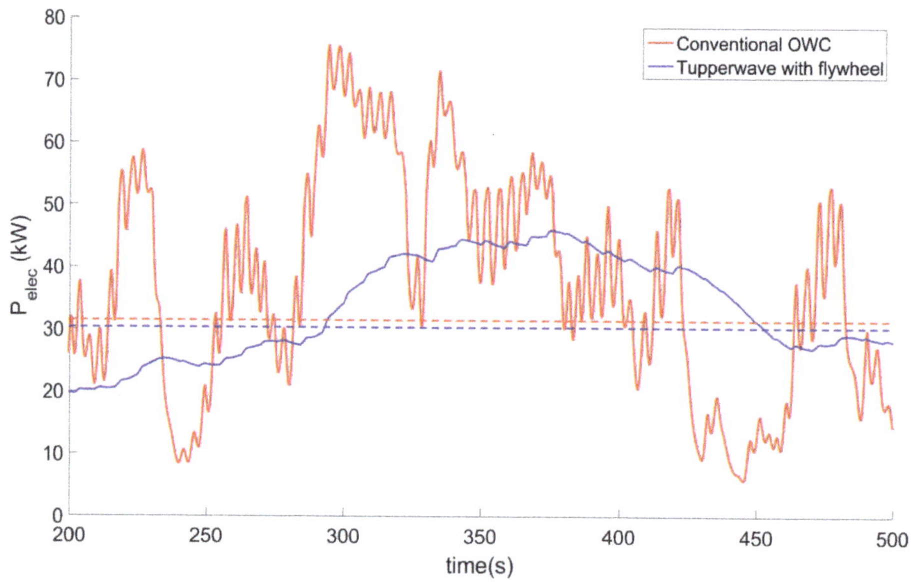

Figure 21. Electrical power produced by the conventional OWC and Tupperwave device (case 2) in sea state ($H_s = 3$ m; $T_p = 9$ s). A flywheel of inertia 36.3 N·m^2 was added to the Tupperwave power-take off (PTO) so that the resulting PTO inertia is the same as in the conventional OWC. Solid lines: Time series; dash lines: Average values.

Finally, Figure 22 displays the average power available at the different stages of the power conversion chain and for the three Tupperwave valves cases displayed in Table 4. It is clear that the efficiency of the Tupperwave device depends on the efficiency of its non-return valves in the conversion from absorbed power to available power at the turbine. The valve efficiencies are of 63.7%, 77.9% and 96.9% for cases 1–3, respectively. The Tupperwave device outperforms the conventional OWC in terms of electrical power production only if the non-return valves efficiency is greater than 80%. Case 3 corresponds to the best possible case tested with valve efficiencies close to 100%. In that case, the Tupperwave device produces 18% more electrical power than the conventional OWC.

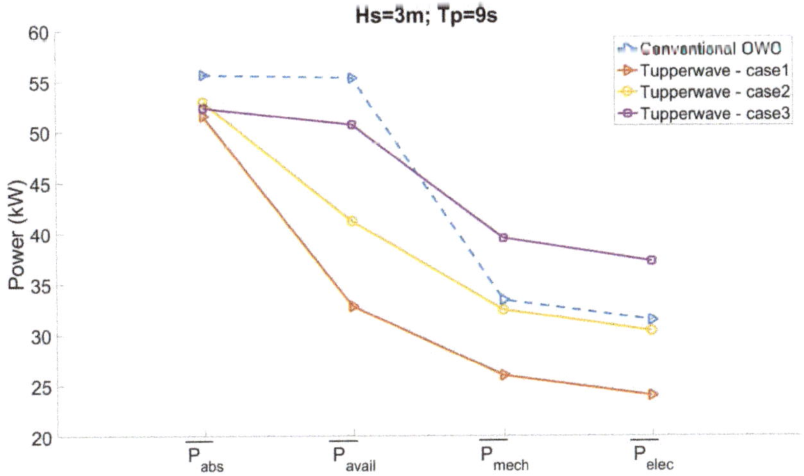

Figure 22. Average power along the power conversion chain of the conventional OWC and Tupperwave device in sea state ($H_s = 3\,\mathrm{m}$; $T_p = 9\,\mathrm{s}$).

5.2. Yearly Power Performance Comparison

A large number of simulations were run to assess and compare the device power performances on the EMEC wave energy test site in a one year time period. The comparison is based on several criteria listed below:

- Annual electrical power production.
- Electrical power fluctuation.
- Use of the security system (bypass valve).

The level of electrical power fluctuation f around its average value is calculated as the standard deviation normalised by the average value:

$$f = \frac{1}{\overline{P}}\sqrt{\frac{1}{t_f}\int_0^{t_f}\left(P(t) - \overline{P}\right)^2 dt} \tag{21}$$

Table 5 compares the devices' performances over a year in the EMEC test site. The performances of the Tupperwave device were assessed with the three different valve characteristics (see Table 4) and also when equipped with a flywheel providing the Tupperwave PTO the same inertia as the conventional OWC PTO.

Table 5. Devices performances over a year in the EMEC test site.

	Tupperwave				Conventional OWC
Valve characteristics	case 1	case 2	case 3	case 2	-
Flywheel inertia (N·m^2)	-	-	-	36.3	-
Annual electrical production (MWh)	70.6	97.6	119.5	99.3	99.9
Average power fluctuation (%)	64.9	55.0	55.1	23.2	70.2
Annual pneumatic energy dissipated in bypass valves (MWh)	0.009	0.051	0.619	0.012	0.582
Bypass valve opening per year	44	287	3209	65	8537

The results show that the Tupperwave device in case 2 produces very similar electrical power on all sea states to the conventional OWC. Figure 23 displays the amount of electrical energy produced by the Tupperwave device (case 2) in each sea state over a year. The results obtained for the conventional OWC being very similar, it is not represented here. As mentioned for the dimensioning of turbine-generator system in Section 3, the most producting sea states over the year are $H_s = 2$–3 m and $T_p = 8$–9 s.

The values of average power fluctuations displayed in Table 5 are obtained for the most productive sea states. As illustrated in Figure 20, the result indicates that the electrical power delivered by the Tupperwave device is smoother than the conventional OWC in all cases. In addition to enabling the turbine to work close to maximum efficiency, the short term pneumatic power storage mechanism using air compressibility in the HP and LP chambers of the Tupperwave device contributes to the enhancement of power quality by mitigating the power fluctuations.

As a result of the smoother electrical power output, the Tupperwave device requires less use of the security system to protect its generator. At equal power production, the bypass valve opens 30 times less often over a year in in the Tupperwave device (case 2) than in the conventional OWC. This naturally leads to greater generator and power electronics longevity and reduces the risks of failure of the security system. Smoother operation of the turbine and generator also result in less fatigue and hence higher reliability of the system. Figure 24 shows on which sea states the use of the bypass valve is required by displaying the amount of pneumatic power dissipated in the bypass valve per year in each sea state for Tupperwave (case 2) and the conventional OWC. To reduce the use of the bypass valve by the conventional OWC, the generator rated power can be increased but this would reduce its efficiency in most sea states, see Figure 5. The total amount of pneumatic power dissipated through the bypass valve is, however, small compared to the yearly production.

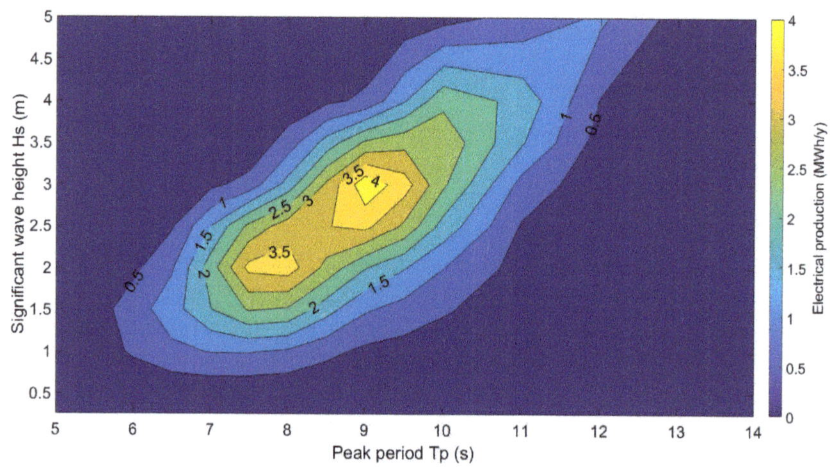

Figure 23. Electrical energy production on all sea states of the EMEC wave energy test site over a year by the Tupperwave (case 2). Quasi-identical figure is obtained for the conventional OWC device.

The addition of a flywheel to the Tupperwave PTO so that its inertia equals the inertia of the conventional OWC PTO (case 2 with flywheel) is very profitable in terms of electrical power quality: The average electrical power fluctuations are considerably reduced and thus, the use of the bypass valve is also reduced. A slight increase in annual electrical production is also observed. This increase is due to the larger production in low energy sea states ($H_s = 1$–2 m) where the device benefits from having a more inertial PTO.

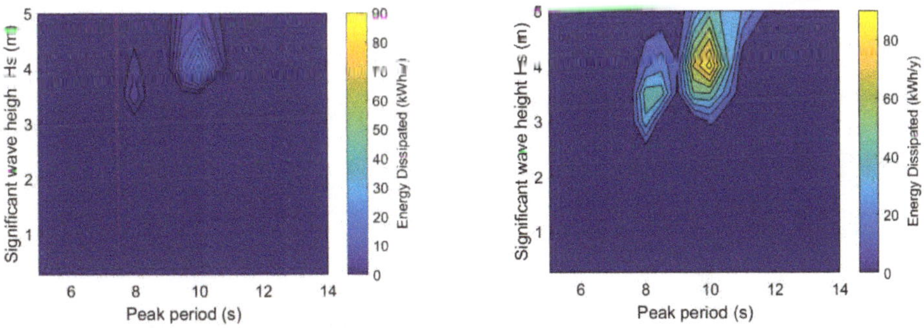

(**a**) Tupperwave: Case 2 (**b**) Conventional OWC

Figure 24. Pneumatic power dissipated in the bypass valve in each sea state per year.

The results show that the Tupperwave device can compete with the conventional OWC provided that the non-return valves are correctly designed to be sufficiently efficient. Low opening pressure and large effective opening area are required. The physical feasibility and practicability of such non-return valves with adequate efficiency, however, remains to be proven.

6. Conclusions

In this research, the time-domain wave-to-wire models of the innovative Tupperwave device and corresponding conventional OWC were built. The two devices use the same floating spar buoy structure. The hydrodynamic equations are based on the linear wave theory and the thermodynamic processes are assumed isentropic. The two devices are equipped with the same floating structure geometry, turbine rotor geometry, generator and control law.

In a previous publication [15], the numerical models from wave to pneumatic power had been validated against tank testing experiments. In this paper, the conversion from pneumatic to electrical power was studied and validated against Hardware-in-the-Loop experiments, completing the validation of the entire power conversion chain. Their power performances in the EMEC wave energy test site were assessed using the validated wave-to-wire models after optimisation of their turbine-generator systems.

The methodology for dimensioning the turbine-generator systems to the wave climate of the test site was detailed and can be applied to similar design problems. The turbine used in the Tupperwave device is half the size of the turbine used in the conventional OWC and is also less mechanically complex. It is therefore likely to be cheaper. However, its high rotational speed requires the use of a mechanical gearbox which is associated with undesirable reliability and maintenance issues. The use of the gearbox could be avoided with a different turbine design.

The comparison of the device power performances allowed the assessment of the Tupperwave device concept against the conventional OWC. Due to its working principle, the Tupperwave device produces smooth pneumatic power across its turbine, allowing this latter to work close to maximum efficiency at all times. In the end, and despite its small PTO inertia, the Tupperwave device produces a smoother electrical power than the conventional OWC, enhancing power quality. This reduces the peak-to-average power ratio and hence requires less use of the security system that protects the generator from power peaks. The smoother operation is likely to have a positive impact on the system reliability. The power quality enhancement becomes particularly remarkable if the Tupperwave PTO inertia is increased to the same level as the conventional OWC PTO inertia. In terms of power production, if the non-return valves dissipate less than 20% of the absorbed wave power, the Tupperwave device can outperform the conventional OWC by up to 20%. Such performance is achievable with sufficiently low opening pressure and a sufficiently large effective opening area of the valves. Further research on the Tupperwave concept should focus on the physical feasibility and reliability of such non-return valves in order to enable definitive conclusions on the relevance of the concept.

The use of the isentropic relationship between air density and pressure for the modelling of the Tupperwave device, in spite of the irreversible thermodynamic processes, represents a limit to the present study. A non-isentropic study of the Tupperwave device will be undertaken in future works to accurately model these processes.

Author Contributions: Conceptualization, P.B.; Data curation, P.B.; Formal analysis, P.B.; Funding acquisition, J.M.; Investigation, P.B. and J.K.; Methodology, P.B. and J.K.; Resources, J.M.; Supervision, J.M. and V.P.; Writing—original draft, P.B.; Writing—review and editing, P.B., J.K., V.P. and J.M.

Funding: The authors would like to acknowledge funding received through OCEANERA-NET European Network (OCN/00028).

Acknowledgments: The authors would like to thanks Bàrbara S. Lopes for kindly sharing the results of her experimental work on the twin-rotor turbine.

Conflicts of Interest: The authors declare no conflict of interest. The funders had no role in the design of the study; in the collection, analyses, or interpretation of data; in the writing of the manuscript, or in the decision to publish the results.

Abbreviations

The following abbreviations are used in this manuscript:

OWC	Oscillating Water Column
HP	High Pressure
LP	Low Pressure

References

1. Czech, B.; Bauer, P. Wave energy converter concepts: Design challenges and classification. *IEEE Ind. Electron. Mag.* **2012**, *6*, 4–16. [CrossRef]
2. Falcão, A.F.O.; Henriques, J.C.C. Oscillating-water-column wave energy converters and air turbines: A review. *Renew. Energy* **2016**, *85*, 1391–1424. [CrossRef]
3. Falcão, A.F.O.; Henriques, J.C.C.; Gato, LMC. Self-rectifying air turbines for wave energy conversion: A comparative analysis. *Renew. Sustain. Energy Rev.* **2018**, *91*, 1231–1241.
4. Falcão, A.F.O.; Gato, L.; Nunes, E. A novel radial self-rectifying air turbine for use in wave energy converters. Part 2. Results from model testing. *Renew. Energy* **2013**, *53*, 159–164. [CrossRef]
5. Lopes, B.S.; Gato, L.M.; Falcão, A.F.O.; Henriques, J.C.C. Test results of a novel twin-rotor radial inflow self-rectifying air turbine for OWC wave energy converters. *Energy* **2019**, *170*, 869–879. [CrossRef]
6. Lopes, B. Construction and Testing of a Double Rotor Self-Rectifying Air Turbine Model for Wave Energy Recovery Systems. Language of Reference: Portuguese. Master's Thesis, Tecnico Lisboa, Lisbon, Portugal, 2017.
7. Borges, J.E. A Three-Dimensional Inverse Method for Turbomachinery: Part II—Experimental Verification. *J. Turbomach.* **1990**, *112*, 355–361. [CrossRef]
8. Masuda, Y.; McCormick, M.E. Experiences in pneumatic wave energy conversion in Japan. In *Utilization of Ocean Waves—Wave to Energy Conversion*; American Society of Civil Engineers: New York, NY, USA, 1986; pp. 1–33.
9. Kofoed, J.P.; Frigaard, P. *Hydraulic Evaluation of the LEANCON Wave Energy Converter*; DCE Technical Reports; Department of Civil Engineering, Aalborg University: Aalborg, Denmark, 2008.
10. Fleming, A.; MacFarlane, G.; Hunter, S.; Denniss, T. Power performance prediction for a vented oscillating water column wave energy converter with a unidirectional air turbine power take-off. In Proceedings of the 12th European Wave and Tidal Energy Conference. EWTEC, Cork, Ireland, 27 August–1 September 2017; p. 1204-1.

11. Benreguig, P.; Murphy, J.; Sheng, W. Model scale testing of the Tupperwave device with comparison to a conventional OWC. In Proceedings of the ASME 2018 37th International Conference on Ocean, Offshore and Arctic Engineering OMAE2018, Madrid, Spain, 17–22 June 2018; American Society of Mechanical Engineers (ASME): New York, NY, USA, 2018.
12. Benreguig, P.; Vicente, M.; Dunne, A.; Murphy, J. Modelling approaches of a closed-circuit OWC wave energy converter. *J. Mar. Sci. Eng.* **2019**, *7*, 23. [CrossRef]
13. Vicente, M.; Benreguig, P.; Crowley, S.; Murphy, J. Tupperwave-preliminary numerical modelling of a floating OWC equipped with a unidirectional turbine. In Proceedings of the 12th European Wave and Tidal Energy Conference (EWTEC), Cork, Ireland, 27 August–1 September 2017.
14. Kelly, J.F.; Wright, W.M.; Sheng, W.; O'Sullivan, K. Implementation and verification of a wave-to-wire model of an oscillating water column with impulse turbine. *IEEE Trans. Sustain. Energy* **2016**, *7*, 546–553. [CrossRef]
15. Benreguig, P.; Pakrashi, V.; Murphy, J. Assessment of Primary Energy Conversion of a Closed-Circuit OWC Wave Energy Converter. *Energies* **2019**, *12*, 1962. [CrossRef]
16. Sheng, W.; Alcorn, R.; Lewis, A. Assessment of primary energy conversions of oscillating water columns. I. Hydrodynamic analysis. *J. Renew. Sustain. Energy* **2014**, *6*, 053113. [CrossRef]
17. Giorgi, G.; Ringwood, J.V. Consistency of viscous drag identification tests for wave energy applications. In Proceedings of the 12th European Wave and Tidal Energy Conference (EWTEC), Cork, Ireland, 27 August–1 September 2017.
18. Falcao, A.F.; Justino, P.A.P. OWC wave energy devices with air flow control. *Ocean Eng.* **1999**, *26*, 1275–1295. [CrossRef]
19. Rajput, R. *A Textbook of Engineering Thermodynamics*; Firewall Media/Laxmi Publications (P) Ltd.: New Delhi, India, 2010.
20. Abdullah, M.A.; Yatim, A.H.M.; Tan, C.W.; Saidur, R. A review of maximum power point tracking algorithms for wind energy systems. *Renew. Sustain. Energy Rev.* **2012**, *16*, 3220–3227. [CrossRef]
21. Faÿ, F.X.; Henriques, J.C.; Kelly, J.; Mueller, M.; Abusara, M.; Sheng, W.; Marcos, M. Comparative assessment of control strategies for the biradial turbine in the Mutriku OWC plant. *Renew. Energy* **2019**, *146*, 2766–2784. [CrossRef]
22. Tedeschi, E.; Carraro, M.; Molinas, M.; Mattavelli, P. Effect of control strategies and power take-off efficiency on the power capture from sea waves. *IEEE Trans. Energy Convers.* **2011**, *26*, 1088–1098. [CrossRef]
23. Penalba, M.; Ringwood, J.V. A reduced wave-to-wire model for controller design and power assessment of wave energy converters. In *Advances in Renewable Energies Offshore, Proceedings of the 3rd International Conference on Renewable Energies Offshore (RENEW 2018), Lisbon, Portugal, 8–10 October 2018*; CRC Press: Boca Raton, FL, USA, 2018; p. 379.
24. Sheng, W.; Alcorn, R.; Lewis, A. A new method for radiation forces for floating platforms in waves. *Ocean Eng.* **2015**, *105*, 43–53. [CrossRef]
25. Duclos, G.; Clément, A.H.; Chatry, G. Absorption of outgoing waves in a numerical wave tank using a self-adaptive boundary condition. *Int. J. Offshore Polar Eng.* **2001**, *11*.
26. *MATLAB*, version 7.10.0 (R2010a); The MathWorks Inc.: Natick, MA, USA, 2010.
27. Falcão, A.F.O.; Gato, L.M.C.; Henriques, J.C.; Borges, J.E.; Pereiras, B.; Castro, F. A novel twin-rotor radial-inflow air turbine for oscillating water column wave energy converters. *Energy* **2015**, *93*, 2116–2125. [CrossRef]
28. Dick, E. *Fundamentals of Turbomachines*; Springer: Berlin/Heidelberg, Germany, 2015; Volume 109.
29. Rea, J.A.; Kelly, J.F.; Alcorn, R.; O'Sullivan, D. Development and operation of a power take off rig for ocean energy research and testing. In Proceedings of the Ninth European Wave and Tidal Energy Conference EWTEC 2011, Southampton, UK, 5–9 September 2011.
30. Kelly, J.F.; Christie, R. Applying Hardware-in-the-Loop capabilities to an ocean renewable energy device emulator. In Proceedings of the Twelfth IEEE International Conference on Ecological Vehicles and Renewable Energies (EVER), Monte Carlo, Monaco, 11–13 April 2017; pp. 1–7.
31. O'Sullivan, D.L.; Lewis, A.W. Generator selection and comparative performance in offshore oscillating water column ocean wave energy converters. *IEEE Trans. Energy Convers.* **2011**, *26*, 603–614. [CrossRef]
32. Falcão, A.F.; Henriques, J.C. The spring-like air compressibility effect in OWC wave energy converters: Hydro-, thermo- and aerodynamic analyses. In Proceedings of the ASME 2018 37th International Conference on Ocean, Offshore and Arctic Engineering, Madrid, Spain, 17–22 June 2018.

33. Blavette, A.; O'Sullivan, D.L.; Lewis, A.W.; Egan, M.G. Impact of a wave farm on its local grid: Voltage limits, flicker level and power fluctuations. In Proceedings of the 2012 IEEE Oceans Yeosu, Yeosu, Korea, 21–24 May 2012; pp. 1–9.
34. O'Sullivan, D.; Salcedo, F.; Blavette, A.; Santos, M.; Lewis, A. Case studies on the benefits of energy storage for power quality enhancement: Oscillating water column arrays. In Proceedings of the 4th International Conference on Ocean Energy (ICOE2012), Dublin, Ireland, 17–19 October 2012; Volume 1719.

energies

MDPI

Article

Assessment of a Hydrokinetic Energy Converter Based on Vortex-Induced Angular Oscillations of a Cylinder

Iro Malefaki and Efstathios Konstantinidis *

Department of Mechanical Engineering, University of Western Macedonia, 50132 Kozani, Greece; iromalefaki@gmail.com

* Correspondence: ekonstantinidis@uowm.gr; Tel.: +30-24610-56754

Received: 30 December 2019; Accepted: 3 February 2020; Published: 7 February 2020

Abstract: Vortex-induced oscillations offer a potential means to harness hydrokinetic energy even at low current speeds. In this study, we consider a novel converter where a cylinder undergoes angular oscillations with respect to a pivot point, in contrast to most previous configurations, where the cylinder undergoes flow-induced oscillations transversely to the incident free stream. We formulate a theoretical model to deal with the coupling of the hydrodynamics and the structural dynamics, and we numerically solve the resulting nonlinear equation of cylinder motion in order to assess the performance of the energy converter. The hydrodynamical model utilizes a novel approach where the fluid forces acting on the oscillating cylinder are split into components acting along and normal to the instantaneous relative velocity between the moving cylinder and the free stream. Contour plots illustrate the effects of the main design parameters (in dimensionless form) on the angular response of the cylinder and the energy efficiency of the converter. Peak efficiencies of approximately 20% can be attained by optimal selection of the main design parameters. Guidelines on the sizing of actual converters are discussed.

Keywords: energy harnessing; energy converter; flow-induced oscillations; vortex-induced vibration; flow–structure interaction; hydrodynamics; vortex shedding; cylinder wake

1. Introduction

Driven by the need to increase the percentage of renewable sources in the energy-production mix during the last decade, much research has focused on the development and advancement in science, technology, and engineering of wave, wind, and current energy converters, primarily in offshore installations [1–4]. Conventional devices harnessing the kinetic energy of water currents are usually based on propeller-like turbines, which require relatively high current speeds of above 1 m/s. Unlike conventional turbines that operate with a single degree of freedom to rotate around a horizontal or vertical axis, a novel concept is to develop hydrokinetic energy converters based on flow-induced oscillatory motions of their power-generating elements. These oscillatory motions have two degrees of freedom, which allows for a more versatile operation; such converters can be designed to exploit high-amplitude and/or high-frequency oscillations to generate significant power, even at currents speeds as low as 0.1 m/s, with minimal disruption of the environment. Thus, novel energy converters based on oscillating structures may, at least in principle, avoid some of the limitations of conventional technologies and expand the availability of resource fields.

The concept of exploiting flow-induced oscillatory motions of bluff bodies for energy harnessing has received considerable attention in recent years. Various aspects and pertinent phenomena have been investigated in several studies with a view to assess the performance and optimize the design of hydrokinetic energy converters [5–21]. Probably, the best-known concept is the VIVACE converter, which was developed by the Marine Renewable Energy Laboratory (MRELab) in the University of Michigan [5,6]. Its operation relies on the fundamental phenomenon of vortex induced vibration: when an elastically-mounted cylinder is placed normal to a cross-flow, it can respond to periodic fluid forcing induced by vortices regularly shed in its after-body wake if the excitation frequency is close to the natural vibration frequency of the structure [22–24]. Vortex-induced vibration has been known for a long time and is mostly unwanted, as it compromises the integrity of equipment, such as the heat-exchanger tubes, oil risers, offshore platforms, transmission cables, cable-stayed bridges, tall chimneys, etc. due to fatigue of the structural components. On the other hand, vortex-induced vibration may be purposefully enhanced in hydrokinetic energy converters where electricity can be produced, e.g., by electromagnetic induction through the oscillating components.

More recently, a hydrokinetic energy converter has been proposed which is essentially a modification of the original configuration of the VIVACE converter; in the new design, the cylinder is attached to a supporting arm so that it can undergo vortex-induced angular oscillations with respect to a pivot point [25]. In this new configuration, the length of the cylinder-supporting arm offers an additional design parameter to control the system's response and, thereby, the efficiency of hydrokinetic energy conversion. This device has been shown to have potential benefits [25]. However, its performance depends on several design parameters, including the dimensions of the oscillating cylinder and its mass moment of inertia, arm length, damping, stiffness, and current speed, whose influences have not been studied in detail and are thus unknown. The objective of the present study is to make a preliminary assessment of the effects of these design parameters on the performance of the hydrokinetic energy converter through the development of a time-domain model to cope with the flow–structure interaction. Based on this preliminary study, a suitable prototype of the converter can be designed and put to test.

2. Physical and Mathematical Models

2.1. Physical System

Figure 1 depicts a three-dimensional schematic of the hydrokinetic energy converter considered in the present study. The system consists of a cylindrical rod, or simply the 'cylinder', which is attached from its top to a supporting arm. The supporting arm can rotate about a fixed pivot bar (pivot point) so that the main cylinder can perform one-degree-of-freedom rotational oscillations about the pivot bar while the supporting arm maintains the cylinder in an upright position. The cylindrical rod is placed perpendicular to a fluid stream with a uniform velocity profile, i.e., the cylinder is considered to be sufficiently far from obstructing solid boundaries. Elastic springs are attached to the supporting arm on one end and to a fixed frame on the other end. The springs provide restoring forces that keep the cylinder at a neutral position with the supporting arm parallel to the incident fluid stream. There are two possible configurations where the pivot point can be positioned either downstream or upstream of the main cylinder. For economy of presentation, we consider in this study only the case where the pivot point is located downstream, since this configuration yields higher efficiencies, as shown in previous experiments [25]. Unsteady hydrodynamic forces due to periodic vortex shedding can induce rotational oscillations of the main cylinder about the pivot point. For a long cylinder with no free-end effects, we assume that the flow and the fluid forces are homogeneous along the cylinder axis, so that it suffices to consider a two-dimensional segment for the analysis of the cylinder motion. In the following, pertinent structural dynamics and hydrodynamics models are developed in order to study the hydrokinetic to mechanical energy conversion.

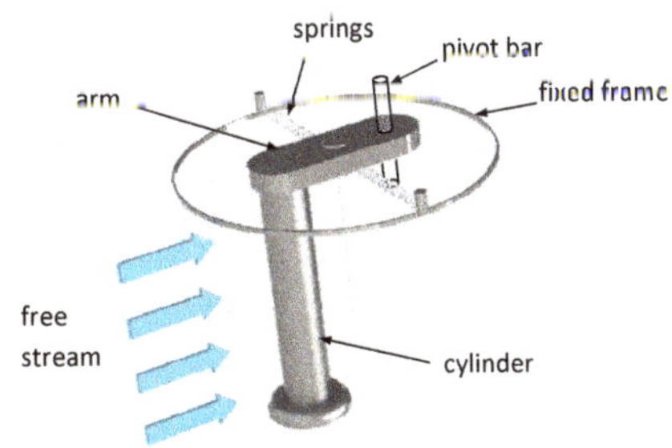

Figure 1. Schematic of the hydrokinetic energy converter under study.

2.2. Kinematics

We start with some equations describing the kinematics of the problem, which are essential in order to formulate the hydrodynamical model, which is based on the relative velocity between the cylinder and the free stream. Figure 2 presents the two-dimensional geometrical model of the problem under consideration. In this configuration, the pivot point is located downstream of the cylinder. The supporting arm has length r and the main cylinder has diameter D. At a random time, the supporting arm forms an angle θ with respect to the free stream of speed U_∞ while the main cylinder moves with linear velocity U_c, arbitrarily taken towards the top. The angle is considered positive in the clockwise direction. The linear velocity of the cylinder is $U_c = r\dot{\theta}$, where $\dot{\theta}$ is the angular velocity (the overdot denotes a derivative with respect to time). The relative velocity between the moving cylinder and the free stream is the vector difference $\vec{U}_{\text{rel}} = \vec{U}_c - \vec{U}_\infty$, where the angle between $\vec{U}_\infty$ and $\vec{U}_c$ is $\alpha + \beta = \pi/2 + \theta$. Thus, the magnitude of the relative velocity is $U_{\text{rel}} = \sqrt{U_\infty^2 + U_c^2 + 2U_\infty U_c \cos(\alpha+\beta)}$. Substituting $U_c = r\dot{\theta}$ and $\cos(\alpha+\beta) = -\sin\theta$ results in

$$U_{\text{rel}} = U_\infty \sqrt{1 + \left(\frac{r\dot{\theta}}{U_\infty}\right)^2 - 2\frac{r\dot{\theta}}{U_\infty}\sin\theta}. \tag{1}$$

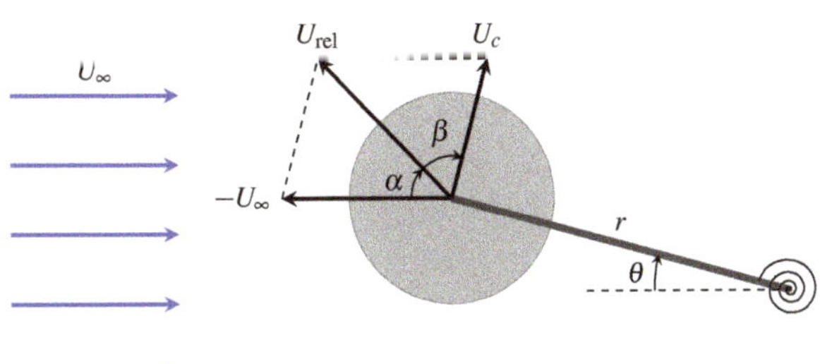

Figure 2. Two-dimensional representation of the cylinder kinematics under study.

The angle β between $\vec{U}_{\text{rel}}$ and $\vec{U}_c$ will be needed in order to project the resolved hydrodynamic forces acting along and normal to the relative velocity vector into the direction of cylinder motion. Using the definitions of internal and external products between vectors, we can express the cosine and sine of β as

$$\cos\beta = \frac{\vec{U}_{\text{rel}} \cdot \vec{U}_c}{\|\vec{U}_c\|\|\vec{U}_{\text{rel}}\|} = \frac{U_c + U_\infty \sin\theta}{U_{\text{rel}}}, \tag{2a}$$

$$\sin\beta = \frac{\vec{U}_{\text{rel}} \times \vec{U}_c}{\|\vec{U}_c\|\|\vec{U}_{\text{rel}}\|\vec{e}_z} = \frac{U_\infty \cos\theta}{U_{\text{rel}}}. \tag{2b}$$

Above, $\vec{e}_z$ is the unit vector along the the axis of the cylinder. Finally, we can express the above equations as functions of the angular displacement θ and the angular velocity $\dot{\theta}$ as

$$\cos\beta = \frac{r\dot{\theta} + U_\infty \sin\theta}{\sqrt{U_\infty^2 + 2r\dot{\theta}U_\infty \sin\theta + \left(r\dot{\theta}\right)^2}}, \tag{3a}$$

$$\sin\beta = \frac{U_\infty \cos\theta}{\sqrt{U_\infty^2 + 2r\dot{\theta}U_\infty \sin\theta + \left(r\dot{\theta}\right)^2}}. \tag{3b}$$

2.3. Structural Dynamics

The main cylinder is assumed to undergo angular oscillations, $\theta(t)$, due to the action of periodic hydrodynamic forcing. The two-dimensional motion of the cylinder is governed by the balance of angular momentum about the pivot point,

$$I_r \ddot{\theta} + c\dot{\theta} + k\theta = rF_\theta, \tag{4}$$

where I_r is the mass moment of inertia about the pivot point, c is the equivalent structural damping factor, k is the stiffness of the torsional spring, r is the length of the supporting arm, and $F_\theta(t)$ is the projection of the total hydrodynamic force in the direction of the instantaneous linear velocity of the cylinder. Note that the linear springs have been replaced by an equivalent torsional spring that provides a restoring force, as shown in Figure 2. It is assumed here that the mechanical restoring force from the springs varies linearly with angular displacement of the supporting arm, and that the structural damping is proportional to the velocity of the arm. By introducing the natural undamped frequency of the system, $f_N = (1/2\pi)\sqrt{k/I_r}$, and the ratio of structural damping to the critical damping, $\zeta = c/2\sqrt{kI_r}$, we can rewrite the equation of cylinder motion as

$$I_r\ddot{\theta} + 4\pi I_r \zeta f_N \dot{\theta} + 4\pi^2 I_r f_N^2 \theta = rF_\theta. \tag{5}$$

The mass moment of inertia of a cylinder rotating about a pivot point can be calculated using the parallel axis theorem, $I_r = I_c + mr^2$, where I_c is the mass moment of inertia of the same body rotating about a parallel axis passing through the center of mass and m is the body mass. For a long homogeneous cylinder, $I_c = \frac{1}{8}mD^2$ and, therefore, $I_r = m\left(r^2 + \frac{1}{8}D^2\right)$. The left-hand side of Equation (5) is an ordinary (linear) differential equation of second order. The non-linearity of the problem rests on the right-hand side of Equation (5), which represents the hydrodynamic forcing.

2.4. Hydrodynamics

The hydrodynamic force per unit spanwise length, F, exerted on the cylinder is decomposed into components, denoted F_R and F_L, respectively along and normal to the instantaneous relative velocity between the oscillating cylinder and the free stream, as shown in Figure 3. It should be noted that at an arbitrary instant, F generally forms an angle with respect to the free stream. According to this model, the longitudinal component F_R always resists the cylinder motion, whereas the normal component F_L acts as an excitation source due to lift in vortex-induced vibration. This hydrodynamical model has been previously tested for the simpler case of rectilinear oscillations of an elastically-mounted circular cylinder transverse to a free stream, i.e., the classical problem of vortex-induced vibration [26,27]. It has also been experimentally verified that F_R always does negative work on the cylinder, whereas F_L primarily does positive work [28].

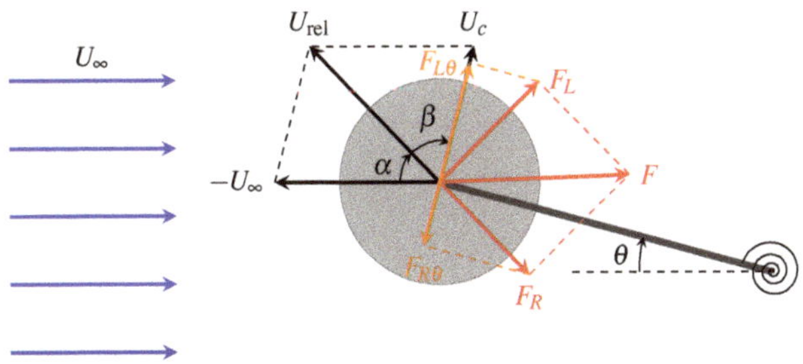

Figure 3. Vector diagram of the hydrodynamic forces acting on a cylinder that performs angular oscillations about the pivot point.

The reaction force incorporates fluid damping due to drag and fluid inertia due to added mass, which can be modeled by the well-known Morison's equation [29], i.e.,

$$F_R(t) = \tfrac{1}{4}\rho\pi D^2 C_A \dot{U}_{\mathrm{rel}} + \tfrac{1}{2}\rho D C_D U_{\mathrm{rel}}^2, \tag{6}$$

where ρ is the density of the fluid, and C_A and C_D are the added mass (inertia) and drag coefficients, respectively. The novelty here is that Morison's equation is used to model the reaction force acting in the direction of the instantaneous relative velocity between the vibrating cylinder and the free stream. A velocity-squared-dependent quasi-steady drag force is applicable for separated flows dominated by the convection of vorticity in the wake [30]. The above reaction force introduces strongly non-linear terms in the total force, as will be seen further below. It should also be noted that F_R depends on the motion of the cylinder; thus, it is time-dependent.

The excitation force is modeled as a periodic function of time

$$F_L(t) = \frac{1}{2}\rho U_\infty^2 D C_L \sin(2\pi f_{vs} t), \tag{7}$$

where C_L is the lift coefficient and f_{vs} is the vortex shedding frequency. That is, vortices that regularly shed in the wake at the shedding frequency induce a periodic force on the cylinder. A major complication in vortex-induced vibration is that the actual vortex shedding frequency depends not only on the Reynolds number as for a fixed cylinder, but also on the frequency of cylinder motion. A large number of studies have been devoted to examining vortex synchronization for cylinders undergoing both forced and free oscillation; e.g., see the review in [23]. Many studies have shown that the frequency of vortex shedding varies between $\frac{1}{2}f_{v0} < f_{vs} < f_{v0}$, where f_{v0} is the vortex shedding frequency from a fixed cylinder at the same Reynolds number, often referred to as the Strouhal frequency. In the following, we will assume that the vortex shedding frequency remains within the above range.

2.5. Flow–Structure Interaction

The next step is to introduce the resolved hydrodynamic forces from Equations (6) and (7) into the equation of cylinder motion (5). Note that $\vec{F}_R$ is collinear with $\vec{U}_{\mathrm{rel}}$ while $\vec{F}_L$ is normal to $\vec{U}_{\mathrm{rel}}$. Thus, we project $\vec{F}_R$ and $\vec{F}_L$ into the direction of $\vec{U}_c$ to obtain $\vec{F}_{R\theta}$ and $\vec{F}_{L\theta}$, respectively. From Figure 3, we see that

$$F_\theta = F_{L\theta} - F_{R\theta} = F_L \sin\beta - F_R \cos\beta, \tag{8}$$

where β is the angle between $\vec{U}_{rel}$ and $\vec{U}_c$. Substitution of the cosine and sine of β from (3a) and (3b), respectively, and of F_R and F_I from (6) and (7), respectively, into F_θ and of the result into the equation of cylinder motion (5), after some lengthy manipulations, leads to

$$\begin{aligned} &\left[I_r + \frac{1}{4}\rho\pi D^2 r^2 C_A \frac{U_\infty^2}{U_{\text{rel}}^2}\left(\frac{r\dot{\theta}}{U_\infty} - \sin\theta\right)^2\right]\ddot{\theta} + \\ &\left[c + \frac{1}{4}\rho\pi D^2 r^2 C_A \frac{U_\infty^2}{U_{\text{rel}}^2}\left(\dot{\theta}\sin\theta - \frac{r\dot{\theta}^2}{U_\infty}\right)\cos\theta + \frac{1}{2}\rho D r^2 C_D U_{\text{rel}}\right]\dot{\theta} + \\ &k\theta - \frac{1}{2}\rho D r C_D U_{\text{rel}} U_\infty \sin\theta = \\ &\frac{1}{2}\rho U_\infty^2 D r \frac{U_\infty}{U_{\text{rel}}}\cos\theta C_L \sin\left(2\pi f_{vs} t\right). \end{aligned} \tag{9}$$

With the aid of the above equation, we can examine the actions of the different terms. The terms inside the square bracket on the first line represent the total inertia of the system. Similarly, the terms on the second line represent the total damping of the system, and the terms on the third line represent the total stiffness of the system. The term on the fourth line, i.e., on the right-hand side of the equation, represents the excitation function. It can be seen that the hydrodynamics modify the properties of the structural system. Terms involving C_A contribute to both the inertia and damping of the system. In fact, fluid inertia introduces strong non-linearity in the equation of motion, as the corresponding terms involve the squared angular velocity. Terms involving C_D contribute to the damping and stiffness of the system. It is interesting to note that the C_A term on the second line of Equation (9) introduces small negative damping on the average (instantaneously takes both positive and negative values). However, the C_D term on the second line of Equation (9) is solely positive and dominates the fluid damping. It should also be noted that for small angles, $\sin\theta \approx \theta$ and the term on the third line of Equation (9) can be expressed as $\left(k - \frac{1}{2}\rho D L C_D U_{\text{rel}} U_\infty\right)\theta$; thus, fluid damping due to drag lowers the total stiffness of the system.

The equation of cylinder motion (9) can be cast in the following non-dimensional form

$$\begin{aligned} &\left[m^*\left(1 + \frac{1}{8L^{*2}}\right) + \left(\frac{L^*\dot{\theta}^*}{U^*} - \sin\theta^*\right)^2 U_{\text{rel}}^{*-2} C_A\right]\ddot{\theta}^* + \\ &\left[4\pi m^*\zeta\left(1 + \frac{1}{8L^{*2}}\right) + \left(\dot{\theta}^*\sin\theta^* - \frac{L^*\dot{\theta}^{*2}}{U^*}\right)\cos\theta^* U_{\text{rel}}^{*-2} C_A + \frac{2}{\pi}U^* U_{\text{rel}}^* C_D\right]\dot{\theta}^* + \\ &4\pi^2 m^*\left(1 + \frac{1}{8L^{*2}}\right)\theta^* - \frac{2}{\pi}\frac{U^{*2}}{L^*}U_{\text{rel}}^* C_D \sin\theta^* = \frac{2}{\pi}\frac{U^{*2}}{L^*}\frac{\cos\theta^*}{U_{\text{rel}}^*}C_L \sin\left(2\pi S_f U^*\tau\right), \end{aligned} \tag{10}$$

where time is normalized with the undamped natural oscillation period of the system, i.e., $\tau = t/T_N = f_N t$, i.e., the dependent variables become $\theta^* = \theta/(1\text{ radian})$, $\dot{\theta}^* = \dot{\theta}/f_N$, and $\ddot{\theta}^* = \ddot{\theta}/f_N^2$. The relative velocity is normalized with the free-stream velocity, $U_{\text{rel}}^* = U_{\text{rel}}/U_\infty$. The resulting equation of motion (10) includes the following dimensionless parameters:

non-dimensional arm length, $L^* = \dfrac{r}{D}$,
mass ratio, $m^* = \dfrac{m}{\frac{1}{4}\pi\rho D^2}$,
damping ratio, $\zeta = \dfrac{c}{2\sqrt{kI_r}}$,
reduced velocity, $U^* = \dfrac{U_\infty}{f_N D}$.

Equation (10) is highly non-linear. Thus, for the purposes of the present study, we numerically solved the equation of cylinder motion using the MATLAB ODE Suite, which provides a powerful yet easy to implement tool for numerical solution of ordinary differential equation initial value problems [31]. The solution of Equation (10) requires as input the hydrodynamic parameters, C_A, C_D, C_L, and S_f, where $S_f = f_{vs}D/U_\infty$, which is the actual Strouhal frequency of vortex shedding driving the motion of the cylinder.

2.6. Power Extraction Efficiency

The instantaneous power $P(t)$ delivered to the energy converter can be calculated from the product of the driving force times the velocity of the cylinder, $P(t) = F_\theta(t)U_c(t) = F_\theta(t)\, r\dot{\theta}(t)$. The term $F_\theta r$ can be obtained from the left-hand side of the equation of motion (5) so that

$$P(t) = \left(\ddot{\theta} + 4\pi\zeta f_N\dot{\theta} + 4\pi^2 f_N^2\theta\right) I_r\dot{\theta}. \tag{11}$$

The average power over many cycles of motion is a practical metric to quantify the performance of the energy converter. Assuming that the motion of the cylinder is periodic with period T, the average power can be calculated from the following integral

$$\langle P\rangle = \frac{1}{T}\int_T P(t)\mathrm{d}t = \frac{1}{T}\int_T \left(\ddot{\theta}(t) + 4\pi\zeta f_N\dot{\theta}(t) + 4\pi^2 f_N^2\theta(t)\right) I_r\dot{\theta}(t)\mathrm{d}t. \tag{12}$$

The first and third term in the integral make a zero net contribution, so that only the second term contributes to the net power. Typically, the average power delivered to the energy converter is normalized by the power of the fluid stream within the frontal area of the cylinder [7], which yields the following efficiency metric:

$$\eta = \frac{\langle P\rangle}{\frac{1}{2}\rho D U_\infty^3} = \frac{2\pi^2 m^*\zeta}{U^{*3}}\left(L^{*2} + \frac{1}{8}\right)\frac{1}{T^*}\int_{T^*}\dot{\theta}^{*2}(\tau)\mathrm{d}\tau. \tag{13}$$

In order to compute the efficiency of the hydrokinetic energy converter, the time series $\dot{\theta}^*(\tau)$ appearing in the integral can be obtained from the numerical solution of Equation (10).

3. Results

In this section, we present results from simulations with the non-linear model developed in the previous section (Equation 10) in order to assess the effect of the basic mechanical parameters of the hydrokinetic energy converter on its efficiency (Equation 13). In particular, we consider variations of the dimensionless arm length L^*, mass ratio m^*, damping ratio ζ, and reduced velocity U^*. A normalized time step of $\delta\tau = 5\times10^{-4}/U^*St$ was employed for the integration of Equation (10) for more than 200 cycles of oscillation, for which the resulting motion typically (but not always) settled to a steady periodic oscillation about the neutral position at zero angle.

The values of the hydrodynamic parameters employed in the present study, which are required as input parameters for the solution of the equation of cylinder motion, Equation (10), are listed in Table 1. These values were selected on the basis of the following considerations. The added mass coefficient takes the theoretical value for inviscid potential flow about a circular cylinder, as discussed in [32]. The drag and lift coefficients are higher than their corresponding values for a non-oscillation cylinder in the same range of Reynolds numbers due to vortex synchronization in the wake of oscillating cylinders [22,23]. The Strouhal number is assumed to have a constant value lower than the corresponding value for a non-oscillation cylinder, as discussed earlier; the selected value is typical around peak amplitude for cylinders undergoing vortex-induced vibration transversely to a free stream [23]. The above choices of the hydrodynamical parameters should be considered appropriate for smooth cylinders placed in uniform free streams at Reynolds numbers for which the boundary layer on the cylinder is laminar and transition to turbulence occurs in the shear layers after flow separation, that is, $Re \approx 10^3 - 5\times10^5$.

Table 1. Hydrodynamic parameters employed in the present study.

C_A	C_D	C_L	S_f
1.00	1.35	1.50	0.155

Sets of simulations were carried out where the reduced velocity U^* is varied at set values of L^*, m^*, and ζ. Due to the large size of the test matrix, initially, we present results in the form of contours in the 2D parameter space of U^* and each one of $\{L^*, m^*, \zeta\}$ for some representative values of the other two parameters from the set. The cylinder response is characterized by the amplitude of angular deflection, denoted θ_o, and the amplitude of displacement transversely to the free stream, $A^* = L^*\theta_o$. It should be noted that θ_0 is dimensionless, since the equation of motion has been solved for the dimensionless angular displacement θ^*, which corresponds to angles in radians normalized with 1 radian. The efficiency of hydrokinetic energy conversion is characterized by the efficiency η.

3.1. Effect of Arm Length

Figure 4 shows contours of θ_o, A^*, and η as functions U^* and L^* at fixed values of $m^* = 5$ and $\zeta = 0.01$. It can be observed that θ_o strongly depends on both U^* and L^*. At a fixed L^* value, θ_o attains a peak level that corresponds to $U^* \approx St^{-1}$ with the precise value being slightly dependent on L^*, i.e., U^* at peak shifts to slightly higher values with L^*. This may be attributable to the decrease in the total stiffness of the system due to the drag term $\left(k - \frac{1}{2}\rho DLC_D U_{\text{rel}} U_\infty\right)\theta$, which increases in proportion to L. The peak level of θ_o gradually decreases with increasing L^*, while a global maximum of 1.26 is attained at the lowest L^* value. The variation of the transverse amplitude A^* as a function of only U^* also displays a peak at the same location as θ_o does, but, conversely to the variation of θ_o, peak levels of A^* increase with L^*, particularly in the range of low L^* values; in the range of high L^* values, A^* is almost independent of L^*. The power efficiency η attains a global maximum value of 7.5% at $(L^*, U^*) = (0.5, 5.6)$, where θ_o also displays a global maximum. However, it should be noted that global maxima in θ_o and η do not coincide in general.

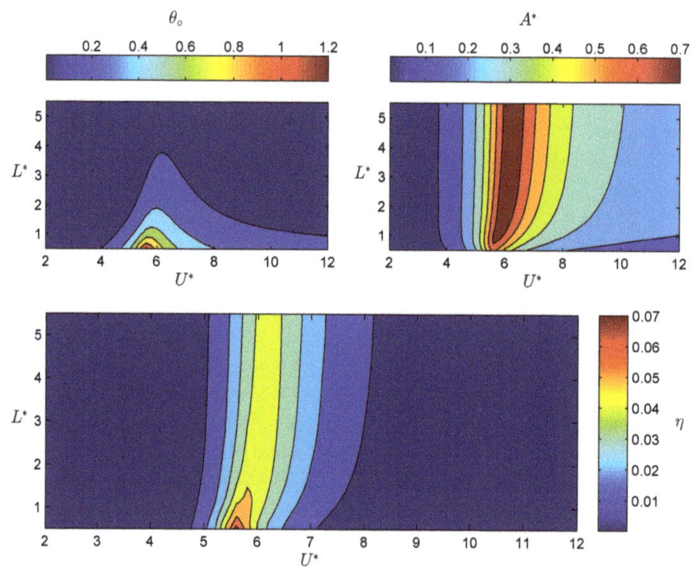

Figure 4. Contours of θ_o and η as functions of L^* and U^* for $m^* = 5$ and $\zeta = 0.01$.

In Figure 5, we present contours of θ_o, A^*, and η as functions U^* and L^* at fixed values of $m^* = 50$ and $\zeta = 0.01$. Both θ_o and A^* display similar variations to those in the previous case with $(m^*, \zeta) = (5, 0.01)$, but their peak levels decreased by approximately half (cf. Figure 4). Yet, the conversion efficiency η approximately doubled for $(m^*, \zeta) = (50, 0.01)$, reaching a global maximum of 18.1% at $(L^*, U^*) = (0.5, 6.4)$. In addition, the peak level of η as a function of U^* at fixed L^* does not depend strongly on L^* any more, unlike at $m^* = 5$. However, the range of appreciable response and high efficiency shrunk by approximately half, now limited in the range $5.5 < U^* < 7.5$ at $m^* = 50$ compared to $4 < U^* < 10$ at $m^* = 5$.

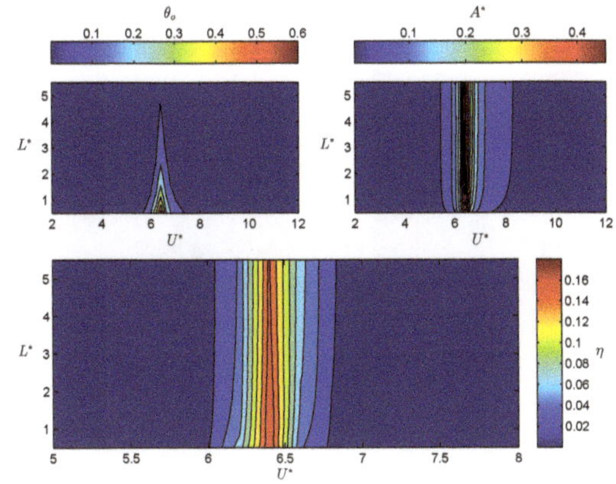

Figure 5. Contours of θ_o and η as functions of L^* and U^* for $m^* = 50$ and $\zeta = 0.01$.

Figure 6 shows contours of θ_o, A^*, and η as functions U^* and L^* at fixed values of $m^* = 5$ and $\zeta = 0.1$, i.e., damping has now been increased. The contours are quite similar to those for $\zeta = 0.01$, but now, peak levels θ_o and A^* have decreased, whereas peak levels of η have increased, reaching a global maximum of 22.1% at $(L^*, U^*) = (0.5, 5.6)$. , Finally, we can see in Figure 7 that using high values of both mass and damping ratios of $(m^*, \zeta) = (50, 0.1)$ results in a reduction in both the cylinder response and the efficiency of the converter comparatively to all previous cases with lower m^* or ζ values.

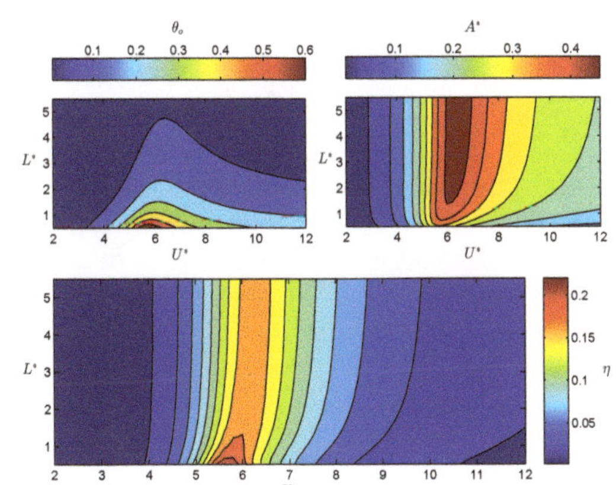

Figure 6. Contours of θ_o and η as functions of L^* and U^* for $m^* = 5$ and $\zeta = 0.1$.

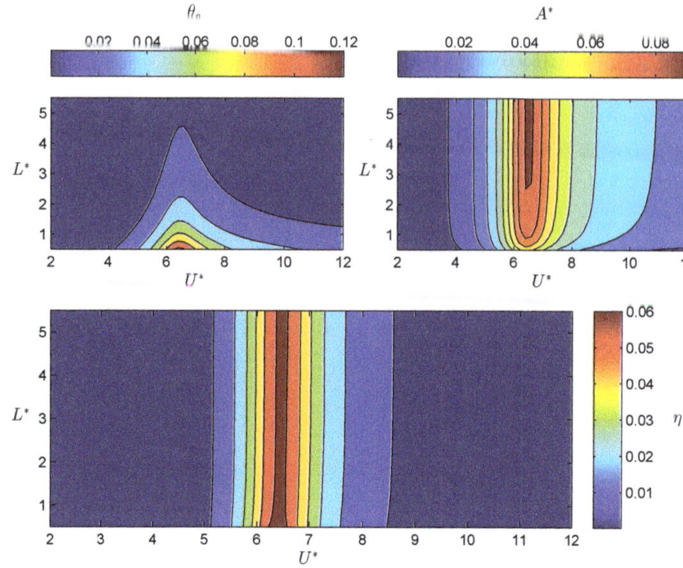

Figure 7. Contours of θ_o and η as functions of L^* and U^* for $m^* = 50$ and $\zeta = 0.1$.

3.2. Effect of Mass Ratio

The results presented in the previous section show that the mass ratio m^* has a marked effect on the response and efficiency of the energy converter. In order to get a more detailed picture of the influence of variations in m^*, we carried out series of simulations with varying m^* at a constant value of ζ and L^*. A value of $L^* = 0.8$ was employed, as the above simulations have shown that the best efficiencies are obtained at the lowest L^* values. Figure 8 shows contours of θ_o and η as functions U^* and m^* at fixed values of $L^* = 0.8$ and $\zeta = 0.01$. It can be seen that θ_o strongly depends on both U^* and L^*. At a fixed m^* value, θ_o peaks at a specific U^* value that depends on m^*. In the lower m^* range, the location of peak θ_o quickly shifts to higher U^* values. Moreover, the U^* range of significant θ_o response is wide but shrinks with increasing m^*. We have observed that irregular oscillations about a non-zero mean deflection angle appear in the higher U^* range; for instance, at $m^* = 1.644$, oscillations about a mean deflection angle appear for $U^* > 7.2$, which might be related to the shift in the location of peak θ_o at low m^* values. The efficiency attains a sharp peak as a function of U^* and a global maximum η of 18.8% is obtained at $(m^*, U^*) = (74, 6.4)$. In addition, a local maximum η of 14.4% is obtained at $(m^*, U^*) = (19.7, 6.2)$. Note that axes in the η contours have been modified to better depict results in the neighborhood of interest. Figure 9 shows corresponding results for $L^* = 0.8$ and $\zeta = 0.1$. It can be observed that a maximum η of 19.5% is now obtained at $(m^*, U^*) = (5.24, 5.8)$, i.e., the optimum m^* in terms of efficiency decreases as ζ increases while keeping L^* and ζ fixed. This can be attributable to competing indirect and direct effects; a nonlinear decrease in angular response θ_o and a linear increase in power efficiency with m^*.

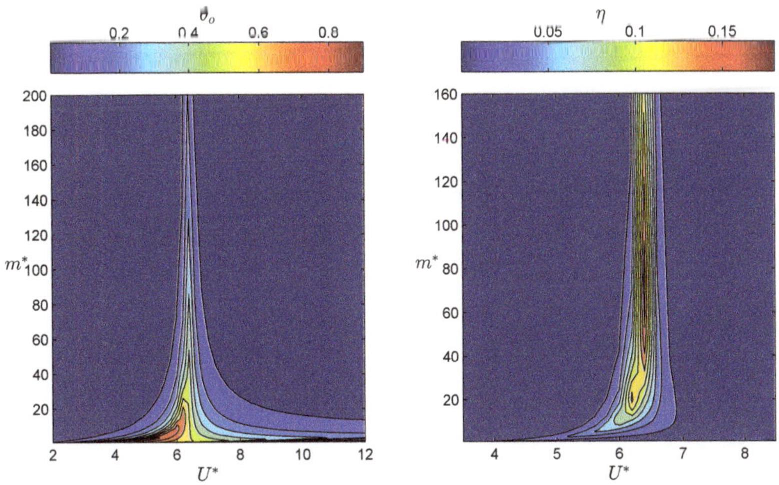

Figure 8. Contours of θ_o and η as functions of U^* and m^* for $L^* = 0.8$ and $\zeta = 0.01$.

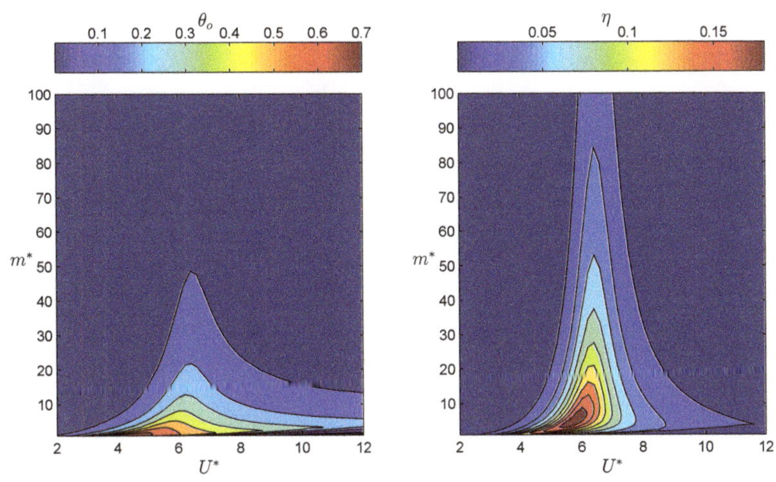

Figure 9. Contours of θ_o and η as functions of U^* and m^* for $L^* = 0.8$ and $\zeta = 0.1$.

3.3. *Effect of Damping Ratio*

In this section, we focus on the effect of the damping ratio ζ on the cylinder response and energy conversion efficiency guided by the results of the foregoing sections. For the first series of simulations, we employ a value of $m^* = 75$, which corresponds to the optimum m^* in terms of efficiency at a low value of the damping ratio of $\zeta = 0.01$ (see Figure 8). Figure 10 shows contours of θ_o and η as functions of U^* and ζ for $(L^*, m^*) = (0.8, 75)$. It can be seen that θ_o decreases with increasing damping ratio as might be expected. However, η attains a global maximum of 19.0% at $(U^*, \zeta) = (6.4, 0.0083)$, which does not correspond to the lowest value of the damping ratio. Nevertheless, the maximum efficiency is attained at a relatively low value of the damping ratio, which is not practical for energy converters where high damping is required in the electromechanical converter (conversion from mechanical to electrical energy). On the other hand, employing a comparatively low value of $m^* = 5$ yields appreciable response along with high efficiencies over a wide range of reduced velocities, as can be seen in Figure 11. A global maximum of η of 19.4% is obtained at $(\zeta, U^*) = (0.10, 5.8)$ for $(L^*, m^*) = (0.8, 5)$.

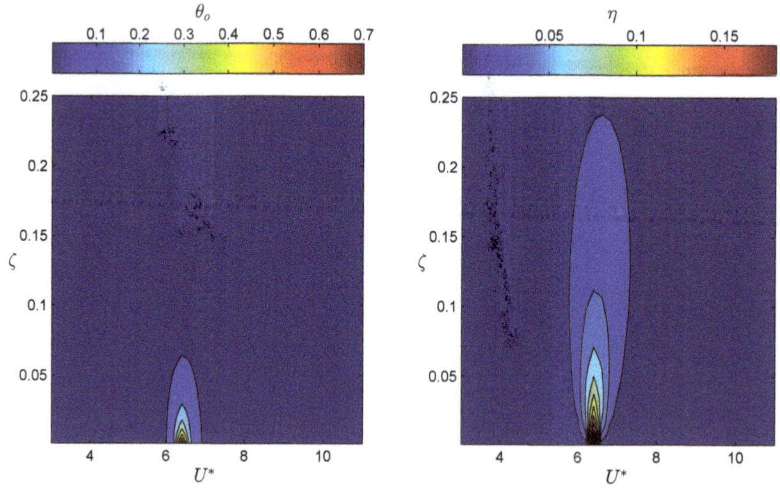

Figure 10. Contours of θ_o and η as functions of U^* and ζ for $L^* = 0.8$ and $m^* = 75$.

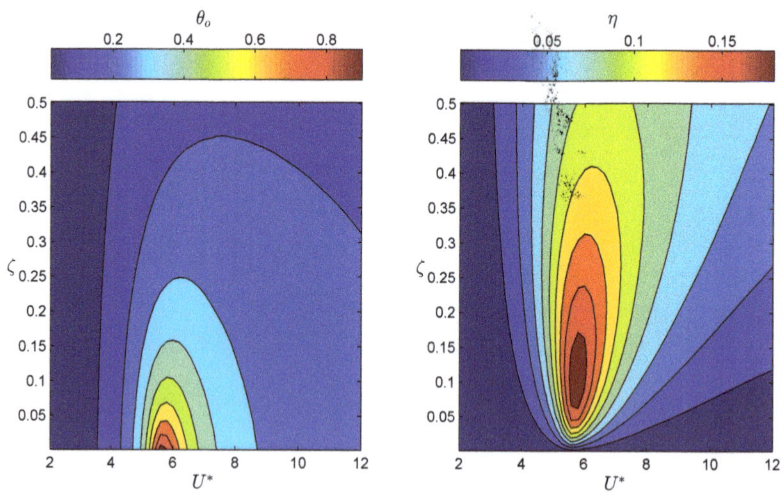

Figure 11. Contours of θ_o and η as functions of U^* and ζ for $L^* = 0.8$ and $m^* = 5$.

4. Discussion

In this section, the results from simulations regarding the performance of the kinetic-energy harvester based on vortex-induced angular oscillations of a circular cylinder are discussed in a broader context. It should be pointed out that the objective of this preliminary work is to assess the effect of the main design parameters related to the conversion of the kinetic energy of a free stream to mechanical energy of the oscillating cylinder (aerodynamic efficiency). The contours of the efficiency η as a function of pairs of the parameters $\{L^*, m^*, \zeta, U^*\}$ show that high aerodynamic efficiencies up to 22.1% can be achieved via appropriate selection of the main design parameters. The estimated aerodynamic efficiency is comparable to that of the original VIVACE based on purely transverse vortex-induced oscillations of the cylinder, which can reach 33.2%. However, we expect that the actual efficiency of the new converter will be higher than estimated here, as we have employed a conservative value for the excitation lift coefficient C_L. Previous experimental assessment of the novel energy harvester yielded a maximum aerodynamic efficiency η of 31.4% [25]. It should be noted that η was not directly measured, but estimated from motion results using a theoretical formula for a micro-electric generator. In that experiment, a rigid cylinder made of polypropylene was tested in a water channel, which is estimated to have a mass ratio of the order of 10. However, the actual values of mass and damping ratios were not reported in [25]. Taking into account the working hypotheses and assumptions employed here, our theoretical study yields comparable results and complements the previous experimental work on the effects of mass and damping ratios, thereby providing useful guidance on the best design of the hydrokinetic energy harvester.

The model results clearly indicate that the highest efficiencies are obtained at the smallest L^* values. Therefore, the swept frontal area of the proposed converter will be smaller by at least 10% (estimated from A^* contours in Figure 6) and the power density of the proposed converter will be increased comparatively to the original VIVACE. The actual performance of the energy harvester might be somewhat different because, for small arm lengths, the motion of the cylinder differs more from a rectilinear oscillation than for large arm lengths. It is known that the frequency of vortex shedding is related to the wake width behind the cylinder, which inevitably depends on the amplitude of transverse oscillation of the cylinder [33]. Although the present hydrodynamic model takes into account only the relative velocity between the moving cylinder and the free stream, it is implicitly assumed—in the selection of the Strouhal number value—that vortex synchronization in the cylinder wake will be similar to that in the case of oscillations of the cylinder purely transverse to the free stream. To our knowledge, this has yet to be studied in the published literature. Nonetheless, we have found that the trends in the contours of the aerodynamic efficiency remain similar for different values of the Strouhal number and, hence, our findings might be expected to hold true in general.

At a given value of the mass ratio m^*, there is an optimal value of the damping ratio ζ. Conversely, at a given ζ value, there corresponds an optimal m^* value. This might suggest that there exists an optimal value of the combined mass–damping parameter $m^*\zeta$ for which the efficiency is maximized. For purely transverse vortex-induced vibrations, the best efficiencies have been obtained at approximately constant $m^*\zeta$ values of 0.2 at $Re = 3800$ and 0.35 at $Re = 10^4$, according to a parametric study in [19]. For our results presented in Figures 8–11, $m^*\zeta$ spans a range between a minimum value of 0.005 to a maximum value of 18.75. Over this very wide range, the efficiency attains peak values of approximately 19% at four points listed in Table 2. All four points fall within a comparatively narrow range of $m^*\zeta$ values with an average value of 0.6. This strongly supports the suggestion of an optimal $m^*\zeta$ value. The optimal value of combined mass–damping and the attained best efficiency in the present study are both higher than in reference [19], which might be attributable to the different configurations considered, i.e., angular vs. purely transverse oscillations.

Table 2. Best efficiency points of the novel hydro-kinetic energy converter for $L^* = 0.8$.

m^*	ζ	$m^*\zeta$	η
74	0.01	0.74	0.188
5.24	0.1	0.524	0.195
75	0.0083	0.6225	0.190
5	0.1	0.5	0.194

In practice, the mechanical energy captured by the harvester will typically be converted into electrical energy by a generator. For a first approximation, this conversion process can be modeled as added damping in the system [6]. Thus, the total system damping is composed of

$$\zeta = \zeta_{\text{struct}} + \zeta_{\text{loss}} + \zeta_{\text{gen}}, \tag{14}$$

where ζ_{struct} is the structural damping of the energy converter (kinetic-to-mechanical energy conversion), ζ_{loss} is the damping associated with power losses in the bearings, belts, etc. of the converter (transmission system), and ζ_{gen} is the damping associated with the electrical loads (mechanical-to-electrical energy conversion). Only the ζ_{gen} part of the damping provides harnessed energy. Therefore, a practical system requires relatively high values of ζ_{gen} combined with minimal values of ζ_{struct} and ζ_{loss}. It should be noted that the efficiencies reported here are based on the total damping, i.e., in the efficiency computation using Equation (13) ζ should be replaced by ζ_{gen}. Therefore, actual efficiencies will be slightly lower than reported. However, this is partly compensated by the underestimation of efficiency due to the use of a conservative excitation coefficient.

Given the considerations regarding damping given in the foregoing paragraph, the total damping of the system—including the electrical generator—has to be high, which necessitates the use of relatively low m^* values. As an example, taking $m^* = 5$, a nearly optimized design will have a total damping of $\zeta = 0.10$. Interestingly, the latter value is close to the optimal one ($\zeta = 0.12$) of the VIVACE converter at its best efficiency point [6]. At lower mass ratios, e.g., $m^* < 3$, the operation of the system might be compromised, as we have noted that small amplitude oscillations about a non-zero mean angle can be induced.

In turn, the total damping ratio of the system should correspond to the optimal one for the given mass ratio ($\zeta \approx 0.1$ for low m^* values). Since the damping ratio depends on the system damping c (including damping in mounting structure, losses, and electric loads) as well as on the stiffness k of the supporting springs, the actual values of c and k have to be jointly determined, taking into account the fact that peak efficiencies are achieved within a narrow band of reduced velocities ($U^* \approx 5.8$). These requirements result in the following relationship:

$$U^*\zeta = \frac{c\pi U_\infty}{kD} \Rightarrow k \approx 5.4\frac{U_\infty}{D}c. \tag{15}$$

Pairs of (c, k) values can be selected on the basis of the above relationship, taking into account the practical limits of these mechanical parameters for the intended size of the equipment. This allows for a versatile mechanical design of the converter without directly affecting the aerodynamic efficiency.

At Reynolds numbers lower than assumed here, i.e., $Re < 10^3$, the aerodynamic efficiency might be expected to be limited by viscous effects. As a consequence, the minimum diameter of the cylinder is determined by the minimum required Reynolds number for a specific fluid (air or water) at a given wind or current speed. In practice, this factor does not pose a substantial constraint even at low speeds. For example, the minimum diameter for a wind speed of 0.5 m/s is 3 cm and that for a water current speed of 0.1 m/s is 1 cm. The weight (mass) of the oscillating structure is determined by the mass ratio requirement discussed in the previous paragraph ($m^* \approx 5$). The requirement for low mass ratios results in very lightweight structures in air, which makes the use of the proposed kinetic-energy converter impractical in this case. Therefore, the use of the converter mostly aims to harness the hydro-kinetic energy of water currents. Based on the above considerations, Table 3 shows calculations of the power generated (per unit length of the cylinder) by the novel hydro-kinetic energy converter at representative current speeds and cylinder diameters.

Table 5. Estimated power output per unit span of the cylinder generated by the novel hydro-kinetic energy converter.

U_∞ (m/s)	D (cm)	m (kg/m)	P (mW/m)
0.1	1	0.39	1
0.1	5	9.82	5
0.1	25	245.4	25
0.5	1	0.39	125
0.5	5	9.82	625
0.5	25	245.4	3125
1.0	1	0.39	1000
1.0	5	9.82	5000
1.0	25	245.4	25,000

Author Contributions: Conceptualization, E.K.; Methodology, I.M. and E.K.; Software, I.M.; Validation, I.M. and E.K.; Investigation, I.M.; Supervision, E.K.; Writing—Original draft preparation (in Greek), I.M.; Writing—Review and editing, E.K. All authors have read and agreed to the published version of the manuscript.

Funding: This research received no external funding.

Conflicts of Interest: The authors declare no conflict of interest.

References

1. Karimirad, M. *Offshore Energy Structures: For Wind Power, Wave Energy and Hybrid Marine Platforms*; Springer: Cham, Switzerland, 2014.
2. Karimirad, M.; Michailides, C. V-shaped semisubmersible offshore wind turbine: An alternative concept for offshore wind technology. *Renew. Energy* **2015**, *83*, 126–143. [CrossRef]
3. Collu, M.; Borg, M. Design of floating offshore wind turbines. In *Offshore Wind Farms*; Ng, C., Ran, L., Eds.; Elsevier Inc.: Amsterdam, The Netherlands, 2016; pp. 359–385. [CrossRef]
4. Kong, F.; Su, W.; Liu, H.; Collu, M.; Lin, Z.; Chen, H.; Zheng, X. Investigation on PTO control of a combined axisymmetric buoy-WEC(CAB-WEC). *Ocean Eng.* **2019**, *188*, 106245. [CrossRef]
5. Bernitsas, M.M.; Raghavan, K.; Ben-Simon, Y.Y.; Garcia, E.M. VIVACE (Vortex Induced Vibration Aquatic Clean Energy): A New Concept in Generation of Clean and Renewable Energy From Fluid Flow. *ASME J. Offshore Mech. Arct. Eng.* **2008**, *130*, 041101. [CrossRef]
6. Lee, J.; Bernitsas, M. High-damping, high-Reynolds VIV tests for energy harnessing using the VIVACE converter. *Ocean Eng.* **2011**, *38*, 1697–1712. [CrossRef]
7. Barrero-Gil, A.; Pindado, S.; Avila, S. Extracting energy from Vortex-Induced Vibrations: A parametric study. *Appl. Math. Modell.* **2012**, *36*, 3153–3160. [CrossRef]
8. Hobbs, W.B.; Hu, D.L. Tree-inspired piezoelectric energy harvesting. *J. Fluids Struct.* **2012**, *28*, 103–114. [CrossRef]
9. Wang, D.A.; Chiu, C.Y.; Pham, H.T. Electromagnetic energy harvesting from vibrations induced by Kármán vortex street. *Mechatronics* **2012**, *22*, 746–756. [CrossRef]
10. Grouthier, C.; Michelin, S.; Bourguet, R.; Modarres-Sadeghi, Y.; de Langre, E. On the efficiency of energy harvesting using vortex-induced vibrations of cables. *J. Fluids Struct.* **2014**, *49*, 427–440. [CrossRef]
11. Liu, J.; Bernitsas, M.M. Envelope of Power Harvested by a Single-Cylinder VIVACE Converter. In Proceedings of the ASME 34th International Conference on Ocean, Offshore, and Arctic Engineering, St. John's, NL, Canada, 31 May–5 June 2015, Volume 9. [CrossRef]
12. Dai, H.L.; Abdelkefi, A.; Yang, Y.; Wang, L. Orientation of bluff body for designing efficient energy harvesters from vortex-induced vibrations. *Appl. Phys. Lett.* **2016**, *108*, 53902. [CrossRef]
13. Zhang, J.; Liu, F.; Lian, J.; Yan, X.; Ren, Q. Flow Induced Vibration and Energy Extraction of an Equilateral Triangle Prism at Different System Damping Ratios. *Energies* **2016**, *9*, 938. [CrossRef]
14. Kim, E.S.; Bernitsas, M.M. Performance prediction of horizontal hydrokinetic energy converter using multiple-cylinder synergy in flow induced motion. *Appl. Energy* **2016**, *170*, 92–100. [CrossRef]
15. Bhattacharya, A.; Shahajhan, S.S.S. Power extraction from vortex-induced angular oscillations of elliptical cylinder. *J. Fluids Struct.* **2016**, *63*, 140–154. [CrossRef]

16. Soti, A.K.; Thompson, M.C.; Sheridan, J.; Bhardwaj, R. Harnessing electrical power from vortex-induced vibration of a circular cylinder. *J. Fluids Struct.* **2017**, *70*, 360–373. [CrossRef]
17. Chandran, V.M.S.; Janardhanan, S.; Menon, V. Numerical Study on the Influence of Mass and Stiffness Ratios on the Vortex Induced Motion of an Elastically Mounted Cylinder for Harnessing Power. *Energies* **2018**, *11*, 2580. [CrossRef]
18. Ding, L.; Zou, Q.; Zhang, L.; Wang, H. Research on Flow-Induced Vibration and Energy Harvesting of Three Circular Cylinders with Roughness Strips in Tandem. *Energies* **2018**, *11*, 2977. [CrossRef]
19. Barrero-Gil, A.; Vicente-Ludlam, D.; Gutierrez, D.; Sastre, F. Enhance of Energy Harvesting from Transverse Galloping by Actively Rotating the Galloping Body. *Energies* **2019**, *13*, 91. [CrossRef]
20. Lu, Z.; Wen, Q.; He, X.; Wen, Z. A Flutter-Based Electromagnetic Wind Energy Harvester: Theory and Experiments. *Appl. Sci.* **2019**, *9*, 4823. [CrossRef]
21. Naseer, R.; Dai, H.; Abdelkefi, A.; Wang, L. Comparative Study of Piezoelectric Vortex-Induced Vibration-Based Energy Harvesters with Multi-Stability Characteristics. *Energies* **2019**, *13*, 71. [CrossRef]
22. Bearman, P.W. Vortex shedding from oscillating bluff bodies. *Ann. Rev. Fluid Mech.* **1984**, *16*, 195–222. [CrossRef]
23. Sarpkaya, T. A critical review of the intrinsic nature of vortex-induced vibrations. *J. Fluids Struct.* **2004**, *19*, 389–447. [CrossRef]
24. Williamson, C.; Govardhan, R. Vortex-Induced Vibrations. *Ann. Rev. Fluid Mech.* **2004**, *36*, 413–455. [CrossRef]
25. Arionfard, H.; Nishi, Y. Experimental investigation of a drag assisted vortex-induced vibration energy converter. *J. Fluids Struct.* **2017**, *68*, 48–57. [CrossRef]
26. Konstantinidis, E. A new forced-excitation model for the prediction of maximum response in vortex-induced vibration. In Proceedings of the 11th International Conference on Flow-Induced Vibration, TNO, The Hague, The Netherlands, 4–6 July 2016.
27. Konstantinidis, E. A physics-based model for VIV analysis. In Proceedings of the ASME 36th International Conference Offshore Mechanics and Arctic Engineering, Trondheim, Norway, 25–30 June 2017; Volume 2, p. OMAE2017-62483. [CrossRef]
28. Konstantinidis, E.; Zhao, J.; Leontini, J.; Lo Jacono, D.; Sheridan, J. Excitation and Damping Fluid Forces on a Cylinder Undergoing Vortex-Induced Vibration. *Front. Phys.* **2019**, *7*, 185. [CrossRef]
29. Morison, J.R.; O'Brien, M.P.; Johnson, J.W.; Schaaf, S.A. The force exerted by surface waves on piles. *AIME Petrol. Trans.* **1950**, *189*, 149–154. [CrossRef]
30. Howe, M.S. *Hydrodynamics and Sound*; Cambridge University Press: Cambridge, UK, 2007.
31. Shampine, L.F.; Reichelt, M.W. The MATLAB ODE Suite. *SIAM J. Sci. Comput.* **1997**, *18*, 1–22. [CrossRef]
32. Konstantinidis, E. Added mass of a circular cylinder oscillating in a free stream. *Proc. R. Soc. Lond. A* **2013**, *469*, 20130135. [CrossRef]
33. Griffin, O.M. A universal Strouhal number for the 'locking-on' of vortex shedding to the vibrations of bluff cylinders. *J. Fluid Mech.* **1978**, *85*, 591–606. [CrossRef]

energies

MDPI

Article

Wave-Turbulence Decomposition Methods Applied to Tidal Energy Site Assessment

Larissa Perez [1,*], Remo Cossu [1], Camille Couzi [2] and Irene Penesis [2]

[1] School of Civil Engineering, University of Queensland, Brisbane 4072, QLD, Australia; r.cossu@uq.edu.au
[2] Australian Maritime College, University of Tasmania, Hobart 7250, TAS, Australia; camille.couzi@utas.edu.au (C.C.); i.penesis@utas.edu.au (I.P.)
* Correspondence: l.perez@uqconnect.edu.au

Received: 7 February 2020; Accepted: 5 March 2020; Published: 7 March 2020

Abstract: High levels of turbulence have been proven to substantially increase the blade loadings on tidal turbines, outlining the need of properly characterizing turbulence parameters in tidal energy sites. The presence of long surface gravity waves may cause a significant bias on the estimation of these parameters, which requires wave-turbulence decomposition methods that are currently missing from guidelines. Here, three techniques of decomposing wave and turbulence are tested: the stopband filter (SB), moving average filter (MA), and synchrosqueezing wavelet transform (SWT). The study site, Banks Strait, Tasmania, is a 16 km wide channel that presents high potential for tidal energy generation. Wave peak periods at the study site were found to vary mostly between 7 and 12 s, with maximum exceeding 15 s. Turbulence intensities (TI), turbulent kinetic energy (TKE), and integral scales are quantified. Our results indicate differences between the estimates obtained from each method. The MA highly underestimates turbulence, resulting in TI values which were nearly 50% lower than those obtained from other decomposition methods. While TI and TKE estimated from the SB and the SWT techniques are quite similar, integral length scales are considerably underestimated by the SB. These findings reveal that the SWT is a more reliable method because of the more accurate estimates of turbulence parameters and indicate the need of establishing guidelines which address wave-turbulence decomposition in tidal stream energy site assessments. Despite having shown to be quite a versatile technique, further investigation of its applicability in data from other prospective tidal energy sites is necessary to fully assess the generality of the SWT technique.

Keywords: tidal energy; site assessment; wave-current interaction; turbulence; integral length scales; wave-turbulence decomposition

1. Introduction

The ocean stores a vast amount of energy that can be harvested, for instance, from tidal currents. Prospective tidal energy sites are potentially highly energetic locations commonly characterized by high levels of turbulence and large waves, which can considerably increase loads on structures [1,2]. Hence, the research effort has been directed toward advancing tidal stream technologies to ensure they are reliable yet cost-effective. For instance, significant loads on a tidal turbine are caused by thrust forces, which tend to cause flapwise bending moments, and inertial forces, which cause chordwise bending moments. This implies that the rotor experiences extremely high dynamic forces because of the passing waves, turbulence, vortex shedding, and velocity shear, resulting in high bending moments [3]. These forces can lead to fatigue of the blade and hence are an important aspect to be considered when designing these devices. As highlighted by Ouro et al. [4] and Vazquez et al. [5], tidal energy LCOE still represents a challenge for it to become a mature technology, being highly related to the turbine's manufacturing and maintenance costs. Nonetheless, turbine developers have been overly conservative because of an uncertainty regarding extreme loads, which significantly increases the levelized cost of energy (LCOE) [6]. Despite having an effect on blade loadings, the presence of fine-scale velocity fluctuations does not increase resource uncertainty substantially, reinforcing the predictability of this source of energy [7].

Laboratory experiments conducted by Blackmore et al. [8] showed that variations in turbulence intensity and length scales affect the turbine power and thrust coefficients by over 10%, highlighting the significance of turbulence to turbine performance. The results obtained from the laboratory work reported by Mycek et al. [9] revealed turbulence intensities did not affect the mean power and thrust coefficients substantially but it increased the standard deviations, which may lead to device fatigue. On a full-scale turbine, Harrold and Ouro [2] investigated the loadings on a 12 m diameter tidal turbine deployed in Ramsey Sound, UK. Observation of low and high frequency variations at turbine loadings indicated the presence of a variety of length scales and led to the conclusion that quantifying turbulence parameters is important to guarantee the device's long-term functionality. These results highlight the need to obtain accurate site-specific estimates of turbulence characteristics to ensure the optimum performance of tidal turbines and avoid unnecessary conservativeness and premature fatigue [10–14].

The presence of long surface gravity waves may result in orbital velocities reaching deeper parts of the water column. As a consequence, velocity fluctuations are no longer governed strictly by turbulence, but also by wave orbital motion, and turbulence parameters may be considerably overestimated [15–18]. In order to ensure more accurate estimates of turbulence parameters, and thus avoiding excessive levels of conservativeness, wave orbital motion and turbulent fluctuations must be decomposed in the total velocity fluctuation. Although the literature presents a variety of methods to perform wave-turbulence decomposition, little has been discussed regarding the practicality and applicability of these methods for characterization of turbulence in a tidal energy site assessment. For instance, the work by Parkinson and Collier [19] using the software Tidal Bladed (DNV GL, Bristol, UK), well-known in the industry, validated the results using acoustic Doppler current profiler (ADCP) data from the Fall of Warness. When addressing wave-turbulence interaction, the authors excluded periods of high wave activity in the analysis. The lack of best practice is further emphasized by the fact that the International Electrotechnical Commission (IEC) guidelines as well as, the European Marine Energy Centre (EMEC) do not include recommendations regarding criteria for turbulence measurements, wave-turbulence interaction, and wave-turbulence decomposition. To the authors' knowledge, the present work is the first to test and compare the wave-turbulence decomposition methods for a tidal energy site characterization and to report high frequency measurements taken over such a long period.

Since turbulent and wave orbital motion are usually overlapped on the same frequency bands, it can be argued that performing this task is not as trivial as band-pass filtering [20,21]. Nonetheless, this approach has still been adopted by several researchers [15,18,22,23]. Alternatively, Soulsby and Humphrey [24] have developed a technique based on separating wave induced motion and turbulent fluctuations within the frequency domain. It is assumed that the area under the energy spectrum is equal to the total velocity variance. By extrapolating the expected inertial subrange −5/3 slope [25], orbital velocity variance can be subtracted from the total variance. This technique, however, relies on the assumption that the energy peak caused by waves happens in the isotropic inertial subrange, which is not always the case. Another method is based on the phase lag between horizontal and vertical velocities at the wave frequencies [20]. However, since this method addresses directly the horizontal and vertical velocities, rather than beam velocities, it is not the most appropriate choice for data collected with ADCPs, as turbulence estimates from these instruments rely on the variance method [26,27]. This decomposition has also been successfully performed with a moving average filter, under the assumption that the filter result is the parcel of fluctuations attributed to waves [18,28]. Smyth et al. [15] have described and compared three methods of decomposing wave and turbulence. The first one was the filter method—applying a band-pass filter to the signal. The second one was based on the estimation of vertical wave velocity by assuming linear wave theory. Lastly, a technique based on turbulence dissipation rate was applied. Resulting profiles of vertical root-mean-square velocities have shown to be very similar for the three methods and that the filter and the linear wave theory methods produce reasonable estimates of the turbulence intensity profiles. Similarly, Bian et al. [21] tested and compared three existing decomposition methods—the coherence, cospectra, and ensemble empirical mode decomposition (EEMD)—as well as introduced the synchrosqueezing wavelet transform (SWT), a signal processing technique based on wavelet transforms combined with the reallocation method [29]. Tests were conducted with acoustic Doppler velocimeter (ADV) data for estimation of Reynolds stresses. Even though all four methods slightly underestimated turbulence, the SWT method performed best.

In this paper we compare three techniques to conduct wave-turbulence decomposition for estimation of turbulence parameters at tidal energy sites: the stopband (SB) method, the moving average (MA) method, and the SWT method. While the first two are well-known filtering techniques, the latter is a novel signal processing method promising to provide accurate frequency estimates. These methods were chosen for comparison based on their different levels of practicality, sophistication, and computational time. The SB and MA are relatively simple, hence easily reproducible, and require little computational resources. Moreover, the debate in the literature regarding the applicability of the SB to wave-turbulence decomposition is still ongoing. Contrastingly, the SWT is a more complex and computationally demanding technique, which has provided more accurate results when compared to other methods [20]. The study site—Banks Strait, Tasmania—has been investigated for the Australian Tidal Energy (AUSTEn) project [30–32]. Analyzed data were collected by two Nortek Signature AD2CPs, deployed between March and July 2018, and a Nortek acoustic wave and current profiler (AWAC), deployed between March and May 2018. Instruments were configured to measure wave parameters, current velocities, and high-frequency velocity variations. A link to the dataset used for this work is available in the Supplementary Materials section. A wave climate and current characterization is conducted initially, followed by the application of the three decomposition methods and the estimation of turbulence intensities, turbulent kinetic energy, and integral scales. Finally, estimates obtained from each method are compared. The findings of this work provide important information to allow for the elaboration of guidelines for wave-turbulence decomposition and turbulence characterization in prospective tidal energy sites.

2. Materials and Methods

2.1. Study Site and Data Collection

Banks Strait is a promising site for tidal stream energy generation in Australia. It is a 15-km wide tidal channel located in the Eastern part of Bass Strait, Northeast Tasmania, between Swan and Clarke Islands. The depth in the area varies mostly between 25 and 60 m, with some deeper areas reaching over 70 m, and it has been highlighted for being suitable for tidal energy converters [30] (Figure 1). Even though this area does not exhibit large tidal ranges, strong tidal currents are prevalent. Power density has been estimated to be higher in the middle to northern side of the channel, reaching over 2000 W/m^2 and presenting stronger currents during ebb tides [31,33]. The dominant tidal harmonics constituents are M2 and N2, which are the principal and large elliptical lunar semidiurnal constituents respectively. Maximum current speeds at the site were found to reach over 2 m/s, with significant variation across the channel. Significant wave heights vary mostly between 1 and 2 m, with maximum values reaching over 5 m. Wave periods in Banks Strait vary dominantly between 7 and 12 s. Because of its geomorphology and dominant wind and current directions, the area is subject to strong wave-current interaction, requiring wave-turbulence decomposition prior to estimation of turbulence parameters [32].

During the deployment period dominant wind direction was West-Northwest. Maximum wind speeds have reached over 60 km/h, with approximately 80% of values below 35 km/h and mean speeds of approximately 20 km/h [34].

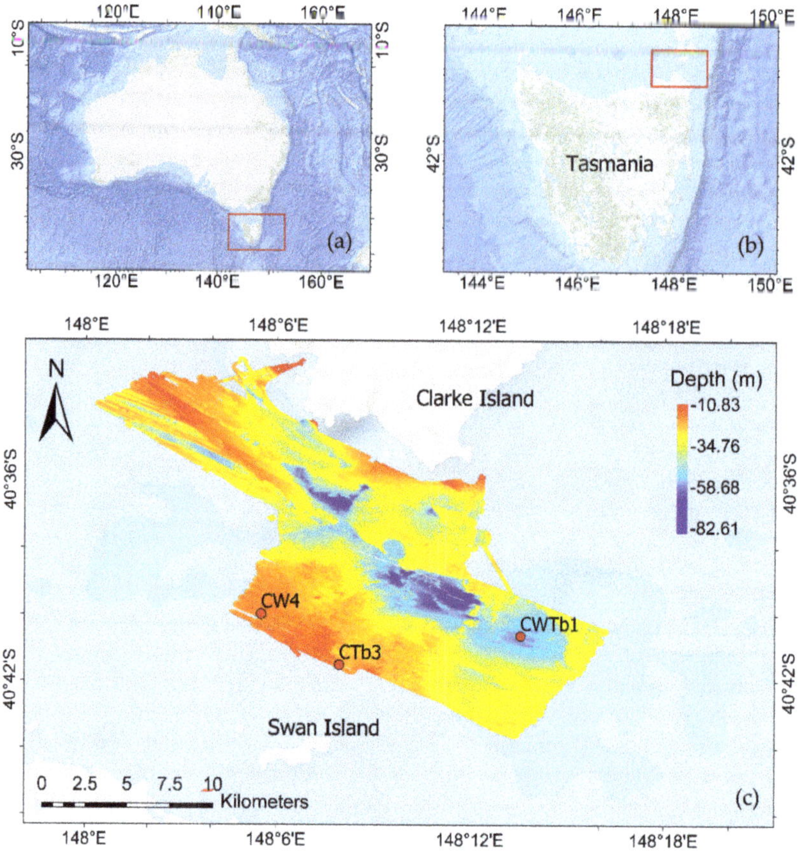

Figure 1. Indication of (**a**) Tasmania in Australia, (**b**) Banks Strait, Northeast Tasmania, and (**c**) measurement sites, located between Clarke and Swan Islands. (Sources: Esri, Garmin, GEBCO, NOAA NGDC and other contributors).

This work analyses data collected with two Nortek AD2CP Signature instruments (500 kHz and 1000 kHz) and a Nortek AWAC 1000 kHz. Investigated depths for turbulence estimation consider as reference a hub-height of 15 m above seabed. Bottom tip and top tip are defined as 10 and 20 m above seabed respectively. Figure 1 shows the deployment location for all three instruments. The signature series is a new generation ADCP that includes a fifth vertical beam. The instrument has a high sampling frequency (up to reach 16Hz depending of chosen configurations) as well as the ability to interleave between two acoustic measurement configurations during a single deployment. Thus, the signature instruments can obtain current velocity, turbulence, and wave data during the same deployment [35].

The Nortek Signature 1000 kHz (CTb3) was deployed in the southern side of the channel (40.5454° S, 148.078° E) at a depth of 33 m, between March and July 2018. In order to ensure data quality, samples which presented low amplitude or high amplitude spikes were removed as well as the top 10% of the water column, to avoid side lobe interference. As per manufacturer's recommendations, samples with beam correlation below 50% were also discarded.

The Nortek Signature 500 kHz (CWTb1) was deployed in the center of the channel (40.6772° S, 148.226° E), in approximately 60 m between March and July 2018. This instrument was set to interleave two different configuration plans. In the first configuration plan, CWTb1 was set to measure the current velocities and turbulence. In the second configuration plan, the instrument was set to measure the current velocities and waves. Configuration for current measurements was the same for the first and second plan. For estimation of directional wave parameters, the instrument measures velocity with its four slanted beams, while using its vertical beam for acoustic surface tracking (AST). Data processing for quality control was done similarly to instrument CTb3.

A Nortek AWAC 1000 kHz (CW4) was set to measure current velocities and waves. The instrument was placed close to CTb3 (40.6664° S, 148.092° E), at a depth of 30 m, and was measuring between March and May 2018. Top of water column was removed in order to avoid side-lobe interference. All instruments were mounted on gravity-based moorings 1 m above the seafloor. Data were corrected for magnetic declination. Instruments configurations are summarized in Table 1.

Table 1. Summary of instruments configurations and deployment details.

Station	CTb3	CWTb1	CW4
Instrument	Signature 1000 kHz	Signature 500 kHz	AWAC 1000 kHz
Depth	33 m	60 m	30 m
Latitude	40.5454° S	40.6772° S	40.6664° S
Longitude	148.078° E	148.226° E	148.092° E
Currents			
Sampling interval	-	150 s	60 s
Averaging interval	-	300 s	300 s
Number of samples	-	210	180
Blanking distance	-	0.5 m	0.4 m
Cell size	-	1 m	1 m
Measurement length	-	108 days	85 days
Standard deviation [1]	-	H: 1 cm/sB: 0.6 cm/s	H: 2.2 cm/sV: 0.7 cm/s
Turbulence			
Burst length	900 s	1200 s	-
Measurement interval	1800 s	3600 s	-
Sampling frequency	8 Hz	4 Hz	-
Beams	4 slanted	Vertical only	-
Blanking distance	0.1 m	0.5 m	-
Cell size	0.5 m	1 m	-
Measurement length	108 days	108 days	-
Standard deviation [1]	H: 3.44 cm/s B: 2.05 cm/s	V: 4.6 cm/s	-
Waves			
Sampling interval	-	2048 s	2048 s
Measurement interval	-	3600 s	3600 s
Sampling frequency	-	1 Hz	1 Hz
Number of samples	-	2048	2048
Measurement length	-	108 days	55 days
Standard deviation [1]	-	H: 1.4 cm/s B: 0.84 cm/s	H: 2.2 cm/s V: 0.7 cm/s

[1] Note: H: horizontal; V: vertical; B: beam. Values obtained from instruments' manufacturer software.

2.2. Theory Background

A velocity signal can be decomposed in:

$$u = \overline{u} + u_t{}' \quad (1)$$

where u represents the total velocity, $\overline{u}$ represents the mean velocity, usually averaged over 5 to 10 min, and $u_t{}'$ represents the total instantaneous velocity fluctuation from the mean. However, in the presence of waves, a component relative to the wave orbital motion must be added. Hence, the decomposition becomes:

$$u = \overline{u} + u' + \widetilde{u} \quad (2)$$

where u' is the turbulent parcel of the velocity fluctuation and $\widetilde{u}$ is the orbital velocity. For a proper characterization of turbulence, the latter must be removed from the signal (2).

Estimation of turbulence parameters from data measured by an ADCP cannot be obtained directly from horizontal and vertical velocities [26]. Because of the geometry of the instrument and the beam spread, it is very likely that opposite beams will be sampling different turbulent eddies. However, the variance technique only relies on statistic features from each beam rather than the instantaneous values. This method requires two major assumptions: (a) spatial homogeneity between beams, which means the velocity fields sampled by opposite beams have the same statistical characteristics (mean velocity and velocity variance), and (b) turbulence is stationary over the averaging period [12,26]. The homogeneity assumption can be checked by estimating the velocity error E, defined as the difference between the two independent vertical velocities, which can be calculated by the two pairs of opposing beams. If the flow field is homogeneous, mean E is expected to be zero and this test is done by comparing E mean to its standard deviation. In a horizontally homogeneous flow, the mean should rarely exceed the standard deviation [12,36]. Therefore, wave-turbulence decomposition must be performed to beam velocities, prior to turbulence parameters estimation, rather than to streamwise, cross-stream, and vertical velocities directly.

TKE is a commonly used metric to characterize turbulence, which indicates the amount of energy contained in turbulent fluctuations per unit mass. It is traditionally defined as:

$$q^2 = \frac{\overline{u'^2} + \overline{v'^2} + \overline{w'^2}}{2} \tag{3}$$

where u', v', and w' indicate the turbulent velocity fluctuations in the streamwise, cross-stream, and vertical directions of the flow. The methodology for estimation of TKE from a 4-beams ADCP is fully described in the literature [12,26,27,37]. Because of the absence of true vertical measurements, an assumption regarding turbulence anisotropy is also necessary. This is done by adopting a value $\xi = \frac{\overline{w'^2}}{2}\frac{1}{q^2}$. Togneri and Masters [12], who have performed a similar estimation in a tidal energy site, followed the work of Nezu and Nakagawa [38], adopting $\xi = 0.1684$. Total TKE can, thus, be estimated from a 4-beams ADCP from:

$$TKE = \frac{\sum \overline{b'^2}_i}{4sin^2\theta(1 - \xi(1 - 2\,cot^2\theta))} \tag{4}$$

where $\overline{b'^2}$ is the beam variance over the averaging period and θ is the inclination angle of each beam from the vertical axis. For the Nortek Signature, instruments used in this study, this is equal to 25°.

Another important parameter for tidal turbine developers is TI, defined as:

$$TI = \frac{\sqrt{\overline{u'^2}}}{\overline{u}} \tag{5}$$

representing the portion of total velocity which is attributed to turbulent fluctuations and often given in percentages. Since TI results are influenced by the mean velocity, it is common practice to present and analyze both parameters together. Moreover, elevated TI are expected to happen during low mean current velocities, as in slack tide conditions, which could potentially lead to a misinterpretation of results.

The integral length scale is the turbulence scale containing the most energetic eddies. It can be estimated from the analysis of a temporal autocorrelation of velocity fluctuations which can be described as:

$$\rho(\tau) = \frac{\overline{u'(t)u'(t+\tau)}}{\overline{u'^2}} \tag{6}$$

where τ is a time increment given in seconds. The integral time scale can then be obtained by integrating $\rho(\tau)$ until the first zero-crossing point. By assuming Taylor's frozen field hypothesis, the integral length scale can be calculated by multiplying the integral time scale by the average velocity [11]. Here, this analysis is performed over 5 min periods. It is important to highlight that it is necessary to assume the flow is homogeneous over the plane at which the streamwise velocity is calculated to estimate integral length scales from an ADCP in the streamwise direction. The distance from a velocity bin to the vertical axis above the instrument is equal to $d = 2zsin(\theta)$, where z is the vertical distance above the instrument [39,40]. Therefore, eddies smaller than d cannot be accurately resolved.

The significant wave height is defined as the height obtained from the wave energy spectrum as:

$$H_S = 4\sqrt{M_o} \tag{7}$$

where M_o is the variance of the water surface elevation, which is approximately equal to the sum of all frequency components of the wave energy spectrum. Given the depth of the channel, it is reasonable, as a first order approximation, to assume linear wave theory for wavelength estimation, defined as:

$$L = \frac{g{T_p}^2}{2\pi} \tag{8}$$

where T_p is the peak wave period, corresponding to the peak frequency for the wave spectrum.

2.3. *Wave-Turbulence Decomposition Methods*

2.3.1. Method 1: SB

The method described here uses a SB filter to remove the frequency bands in which waves are dominant in the velocity signal. To do so, an observation of the original energy spectrum is necessary in order to properly select the correct frequencies to be filtered. Data collected in Banks Strait indicated frequencies between 0.09 Hz and 0.16 Hz included orbital velocities. Hence these frequencies were selected as minimum and maximum thresholds in the second-order Butterworth filter which was applied to each beam velocity. This step was performed using the butter function in MATLAB (2018b, MathWorks, Natick, MA, USA). Even though recent works have pointed out the limitations of applying high-pass or bandpass filters in order to remove wave interference from the velocity signal, this method has successfully been applied in many studies, providing good estimates of TI [15,17,22]. In addition, this technique presents the benefit of requiring little computational resources.

2.3.2. Method 2: MA

The MA filter method is a technique that has been adopted by Williams et al. [28]. A simple MA filter is applied to the velocity signal in order to remove high frequency fluctuations. Therefore, the resulting signal is the portion of the signal composed strictly by low frequency fluctuations, which are assumed to be the wave orbital velocities. To perform the decomposition, the orbital velocities are subtracted from the original signal. The use of previous values in the moving average filter induces a slight time shift between the original and filtered signal. Therefore, to ensure the average is central to the data points, here the moving average filter is adapted to a two-sided moving average, defined as:

$$B_t = \frac{1}{2N+1}\sum_{i=-N}^{N} b_{t+i} \tag{9}$$

where B_t is the forecast value at instant t, N is the amount of values before and after the centre value to be averaged over and b_t is the value in the original signal at instant t. The filter is applied in order to obtain averaged values over a time period. A sensitivity analysis using periods of 1, 2, and 3 s was performed in order to identify the best time interval to be used. This indicated using 1 s provided the best result, reinforcing the choice made by Williams et al. [28]. The resulting filtered signal is expected to be approximate to the wave-induced motion and should then be subtracted from the original signal.

2.3.3. Method 3: SWT

The SWT is a time-frequency signal analysis method that has proved to be useful in extracting independent components from nonlinear and nonstationary signals. It uses a combination of a wavelet analysis and the reallocation method, reducing the energy spread and resulting in a more concentrated time-frequency analysis, which allows for the extraction of instantaneous frequencies [29]. This processing method has been applied for the first time with the purpose of decomposing wave and turbulence and described in detail by Bian et al. [21]. Here we have used a Morlet wavelet as performed by Daubechies et al. [29]. Daubechies et al. [29] and Thakur et al. [41] have shown the SWT is highly invariant to the mother wavelet shape and more affected to its concentration in time and frequency. Some other parameters that can be adjusted during the analysis are the number of voices per octave (recommendation is 32 or 64) and the number of curves to be extracted from the original signal. A more detailed description of the adjustable parameters can be found in Thakur et al. [41].

Initially, the original signal is decomposed in beam velocity mean and fluctuation. The SWT of the velocity fluctuation is then calculated in order to obtain the instantaneous frequencies of the signal. Wave dominant frequencies can then be identified as the most energetic frequencies and orbital velocity can be estimated by reconstructing the velocity fluctuation signal and summing it at the wave-dominant frequencies. To estimate turbulent fluctuations, the new reconstructed orbital velocity time-series is subtracted from the total velocity fluctuations. The MATLAB functions used to perform the described steps are part of the Synchrosqueezing Toolbox developed and made publicly available by Thakur et al. [41].

3. Results

3.1. Wave Climate Characterization

Table 2 shows the statistical results obtained from the analysis of the significant wave height and the period of the waves corresponding to the peak frequency for the wave spectrum. Analysis of wave data from instruments CWTb1 and CW4 indicates a considerable variation in significant wave height during the deployment period, with values ranging from approximately 0.5 m to over 5 m. Mean significant wave height observed in measurement locations CWTb1 and CW4 are 1.75 m and 1.42 m respectively, with approximately 75% of results below 2 m. Measured peak periods reveal dominance of long waves in Bank Straits, with maximum periods varying between 13 s and 15 s and mean values 8.05 s and 8.76 s at CWTb1 and CW4 respectively.

Table 2. Summary of statistical results of significant wave height and peak wave period from measurement locations CWTb1 and CW4. 1296 data points were included in the analysis for CWTb1 and 1342 data points for CW4.

	Statistical Variable	CWTb1	CW4
H_{Mo} (m)	Mean	1.75	1.42
	Standard deviation	0.78	1.03
	Maximum	5.17	5.26
	Minimum	0.49	0.04
	50th percentile	1.62	1.05
	75th percentile	2.25	2.01
T_P (s)	Mean	8.05	8.76
	Standard deviation	1.75	2.49
	Maximum	13.28	15.80
	Minimum	2.63	5.20
	50th percentile	7.97	7.79
	75th percentile	9.20	10.52

Figure 2 shows the significant wave height, peak wave period, and estimated wavelength time-series for the full deployment period. Results for 50th and 75th percentiles are illustrated by red and blue lines, with dashed lines corresponding to instrument CWTb1 and dotted lines to instrument CW4. Time-series agree well for both locations during the deployment, which indicates that waves exhibit similar characteristics across the entire channel in Banks Strait.

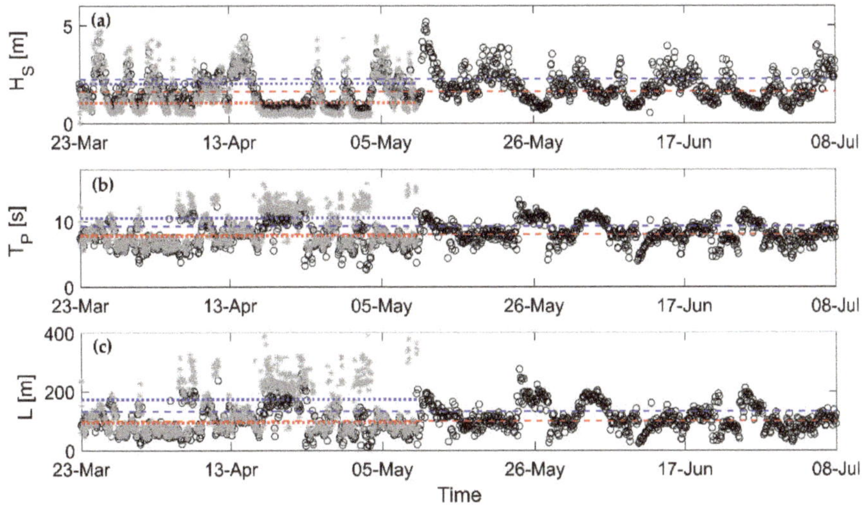

Figure 2. Time-series for (**a**) significant wave height, (**b**) peak wave period, and (**c**) estimated wavelength. Red and blue lines indicate 50th and 75th percentiles respectively. Black circles and dashed lines correspond to instrument CWTb1 and grey stars and dotted lines, to instrument CW4.

In order to characterize for the dominant wave direction, sea surface variance directional wave spectral densities obtained during the full deployment period were averaged (Figure 3). Reported directions correspond to the direction waves are coming from and peak direction occurrence is also presented in percentages in Table 3. The peak direction is related to the peak period obtained from the spectra. At both locations, waves come predominantly from West-Northwest (270°–315°) and East-Southeast (90°–135°) but at the measurement location CWTb1, waves exhibit a higher directional change.

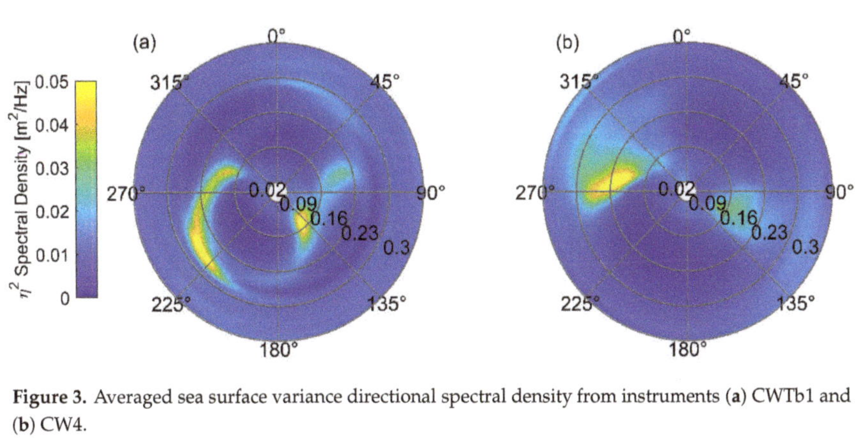

Figure 3. Averaged sea surface variance directional spectral density from instruments (**a**) CWTb1 and (**b**) CW4.

Table 3. Occurrence of peak wave direction from each station. Occurrence is presented in percentages and directions represent the direction the waves are coming from.

Peak Direction	CWTb1	CW4
0 < Dir ≤ 45°	1.00	2.83
45° < Dir ≤ 90°	10.65	6.33
90° < Dir ≤ 135°	22.92	33.08
135° < Dir ≤ 180°	16.28	0.60
180° < Dir ≤ 225°	3.63	0.07
225° < Dir ≤ 270°	8.80	0.22
270° < Dir ≤ 315°	35.65	52.16
315° < Dir ≤ 360°	1.08	4.69

3.2. Current Velocity Characterization

Table 4 summarizes the statistic results for current speeds measured with instruments CWTb1 and CW4. The 30% lowest values were removed from the data set so that slack tide would not be included. Current speeds have reached a maximum of 1.86 m/s and 2.35 m/s during ebb tide at locations CWTb1 and CW4 respectively. During flood tide, maximum currents reached 1.42 m/s and 2.23 m/s. Mean velocities were 0.85 m/s and 1.21 m/s during ebb tides and 0.67 m/s and 1.21 m/s during flood, indicating stronger currents during ebb tides at location CWTb1. Speed asymmetry here is defined as $|Ebb_{mean}/Flood_{mean}|$ to better understand the differences in current speeds during the two different stages of the tidal cycle.

Table 4. Summary of statistic results of current speed measured with instruments CWTb1 and CW4.

Statistic Result		CWTb-1		CW-4	
		Ebb	Flood	Ebb	Flood
Maximum speed	(m/s)	1.86	1.42	2.35	2.23
Mean speed	(m/s)	0.85	0.67	1.21	1.21
Standard deviation	(m/s)	0.28	0.17	0.31	0.30
Speed asymmetry	-	1.26		1.00	
Mean direction	(°)	117.13	280.91	104.79	292.16
Directional asymmetry	(°)	16.22		7.37	
Standard deviation	(°)	12.75	19.95	15.25	11.97

During ebb tides currents flowed dominantly to approximately 117.13° and 104.79° degrees from North, while during floods, these values are 280.91° and 292.16°, at locations CWTb1 and CW4 respectively. Figure 4 presents the current speed and direction for both stations and shows that current direction agrees well with the dominant wave direction. Asymmetry is defined as $|\overline{\theta}_{ebb} - \overline{\theta}_{flood} - 180°|$. Higher directional asymmetry—approximately 16.22°—is observed from data measured with instrument CWTb1, which was located in the Eastern part of the channel. Location CW4 presents a directional asymmetry of 7.37°.

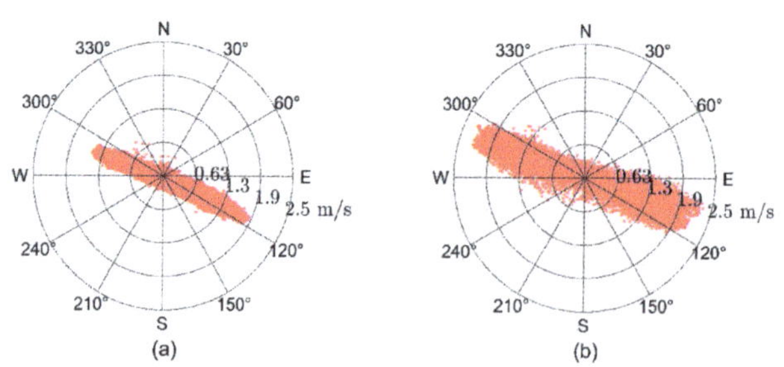

Figure 4. Current speed and direction from instruments (**a**) CWTb1 and (**b**) CW4.

3.3. Wave-Turbulence Decomposition Methods

3.3.1. Southern Part of the Channel

The original turbulent kinetic energy spectrum from a single 15-min burst measured with instrument CTb3 is shown in Figure 5. The burst was measured during fully developed spring flood tide, beginning at 6:30 PM on the 26th April 2018, when mean streamwise velocity was 1.26 m/s. A clear peak occurs at approximately 0.13 Hz. This frequency is consistent with wave peak frequencies measured at CW4. The peak is followed by the onset of the −5/3 slope, indicating that waves likely occur on the same frequency as larger turbulent eddies.

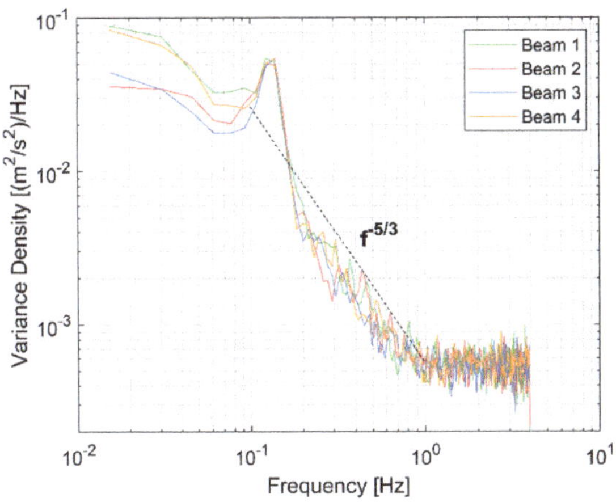

Figure 5. Original turbulent kinetic energy spectrum from velocities measured with each beam of instrument CTb3.

Results from wave-turbulence decomposition with the SB, MA, and SWT methods are shown in Figure 6. The column on the left-hand side of the figure shows a velocity time-series from beam 2 of the first 40 s of the same burst shown in Figure 5. A short period of the burst was selected to better visualize the low frequency orbital motion (Figure 6a,c,e). The right hand side column shows the resulting turbulent kinetic energy spectrum of the burst. The time-series show the original velocity fluctuation, which is a sum of orbital velocities and turbulence, the estimated orbital velocity and the turbulent fluctuations, obtained after the removal of orbital motion from the original signal.

Although an initial comparison of the original velocity fluctuation and the estimated orbital velocity time-series shows good agreement, the spectra obtained from each method highlight the difference between these techniques. The spectrum resulting from the SB filter method presents a pronounced drop in energy over the frequencies in which wave interference was previously dominant, which yields a consequent underestimation of turbulent energy (Figure 6b). Likewise, a considerable drop in energy is observed from the MA filter method in the spectrum shown in Figure 6d, in which the −5/3 slope has vanished. In addition, a peak at the wave dominant frequencies is still present, indicating that this method was not sufficient in fully removing the orbital velocities from the analyzed signal. Looking at the orbital velocity time-series originated from this method (Figure 6 c), high frequency fluctuations are still visible, which justifies the significant decrease in turbulence. Finally, the application of the SWT method resulted in a slight drop in energy at the wave dominant frequencies, which can be observed in all beams but is more pronounced in beam 4 (Figure 6f). Generally, the resulting spectrum is in good agreement with the theoretical −5/3 slope, indicating less accentuated underestimation of turbulence. Moreover, orbital velocity time-series agrees well with the low frequency motion present in the original signal, suggesting a satisfactory performance.

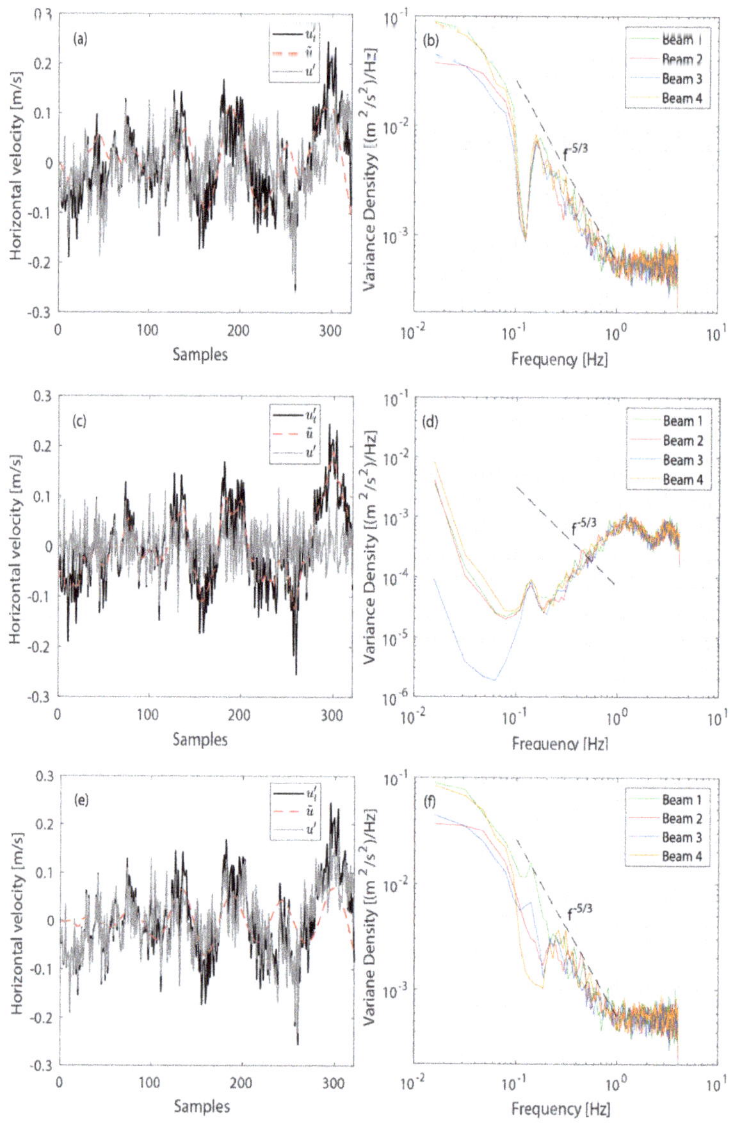

Figure 6. Original and decomposed velocity fluctuations time series from beam 2 and turbulent kinetic energy spectrum and after wave-turbulence decomposition performed in beam coordinates with (**a**,**b**) the stopband (SB) method, (**c**,**d**) moving average (MA) method, and (**e**,**f**) synchrosqueezing wavelet transform (SWT) method. Velocity time series obtained from beams 1, 3, and 4 presented similar results as beam 2. Mean streamwise velocity during this burst measurement was 1.26 m/s.

The performance of the three methods is also compared in Table 5, which shows averaged streamwise turbulence intensity and total TKE obtained from the same burst depicted in Figure 6. As expected from the analysis of the energy spectra, both turbulence intensity and TKE values are significantly underestimated by the MA method. While the SB and the SWT methods presented average turbulence intensities of approximately 11.50%, the MA method resulted in 6.02%. Similarly, average total TKE from this method is equal to nearly one-third of the result obtained from the other two methods. In contrast to the spectral densities, the SB and the SWT decomposition methods reveal similar turbulence intensities (difference of 0.27%) and TKE (difference of 0.0001 m^2/s^2).

Table 5. Average turbulence intensity and total turbulent kinetic energy (TKE) from a 15-min long burst collected on the 26th April 2018 resulting from different wave-turbulence decomposition methods.

Method	Mean Streamwise TI (%)	Mean TKE (m^2/s^2)
SB	11.62	0.0162
MA	6.02	0.0048
SWT	11.35	0.0161

A period of approximately 12 h during spring tide on the 26th April 2018 was selected for the estimation of turbulence intensities and TKE in a complete tidal cycle using the three decomposition methods. Figure 7 shows turbulence intensity and velocity for the three depths at measurement location CTb3, estimated using each method. Results obtained from the SB filter and SWT methods do not differ significantly. Nonetheless, the MA method presents results considerably lower, reaching nearly half of TI and 1/4 of TKE obtained with the other methods. Generally, all methods reveal streamwise turbulence intensities that do not vary considerably between bottom tip, hub, and top tip heights.

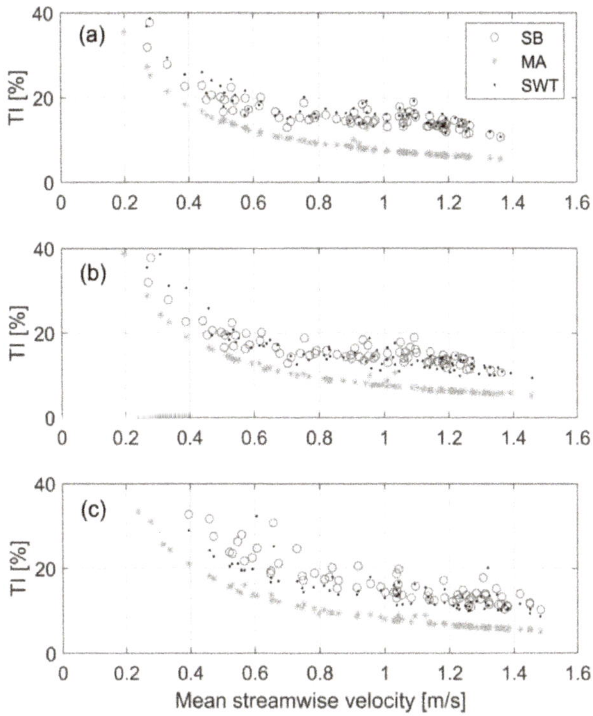

Figure 7. Streamwise turbulence intensities and mean velocity obtained with each method at (**a**) bottom tip height (10 m above seabed), (**b**) hub-height (15 m above seabed), and (**c**) top tip (20 m above seabed) from instrument CTb3.

The total TKE was also estimated at each depth using the three different decomposition methods. The resulting time-series for the full 12-h period is shown in Figure 8. As previously, the time-series obtained from the MA method significantly underestimates TKE, with values being approximately four times smaller than the results obtained with the other two methods. Differences between bottom tip, hub height, and top tip are more prominent in the results from the SB method, with TKE slightly higher nearer the surface (top tip, in Figure 6a). Results originated from the SWT method do not seem to vary significantly between the three selected depths. Similarly to turbulence intensities estimates, total TKE obtained with the SB and the SWT methods do not differ substantially. Linear regression fit between TKE at hub-height from the two methods indicate highly similar results, with a regression coefficient of over 0.92 and R^2 of approximately 0.87 (Figure 9).

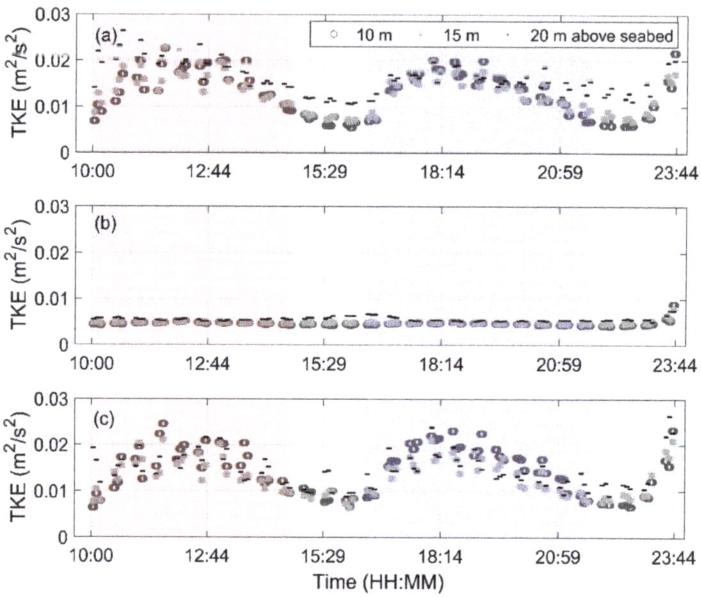

Figure 8. Total TKE estimated from (**a**) the SB, (**b**) MA, and (**c**) SWT decomposition methods at bottom tip height (10 m above seabed), hub-height (15 m above seabed), and top tip (20 m above seabed). Time-series show an approximately 12-h period, measured on the 26th April 2018 at measurement location CTb3. Red area shows ebb tide and blue area shows flood tide.

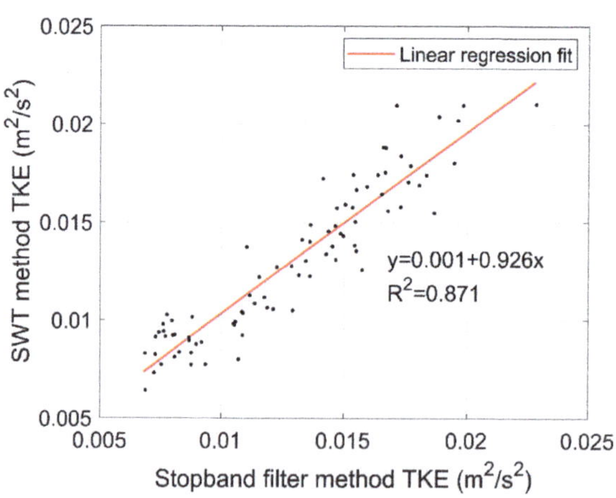

Figure 9. Linear regression fit between total TKE estimates at hub-height from instrument CTb3 obtained with the SB filter and the SWT methods.

3.3.2. Deep Pocket in the Center of the Channel

Because of the vertical beam measurements taken with the Signature 500 kHz (CWTb1), vertical TKE based on true vertical velocities could be estimated. Time-series of nearly the same times shown in Figure 8 are plotted in Figure 10. It is evident that values obtained after applying the SB and the SWT methods agree well, while results from the MA seem to considerably underestimate the TKE quantities. Linear regression fit for vertical TKE at hub-height demonstrates that the two methods perform similarly well (Figure 11).

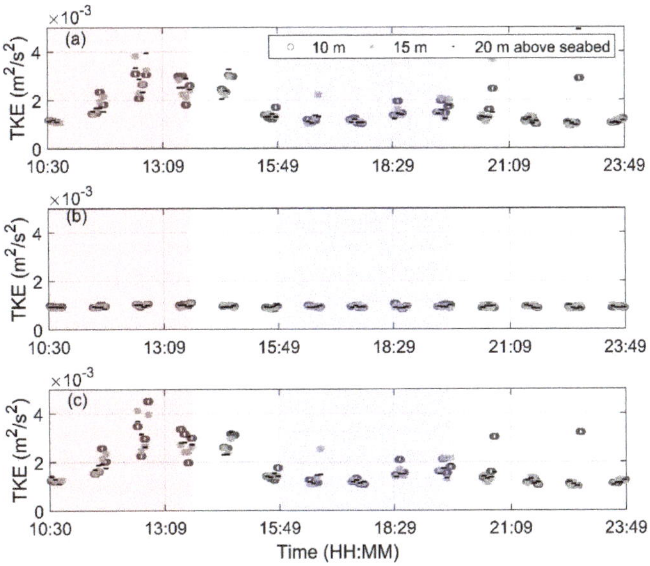

Figure 10. Vertical TKE estimated from vertical beam measurements using (**a**) the SB, (**b**) MA, and (**c**) SWT decomposition methods at bottom tip height (10 m above seabed), hub-height (15 m above seabed), and top tip (20 m above seabed). Time-series show an approximately 12-h period measured on the 26th April 2018 at measurement location CWTb1.

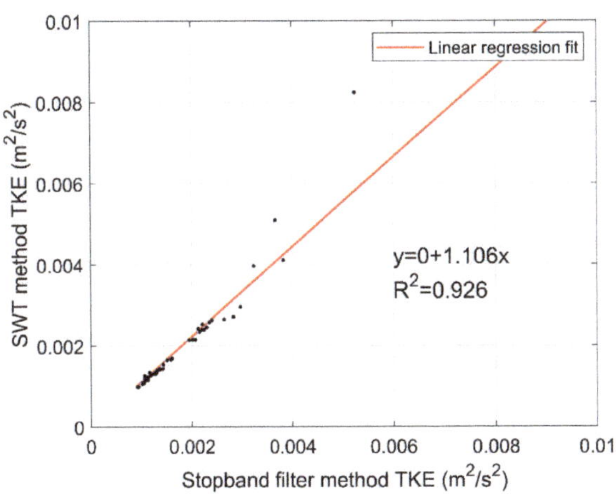

Figure 11. Linear regression fit between vertical TKE estimates at hub-height from instrument CWTb1 obtained with the SB filter and the SWT methods.

3.3.3. Integral Scales

Streamwise mean and maximum integral time and length scales from CTb3 for ebb and flood tides were also estimated for the two most successful methods (SB and SWT) (Table 6). Unlike TI and TKE estimates, results reveal a larger difference. For instance, for the top blade tip depth (20 m above the seabed) during ebb tides, mean integral length scale after the application of the SB was 2.13 m, while the value obtained from the SWT decomposed signal was 4.95 m. Generally, larger integral length scales are observed during ebb tide.

Table 6. Integral time and length scales from CTb3 estimated for ebb and flood tides based on streamwise velocity fluctuations obtained after the SB and the SWT methods.

Height		Tide	Integral Time Scale (s)		Integral Length Scale (m)	
			Mean	Max	Mean	Max
Bottom tip	SB	Ebb	9.12	24.57	9.56	26.91
		Flood	5.76	15.63	5.82	20.79
	SWT	Ebb	8.96	22.23	9.30	25.49
		Flood	4.89	14.35	4.97	19.07
Hub Height	SB	Ebb	5.67	20.41	6.56	25.93
		Flood	4.50	10.81	4.73	12.33
	SWT	Ebb	7.09	22.68	8.21	28.82
		Flood	4.79	14.74	4.87	12.25
Top tip	SB	Ebb	1.77	11.22	2.13	13.25
		Flood	2.75	9.21	3.09	9.78
	SWT	Ebb	4.08	17.51	4.95	20.74
		Flood	4.08	12.41	4.42	12.85

4. Discussion

Our data provide an overview of different techniques to decompose wave-current interaction during turbulent flow conditions in Banks Strait which is a candidate site for commercial development. Wave directions agree well with current direction, mostly at site CW4, as can be seen in Figures 3 and 4. Wave data collected with the Signature 500 kHz at CWTb1 indicate a higher directional spread. However, turbulent energy spectra from both sites reveal a pronounced peak at frequencies ranging from 0.1 Hz to 0.14 Hz, consistent with wave periods observed at both CWTb1 and CW4. This supports our findings of strong wave-current interaction at the sites, which is further corroborated by observations of low frequency fluctuations in the beam velocities.

Initial observation of the resulting turbulent energy spectra obtained after performance of wave-turbulence decomposition with the three tested methods indicates that the SWT method was the most successful. It induces a drop in energy at the wave-dominant frequencies, but this drop is not as accentuated when compared to the spectrum obtained with the SB method and agrees well with the expected −5/3 slope (Figure 6). Bian et al. [21] obtained similar results with an energy peak occurring at approximately 0.1 Hz, which is comparable to the peak frequency observed in Banks Strait. Variations in peak wave frequency could justify the poor performance of the MA method. This simple decomposition method has been introduced by Williams et al. [28] and successfully reproduced by Zhu et al. [18] but in both studies peaks occurred at higher frequencies in the spectrum (approximately 0.3 Hz). We hypothesize that this method reveals to be more suitable for higher frequency surface waves, which restricts its application for tidal energy sites exhibiting longer wave periods such as Banks Strait. The SWT was initially introduced by Daubechies et al. [29] and has been successfully applied to various fields of research over different frequency ranges (e.g., [29,41]). Nonetheless, this technique has never been tested in a tidal energy site and our findings show that this method present satisfactory results when performed to decompose wave and turbulence. More importantly, because of its proven versatility over a large wave frequency range it can be concluded that the SWT method is also applicable at other tidal energy sites, which might have different wave-dominant frequencies compared to the data set presented here. Despite its robust and satisfactory performance, the SWT is also computationally expensive and, hence, time consuming, especially for analysis of large data sets from long deployments. For instance, the decomposition of a 12-h period of an AD2CP dataset with a sampling rate of 8 Hz took over 16 times the time required by the SB and the MA in the same computer. Therefore, applying this method to large data sets for long-term characterization and evaluation of seasonality is not trivial. In contrast, the SB method is computationally simple and easily replicable. The comparison between the SWT and SB methods has given satisfactory results in TI and TKE estimates. Interestingly, the application of this method for wave-turbulence decomposition has not considerably underestimated these turbulence parameters, when compared to the results obtained from the SWT method. Despite the difference in R^2 values obtained from the linear regression fit of TKE shown in Figures 9 and 11 (0.871 and 0.926 respectively) both have provided reasonable results. The slightly lower value obtained from total TKE from CTb3 is likely attributed to the fact that estimation of total TKE required the decomposition of the four beam velocity time-series, while vertical TKE estimation from CWTb1 only required the decomposition of a single velocity time-series. Since each beam velocity is decomposed separately, it is expected that four beams will yield a higher difference in TKE results versus a single beam. Both methods have provided turbulence intensities that are expected at tidal sites (10–15%) [10,42]. However, to ensure minimal turbulence underestimation when conducting the analysis to a long period, the SB method requires good knowledge of the wave dominant frequencies.

Integral length scales have been used to discuss tidal turbine performance (e.g., [4,8]) and poor estimation of this quantity represents a considerable disadvantage for tidal site assessments. Estimates of streamwise integral time and length scales obtained from the SB and SWT methods seem to yield a larger difference than that obtained for TI and TKE. With the exception of results produced during flood tides at bottom blade tip depth, the SB method produced smaller values, which led to the underestimation of turbulence. This difference becomes even more significant closer to the surface, and is likely attributed to the increase in distance between the bins of opposing beams. Estimation of integral length scales from ADCP measurements in the stream-wise direction can only be done by assuming flow homogeneity. In addition, the presence of eddies, which are larger than the water depth, demonstrates inherent anisotropy, since isotropic eddies are limited by the site depth [8]. Since the energy spectrum presents the peak right before the isotropic "−5/3" slope, waves are likely on the same frequency as these identified anisotropic large eddies. Our results suggest that the SB method should be used with caution while the SWT method is able to provide the tidal energy community more accurate estimates of integral scales. These findings also reinforce the results of other studies that the estimation of turbulence parameters can be biased by the presence of surface waves. For instance, Thiébaut et al. [43] has estimated turbulence dissipation rates and integral scales using ADCP data from Alderney Race, France. Their findings reveal estimated parameters were sensitive to wave-induced motion.

It is significant to highlight that the comparison between the wave-turbulence decomposition methods was performed in two measurement locations which present considerably different characteristics. While CTb3 is located in a 30-m deep area and presents strong current speeds, CWTb1 is located in a deep pocket in the center of the channel, being 60 m deep and presenting lower current speeds. This reveals that the presented results are valid under varying conditions.

5. Conclusions

The main goal of the present work was to test the decomposition methods for wave-turbulence interaction to provide turbine developers with information regarding blade loads for device design, ensuring long-term turbine integrity and facilitating tidal stream energy generation. In utilizing data collected from two Nortek Signature AD2CPs and a Nortek AWAC, wave and current characteristics were described and three wave-turbulence decomposition methods—the SB, the MA, and the SWT methods—were compared. While the first two techniques are well-known and easily reproduceable, the third is a more complex and detailed signal processing tool. Mean wave periods were found to be nearly 8 s, but maximum wave periods exceeding 15 s extend the wave period spectrum significantly. Wave direction has shown to agree well with the current direction, reinforcing the previous findings of wave-current interaction at the site [32]. Findings reveal that poor choice of a wave-turbulence decomposition method likely leads to underestimation of fatigue loads, considerably putting long-term turbine integrity at risk.

Based on the estimated energy spectra, the MA method has highly underestimated the turbulence when compared to the other two methods. Streamwise turbulence intensities, turbulent kinetic energy, and streamwise integral time and length scales were estimated from the two most successful methods. Comparison between TI and TKE obtained after conducting the SB method has shown good agreement with those obtained with the SWT. A linear regression between total TKE at hub-height estimated from instrument CTb3 from the two methods indicated similar results. The same analysis was done for vertical TKE estimated from instrument CWTb1. Nonetheless, estimates of streamwise integral scales from the SB and the SWT methods have resulted in a larger difference, especially for the top blade tip depth. Since variations in turbulence length scales have been shown to affect the tidal turbines performance, the consequent underestimation of this parameters from the SB filter method represents a considerable disadvantage when performing a tidal energy site assessment. These diverging turbulence estimates obtained from different methods highlight the need to implement assessment guidelines which address the issue of wave-turbulence interaction at tidal energy sites.

The data collected in Banks Strait help advancing the knowledge about suitable decomposition techniques, but further investigation is necessary to assess the performance of these decomposition methods at sites that are home to different wave climates. Despite being computationally expensive and highly time consuming the SWT has also shown to provide the most accurate estimates of turbulence parameters, facilitating improved quantification of estimated fatigue loads and as such advancing the design and longevity of tidal turbine devices.

Supplementary Materials: The datasets used in this research are publicly available at: https://doi.org/10.14264/ab5f3c0.

Author Contributions: L.P. worked on the conceptualization of this paper, its methodology, data processing and analysis, as well as writing the original draft. R.C. and I.P. have substantially contributed to this paper through funding acquisition, project administration, data collection, improvements on data analysis, as well as supervising and reviewing. C.C. has added to this work through data collection and processing. All authors have read and agreed to the published version of the manuscript.

Funding: The AUSTEn Project is co-funded by the Australian Renewable Agency (ARENA) through the Advancing Renewables Programme (grant G00902), the Australian Maritime College (University of Tasmania), the University of Queensland, CSIRO Oceans and Atmosphere, CSIRO Energy, our industry partners, SIMEC Atlantis Energy and MAKO Tidal Turbines.

Acknowledgments: The authors would like to thank Professor Alex Hay from Dalhousie University, Canada.

Conflicts of Interest: The authors declare no conflict of interest.

Abbreviations

ADCP	Acoustic Doppler Current Profiler
AD2CP	New generation 5-beam Nortek ADCP
ADV	Acoustic Doppler Velocimeter
AUSTEn	Australian Tidal Energy Project
AWAC	Nortek Acoustic Wave and Current Profiler
LCOE	Levelized Cost of Energy
MA	Moving Average filter
SB	Stopband filter
SWT	Synchrosqueezing Wavelet Transform
TI	Turbulence Intensity

TKE	Turbulence Kinetic Energy
u	Total velocity
$\overline{u}$	Mean velocity
u_t'	Total velocity fluctuation
u'	Turbulent fluctuation
$\widetilde{u}$	Wave orbital velocity
q^2	Turbulent kinetic energy
b'_i	ith beam velocity fluctuation
θ	ADCP beam slanted angle
ξ	Constant for anisotropy assumption
ρ	Autocorrelation function
τ	Time increment in the autocorrelation function
H_S	Significant wave height estimated from wave spectrum
M_0	Variance in the water surface elevation
L	Wavelength
T_p	Peak wave period

References

1. Copping, A.; Battey, H.; Brown-Saracino, J.; Massaua, M.; Smith, C. An international assessment of the environmental effects of marine energy development. *Ocean Coast. Manag.* **2014**, *99*, 3–13. [CrossRef]
2. Harrold, M.; Ouro, P. Rotor Loading Characteristics of a Full-Scale Tidal Turbine. *Energies* **2019**, *12*, 1035. [CrossRef]
3. Milne, I.A.; Day, A.H.; Sharma, R.N.; Flay, R.G.J. The characterisation of the hydrodynamic loads on tidal turbines due to turbulence. *Renew. Sustain. Energy Rev.* **2016**, *56*, 851–864. [CrossRef]
4. Ouro, P.; Stoesser, T. Impact of Environmental Turbulence on the Performance and Loadings of a Tidal Stream Turbine. *Flow Turbul. Combust.* **2019**, *102*, 613–639. [CrossRef]
5. Vazquez, A.; Iglesias, G. A holistic method for selecting tidal stream energy hotspots under technical, economic and functional constraints. *Energy Convers. Manag.* **2016**, *117*, 420–430. [CrossRef]
6. Li, W.; Zhou, H.; Liu, H.; Lin, Y.; Xu, Q. Review on the blade design technologies of tidal current turbine. *Renew. Sustain. Energy Rev.* **2016**, *63*, 414–422. [CrossRef]
7. Lewis, M.; Mcnaughton, J.; Márquez-Dominguez, C.; Todeschini, G.; Togneri, M.; Masters, I.; Allmark, M.; Stallard, T.; Neill, S.; Goward-Brown, A.; et al. Power variability of tidal-stream energy and implications for electricity supply. *Energy* **2019**, *183*, 1061–1074. [CrossRef]
8. Blackmore, T.; Myers, L.E.; Bahaj, A.S. Effects of turbulence on tidal turbines: Implications to performance, blade loads, and condition monitoring. *Int. J. Mar. Energy* **2016**, *14*, 1–26. [CrossRef]
9. Mycek, P.; Gaurier, B.; Germain, G.; Pinon, G.; Rivoalen, E. Experimental study of the turbulence intensity effects on marine current turbines behaviour. Part I: One single turbine. *Renew. Energy* **2014**, *66*, 729–746. [CrossRef]
10. Thomson, J.; Polagye, B.; Durgesh, V.; Richmond, M.C. Measurements of Turbulence at Two Tidal Energy Sites in Puget Sound, WA. *IEEE J. Ocean. Eng.* **2012**, *37*, 363–374. [CrossRef]
11. McCaffrey, K.; Fox-Kemper, B.; Hamlington, P.E.; Thomson, J. Characterization of turbulence anisotropy, coherence, and intermittency at a prospective tidal energy site: Observational data analysis. *Renew. Energy* **2015**, *76*, 441–453. [CrossRef]
12. Togneri, M.; Masters, I. Micrositing variability and mean flow scaling for marine turbulence in Ramsey Sound. *J. Ocean. Eng. Sci.* **2016**, *2*, 35–46. [CrossRef]
13. Hay, A.E.; McMillan, J.; Cheel, R.; Schillinger, D. Turbulence and Drag in a High Reynolds Number Tidal Passage Targetted for In-Stream Tidal Power. In Proceedings of the 2013 OCEANS-San Diego, San Diego, CA, USA, 23–27 September 2013; pp. 1–10.
14. Guerra, M.; Thomson, J. Turbulence Measurements from Five-Beam Acoustic Doppler Current Profilers. *J. Atmos. Ocean. Technol.* **2017**, *34*, 1267–1284. [CrossRef]
15. Smyth, C.; Hay, A.E.; Zedel, L. Coherent Doppler Profiler measurements of near–bed suspended sediment fluxes and the influence of bed forms. *J. Geophys. Res. Oceans* **2002**, *107*, 19-1–19-20. [CrossRef]

16. Filipot, J.-F.; Prevosto, M.; Maisondieu, C.; Le Boulluec, M.; Thomson, J. Wave and turbulence measurements at a tidal energy site. In Proceedings of the 2015 IEEE/OES Eleventh Current, Waves and Turbulence Measurement (CWTM), St Petersburg, FL, USA, 2–6 March 2015; pp. 1–9.
17. Newgard, J.P.; Hay, A.E. Turbulence intensity in the wave boundary layer and bottom friction under (mainly) flat bed conditions. *J. Geophys. Res. Oceans* **2007**, *112*. [CrossRef]
18. Zhu, Q.; van Prooijen, B.C.; Wang, Z.B.; Ma, Y.X.; Yang, S.L. Bed shear stress estimation on an open intertidal flat using in situ measurements. *Estuar. Coast. Shelf Sci.* **2016**, *182*, 190–201. [CrossRef]
19. Parkinson, S.G.; Collier, W.J. Model validation of hydrodynamic loads and performance of a full-scale tidal turbine using Tidal Bladed. *Int. J. Mar. Energy* **2016**, *16*, 279–297. [CrossRef]
20. Bricker, J.D.; Monismith, S.G. Spectral wave-turbulence decomposition. *J. Atmos. Ocean. Technol.* **2007**, *24*, 1479–1487. [CrossRef]
21. Bian, C.; Liu, Z.; Huang, Y.; Zhao, L.; Jiang, W. On Estimating Turbulent Reynolds Stress in Wavy Aquatic Environment. *J. Geophys. Res. Oceans* **2018**, *123*, 3060–3071. [CrossRef]
22. Foster, D. *Dynamics of the Nearshore Wave Bottom Boundary Layer*; ProQuest Dissertations Publishing: Washington, DC, USA, 1997.
23. Meirelles, S.; Henriquez, M.; Horner-Devine, A.R.; Souza, A.J.; Pietrzak, J.; Stive, M. Bed Shear Stress on the middle shoreface of the South-Holland Coast. In Proceedings of the Coastal Sediments 2015, San Diego, CA, USA, 11–15 May 2015.
24. Soulsby, R.L.; Humphrey, J.D. Field observations of wave-current interaction at the sea bed. In *Water and Wave Kinematics*; Springer: Dordrecht, The Netherlands, 1990; Volume 178, pp. 413–428.
25. Kolmogorov, A.N. The Local Structure of Turbulence in Incompressible Viscous Fluid for Very Large Reynolds' Numbers. *Dokl. Akad. Nauk* **1941**, *30*, 301–305. [CrossRef]
26. Stacey, M.; Monismith, S.; Burau, J. Measurements of Reynolds stress profiles in unstratified tidal flow. *J. Geophys. Res. Oceans* **1999**, *104*, 10933–10949. [CrossRef]
27. Lu, Y.; Lueck, R. Using a broadband ADCP in a tidal channel. Part II: Turbulence. *J. Atmos. Ocean. Technol.* **1999**, *16*, 1568–1579. [CrossRef]
28. Williams, J.J.; Bell, P.S.; Thorne, P.D. Field measurements of flow fields and sediment transport above mobile bed forms. *J. Geophys. Res. Oceans* **2003**, *108*. [CrossRef]
29. Daubechies, I.; Lu, J.; Wu, H.-T. Synchrosqueezed wavelet transforms: An empirical mode decomposition-like tool. *Appl. Comput. Harmonic Anal.* **2011**, *30*, 243–261. [CrossRef]
30. Penesis, I.; Hemer, M.; Cossu, R.; Hayward, J.; Nader, J.-R.; Rosebrock, U.; Grinham, A.; Sayeef, S.; Osman, P.; Marsh, P. Tidal energy in Australia—Assessing resource and feasibility to Australia's future energy mix. In Proceedings of the Asian Wave and Tidal Energy Conference, Taipei, Taiwan, 9–13 September 2018.
31. Marsh, P.; Penesis, I.; Nader, J.-R.; Couzi, C.; Cossu, R. Assessment of tidal current resources in Banks Strait, Australia. In Proceedings of the European Wave and Tidal Energy Conference, Naples, Italy, 1–6 September 2019.
32. Perez, L.; Cossu, R.; Penesis, I.; Grinham, A.; Nader, J.-R.; Couzi, C. A case study of high frequency AD2CP measurements for tidal site characterization in Banks Straits, Tasmania, Australia. In Proceedings of the European Wave and Tidal Energy Conference, Naples, Italy, 1–6 September 2019.
33. Rahimi, R.; Penesis, I.; Hemer, M.; Mason, L.; Thomas, G.J.R.E. Characterization of the tidal current resource in Tasmania. In Proceedings of the Asian Wave and Tidal Energy Conference, Tokyo, Japan, 28–30 July 2014.
34. Bureau of Meteorology (BOM). Climate data online—Swan Island (092123). Available online: http://www.bom.gov.au/climate/ (accessed on 20 April 2019).
35. Shcherbina, A.Y.; D'Asaro, E.A. Observing Finescale Oceanic Velocity Structure with an Autonomous Nortek Acoustic Doppler Current Profiler. *J. Atmos. Ocean. Technol.* **2018**, *35*, 411–427. [CrossRef]
36. Lu, Y.; Lueck, R.G. Using a broadband ADCP in a tidal channel. Part I: Mean flow and shear. *J. Atmos. Ocean. Technol.* **1999**, *16*, 1556–1567. [CrossRef]
37. Togneri, M.; Jones, D.; Neill, S.; Lewis, M.; Ward, S.; Piano, M.; Masters, I. Comparison of 4- and 5-beam acoustic Doppler current profiler configurations for measurement of turbulent kinetic energy. *Energy Procedia* **2017**, *125*, 260–267. [CrossRef]
38. Nezu, I.; Nakagawa, H. *Turbulence in Open-Channel Flows*; A. A. Balkema Publishers: Rotterdam, The Netherlands, 1993.

39. Greenwood, C.; Vogler, A.; Venugopal, V. On the Variation of Turbulence in a High-Velocity Tidal Channel. *Energies* **2019**, *12*, 672. [CrossRef]
40. Bouferrouk, A.; Hardwick, J.P.; Colucci, A.M.; Johanning, L. Quantifying turbulence from field measurements at a mixed low tidal energy site. *Renew. Energy* **2016**, *87*, 478–492. [CrossRef]
41. Thakur, G.; Brevdo, E.; Fučkar, N.S.; Wu, H.-T. The Synchrosqueezing algorithm for time-varying spectral analysis: Robustness properties and new paleoclimate applications. *Signal Process.* **2013**, *93*, 1079–1094. [CrossRef]
42. Milne, I.A.; Sharma, R.N.; Flay, R.G.J.; Bickerton, S. Characteristics of the turbulence in the flow at a tidal stream power site. *Philos. Trans. R. Soc. A* **2013**, *371*. [CrossRef] [PubMed]
43. Thiébaut, M.; Filipot, J.F.; Maisondieu, C.; Damblans, G.; Duarte, R.; Droniou, E.; Chaplain, N.; Guillou, S. A comprehensive assessment of turbulence at a tidal-stream energy site influenced by wind-generated ocean waves. *Energy* **2020**, *191*, 116550. [CrossRef]

energies

MDPI

Article

Parametric Study for an Oscillating Water Column Wave Energy Conversion System Installed on a Breakwater

Hsien Hua Lee * and Cheng-Han Chen

Department of Marine Environment and Engineering, National Sun Yat-Sen University, Kaohsiung 804, Taiwan; magic1990920@gmail.com

* Correspondence: hhlee@mail.nsysu.edu.tw

Received: 15 March 2020; Accepted: 11 April 2020; Published: 14 April 2020

Abstract: This study focuses on the analysis of the parameters of an oscillating water column (OWC) wave energy conversion system and wave conditions. Interactions between the dimensions of the OWC chambers and wave conditions are all taken into account to design an alternative OWC converter, called caisson-based OWC type wave energy converting system. A numerical method using an unsteady Navier-Stokes equations theorem in conservation form is used to analyze the proposed analytical model. The objective of this study is to try to apply an OWC wave energy converter to a caisson breakwater, which has been constructed in a harbor. The structure proposed in this study is a series of sets of independent systems, in which each set of converters is composed of three chambers to capture the wave energy, while better ensuring the safety of the caisson breakwater. Responses to be analyzed related to the conversion efficiency of the caisson-based OWC wave energy converting system include the airflow velocity from the air-chamber, the pneumatic power and the conversion efficiency in terms of a ratio between the pneumatic power and the energy of the incident waves. Parameters examined in this study include the dimensions of the OWC chamber features such as the orifice of the air-chamber allowing airflow in/output, the chamber length along the direction of incident waves, the size of the opening gate for incident waves and the submersion depth of the air-chamber. As found from the results, a best conversion efficiency from incident waves of 32% can be obtained for the extreme case where the orifice is very small, but for most other cases in the study, the best efficiency is about 15%.

Keywords: OWC; wave energy; wave power converting system; parametric study; caisson breakwater application

1. Introduction

Due to the intensive development of new high technology products and expanding consumer demand for products such as clothes, shoes and electronic devices industrial power supply demand is greater than ever. Besides the industry requirements, the power supply required for society to sustain a more comfortable lifestyle has also increased massively, particularly in newly developed countries. Taiwan also faces these facts, as the growth in the extension and number of industrial parks producing high technology products has forced the government to build more thermal power plants to fulfill the resulting power demand. As a result, other serious issues arise like massive air pollution and the need to treat combustion wastes. Those environmental problems not only bother the local people, affecting their health, but moreover, can eventually cause dramatic climate changes, as evidenced by many studies [1,2]. Therefore, alternative power sources that do not cause such a severe environmental impact are much sought after, such as power from renewable natural resources like solar energy, wind energy, ocean energy or other forms of non-fossil combustion energy.

Ocean energy has been studied for a long time. Among all kinds of ocean energy, wave energy is the most thoroughly studied form because it is widely distributed in oceans around the world and it is an abundant type of exploitable energy. Diverse wave energy conversion systems have been extensively studied, among which oscillating water column (OWC) wave energy conversion systems are the ones with relatively higher efficiency. Studies on OWC wave energy conversion systems are still very active. Some studies are on how to improve of the energy harvesting mechanism, for example by changing the shape of the traditional OWC system into a U-OWC [3,4] or a so called BBDB-OWC using a backward-bent duct buoy [5] or by applying a double-chamber to improve the energy harvesting ability in deep waters [6] or by bending the front wall of the chamber to study its influence on the energy conversion [7]. Research focused on the efficiency of turbine performance for outflow and inflow motions is also performed [8], while a recently released similar study [9] also included a turbine system, where an axial-flow impulse turbine was installed on an OWC model to replace the traditional one and a model was built and placed in a wave flume for experimental tests under regular wave conditions. Investigations, both experimental and numerical, on the wave-height and power take-off damping effect have also been carried out [10]. Some are focused on the performance of the air-chamber [11], where it was shown that the effect of neglecting the air compressibility in an experimental test model scaled down to 1/50 may result in an overestimation by up to 15% for the air pressure in the OWC chamber. This indicates that attention must be paid to any scaled-down model tested in an experimental laboratory, while a full-scale model test will be more accountable. A recent study goes even more deeply inside the wave theorem in considering the viscosity influence of wave properties on the OWC device [12]. By introducing artificial viscous terms into the dynamic free surface boundary conditions and the Bernoulli equation, the authors built a fully nonlinear numerical model based on a higher-order boundary element method (HOBEM) to model the wave dynamics of an OWC device. Application of an OWC wave energy conversion system to buoys for sensors has also been studied [13].

The parameters of an OWC wave energy conversion system include many aspects, such as the dimensional effects of the OWC structure, including both the size and shape of the structure, the associated turbine system and the environmental conditions of the location of the OWC system that may cover geology and wave conditions. As shown in a contemporary paper [14] describing a parametric study of a wave energy converter designed for the Caspian Sea, where a numerical simulation using the commercial software Flow-3D to model the motion of the energy converter for different wave heights was applied for output flow rates in different seasons, extractable wave power and output power, the parameters of an OWC conversion system are strongly related to location factors. Studies on the interactions for different wave periods, wave heights and pneumatic damping factors of OWCs were also performed lately [10,15]. An approach based on artificial neural networks (ANNs) was utilized to develop a virtual laboratory to determine the pneumatic efficiency of an OWC under specific conditions of wave height and period, tidal level and turbine damping [16] before a 1/25 scaled-down physical model similar in the geometry of the prototype designed for an OWC plant to be constructed at the port of A Guarda, Spain was tested in the laboratory. Some designs associated with a breakwater structure were proposed, such as the one proposed by Boccotti [17,18], where a new kind of OWC called U-OWC was considered. According to this report a U-OWC has some advantages for the waves of long periods such as swells or storm waves compared to conventional OWCs.

There seems to be a trend where more and more countries have considered building OWC-like wave energy conversions system on their shorelines, especially in local harbors, as mentioned previously [14,16,19], but construction of an OWC wave energy conversion system usually costs a tremendous amount of money. A thorough study to avoid any uncertainty and safety risks prior to the construction is usually required besides any functioning considerations. In order to reduce the construction cost for an OWC wave energy converting system, the combination of an OWC wave energy converting structure with a breakwater is a cost-reduction idea. However, real in-field experimental data for breakwater-combined OWC wave energy converting systems are still very rare. An OWC system combination with a breakwater was firstly studied in Japan [20,21]. In those advanced studies, not only were experimental tests performed, but also an on-site full-scale structure was built and tested. Many valuable data were acquired and some important conclusions were reached, including that the power efficiency that can be obtained from the conversion system, the estimated direct cost for a system of comparable size and scale, a range of dimensional ratios between the air chamber and the wavelength and the most important one, that the combination of a breakwater with a wave-power conversion system would not affect the function of the breakwater but rather reinforce it. Unfortunately, even though the study of Goda et al. [20,21] is invaluable and pioneering, it is too hard and too costly to be repeated for an on-field experiment even though a small field experiment was performed later on [18]. It is also too confined by many local factors to apply it to caisson breakwaters in other different locations.

The objective of this study is on the application of a conventional OWC wave energy conversion system to a caisson breakwater while upgrading the safety and function of the caisson-based system. However, any study trying to cover all of the parameters that may influence an OWC wave energy converting system would be a big task, if not impossible. In this study, by following the findings of Goda et al. and other contemporary studies, a series of studies were performed by adopting and upgrading a similar system, and thus constituting a caisson-based OWC type wave energy converter for green energy development. This study as a part of a series of studies that will focus on the energy conversion performance and where parameters related to OWC device dimensions and parameters for local wave conditions are both taken into consideration, while a previous study was focused on the structural safety when combined with additional OWC devices. A numerical method using an unsteady Navier-Stokes equations theorem in conservation form is used to analyze the proposed structural model, which has been verified [22] with the on-field experimental data presented by Goda [20,21].

The OWC wave energy converter proposed in this study is a series of sets of independent systems, in which each set of converters is composed of three chambers to capture the wave energy. This is because of the safety considerations for the caisson breakwater since a larger chamber will suffer a larger impact from wave forces. The vertical walls installed between chambers may provide additional support for the chamber and also protect the caisson-based structure. The analysis will focus on the conversion efficiency of the caisson-based OWC wave energy conversion system that includes the airflow velocity from the air-chamber, the pneumatic power and the conversion efficiency in terms of a ratio between the pneumatic power and the energy of the incident waves. Parameters to be examined in this study include the dimensions of the OWC chamber such as the orifice of the air-chamber allowing airflow in/output, the chamber length along the direction of incident waves, the size of the opening gate for incident waves and the submersion depth of the air-chamber. All of these parameters are presented in dimensionless form, while a range of wave-height and wave period variations are considered.

2. Analytical Model and Theorem for the Study

2.1. Analytical Model

The typical OWC wave energy conversion system proposed for this study is presented in a schematic drawing (Figure 1), where the air-chamber of the OWC is divided into three chambers by two inner walls, and where the air is allowed to flow through channels connected to each other near the top ceiling-slab while only one orifice is designed for the in/output air. This kind of design is deliberate as the intention is to build the OWC conversion system as part of a caisson breakwater structure and therefore the OWC chamber must be strong enough to sustain not only regular incident wave but waves from extreme 50 year return period storms, otherwise it might jeopardize the function and even the safety of the breakwater structure. The walls installed to separate the chambers will be also act as reinforcements to support the ceiling slab member, which will suffer the air-pressure induced from the heaving wave motion and also help to sustain the impact of surge motion from the incident waves against the front curtain wall above the open gate of the chamber. Therefore, the application of OWC converting system to the traditional caisson breakwater structure will be in a series of sets of three-chamber OWC converters connected to each other along a breakwater structure.

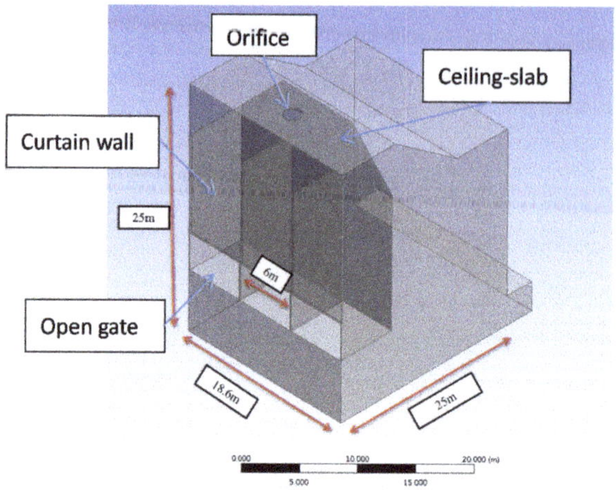

Figure 1. Schematic drawing for a caisson based OWC system.

The basic dimensions of the OWC system are also presented in Figure 1, where based on a local environment such as a water depth of 15 m and wave conditions, a basic design is realized for the dimensions of one set of 3-chamber OWCs as follows: 25 m high, $3 \times 6 = 18$ m wide and 25 m long.

2.2. Theorem of Fluid Mechanics Applied in the Study

The theorem applied in this study includes two parts. The first part is for the calculation of the fluid and wave motions and their influence on the airflow in the OWC chamber such as the velocity of the airflow through the opening orifice. The second part is about the estimation of the efficiency of the wave-energy power conversion by the OWC system to pneumatic power that can drive a turbine generator installed on the OWC system.

2.2.1. Basic Theorems for Fluid

In this study an unsteady Navier-Stokes equations [23,24] theorem in conservation form consisting of continuity equations, momentum equations and turbulence dynamics equations for a fluid with density ρ and velocity U are applied as follows:

Continuity equation:

$$\frac{\partial \rho}{\partial t} + \nabla \cdot (\rho U) = 0 \tag{1}$$

Equation of momentum:

$$\frac{\partial \rho U}{\partial t} + \nabla \cdot (\rho U \times U) - \nabla \cdot \left(u_{eff} \nabla U\right) = \nabla \cdot p' + \nabla \cdot \left(u_{eff} \nabla U\right)^T + B \tag{2}$$

where B is the sum of body force, μ_{eff} is the effective viscosity, $p\prime$ is the revised pressure. The effective viscosity and the revised pressure can be presented as:

$$\mu_{eff} = \mu_t + \mu \tag{3}$$

$$p' = p + \frac{2}{3}\rho k \tag{4}$$

It is also noticed that μ_t is the viscosity of the turbulence, which according to the assumption of k-ε model, is related to the dynamic energy and the dissipation of the dynamic energy as presented as:

$$\mu_t = C_\mu \rho \frac{k^2}{\varepsilon} \tag{5}$$

where k, are obtained directly from the equation of dynamic energy and equation of energy dissipation presented as follows:

$$\frac{\partial(\rho k)}{\partial t} + \nabla\cdot(\rho U k) = \nabla\cdot[\left(\mu + \frac{\mu_t}{\sigma_k}\right)\nabla k] + P_k - \rho\varepsilon \tag{6}$$

$$\frac{\partial(\rho\varepsilon)}{\partial t} + \nabla\cdot(\rho U \varepsilon) = \nabla\cdot\left[\left(\mu + \frac{\mu_t}{\sigma_\varepsilon}\right)\nabla\varepsilon\right] + \frac{\varepsilon}{k}(C_{\varepsilon 1}P_k - C_{\varepsilon 2}\rho\varepsilon) \tag{7}$$

where $C_{\varepsilon 1}$, $C_{\varepsilon 2}$, σ_k, σ_ε are constant parameters while P_k is related to viscosity and floating force and can be presented as:

$$P_k = \mu_t \nabla U\cdot\left(\nabla U + \nabla U^T\right) - \frac{2}{3}\nabla\cdot U(3\mu_t \nabla\cdot U + \rho k) + P_{kb} \tag{8}$$

2.2.2. Conversion Efficiency Estimation

The estimation of conversion efficiency is based on the ratio of the power generated by the airflow from the orifice of the OWC chamber and the power induced from the incident waves. The power of the incident waves will include both the potential and the kinetic energy. The gradient of potential for a differential wavelength dx can be presented as follows [25] and shown in Figure 2, where all coefficients for a propagating wave are also presented such as wave-height H, elevation of water surface η, water depth h and wavelength L:

$$d\Phi = \rho g \frac{(h+\eta)^2}{2} dx \tag{9}$$

Then the potential of one wavelength for a unit area A_w of water surface can be obtained [26] as

$$\Phi = \frac{1}{L}\int_x^{x+L} d\Phi = \frac{\rho g}{2\mathrm{L}}\int_x^{x+L} (h+\eta)^2 dx = \frac{\rho g}{2}\mathrm{h}^2 + \frac{\rho g}{16}H^2 \tag{10}$$

When the potential of waves up to still water level is reduced the potential for a unit wavelength of wave propagating with H wave-height is

$$\Phi_H = \Phi - \Phi_{H=0} = \frac{\rho g}{16}H^2 \tag{11}$$

The kinetic work K from a unit horizontal area of propagating wave can be obtained from a wave with velocity of horizontal component u and vertical component w as:

$$K = \int_x^{x+L}\int_{-h}^{\eta} \frac{\rho}{2}\left(u^2 + w^2\right) dxdz = \frac{\rho g}{16}H^2 \tag{12}$$

The total energy E from a unit horizontal area of an incident wave including both the potential and kinetic work will be as:

$$E = K + \Phi = \frac{\rho g}{8}H^2 \tag{13}$$

The kinetic work from the airflow with density ρ_a and velocity v_a through a cross-section area A_a with a wave-period T can be presented as:

$$\mathrm{E_a} = \frac{1}{2}m{v_a}^2 = \frac{1}{2}\rho_a A_a {v_a}^3\cdot T \tag{14}$$

Then the efficiency of the power converted from a wavelength of incident wave into power of an OWC in a form of airflow velocity can be presented as:

$$\frac{E_a}{E} = \frac{\frac{1}{2}\rho_a A_a v_a{}^3 \cdot T}{\frac{\rho g}{8} H^2 A_w} \tag{15}$$

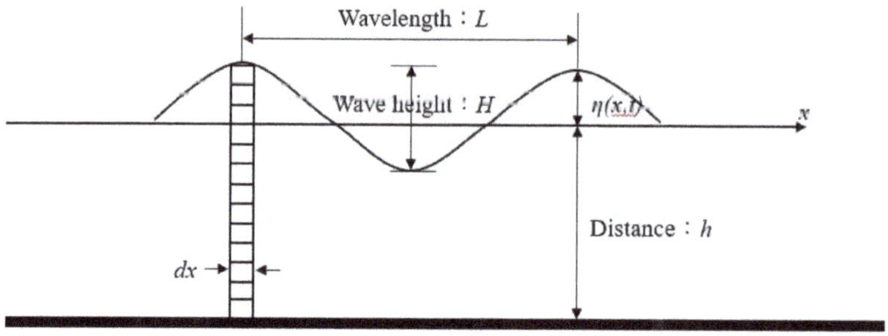

Figure 2. Schematic drawing showing the coefficients of a wave.

3. Dimensional Parameters Applied to the Study and Response Analysis

The parameters chosen in the analysis basically are related to the dimensions of the air-chamber for the OWC converter such as the area ratio between the opening of orifice (A) and the water surface (A_w) in the chamber presented as R_A = A/A_w, the ratio of length of air-chamber (B) in the direction of incident waves to the wave-length (L) presented as R_B = B/L, the ratio of the open-gate (O) in the front side facing the incident wave to the water depth (h) presented as R_O = O/h and the ratio of submerged depth (Z) of the gate of OWC converter to the water depth (h) presented as R_Z = Z/h under a condition that during the operation the whole device is submerged in the water. Figure 3 shows a schematic drawing of the side view of the caisson-based OWC converter. The dimensions of the device related to the parameters to be examined are all marked in the figure and shown in variables as A, B, O and Z. A listing of the variations of these parameters is presented in Table 1. It is noted that because too many parameters are to be analyzed when one parameter is a variable the other parameters will be set as constant. Such a referenced model is set as: R_A = 0.02 (A = 2.16 m^2), R_O = 25%, R_Z = 33% and R_B = a ratio of 6-m divided by wavelength applied for each case of analysis, which are highlighted in Table 1.

Table 1. Parameters utilized in the analysis.

Parameter	Code	Variables
Area-ratio of orifice	R_A	0.002, 0.004, 0.006, 0.008, 0.01, **0.02**, 0.04, 0.06, 0.08, 0.10
Length-ratio of chamber	R_B	2m, 4m, **6m**, 8m, 10m, 12m / wavelength
Opening-ratio of gate	R_O	10%, **25%**, 33%, 50%, 66%
Submerge-ratio of chamber	R_Z	0.20, **0.33**, 0.47, 0.60, 0.73

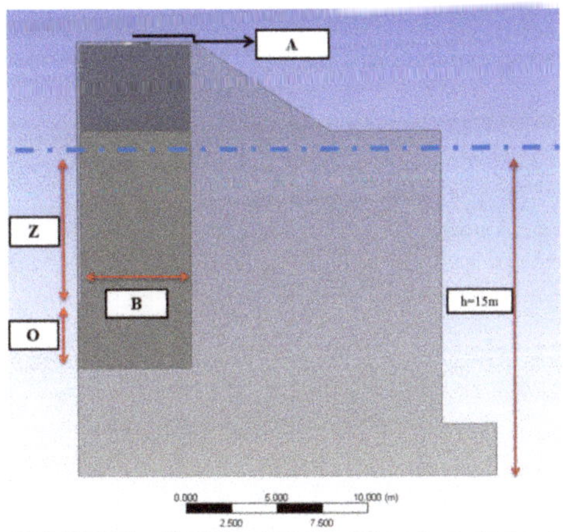

Figure 3. Parameters to be examined in the study.

Except for the variable dimensional parameters, some other variables are the input wave conditions including various wave heights and periods. The wave heights applied in this study are 1.0, 2.0 and 3.0 m while the wave periods are 6.0, 8.0 and 10.0 s. During the analysis, the responses of the OWC system for the converting efficiency to be discussed include the velocity of the airflow through the orifice, the power produced by the airflow and the conversion efficiency in terms of the ratio of pneumatic power and power of incident waves. Comparisons of these responses corresponding to the given wave conditions are presented and discussed in the following sections in terms of parameters like the area-ratio of orifice opening R_A, ratio of chamber-length to the wavelength R_B, opening ratio of OWC gate to the water depth Ro and submersion depth of the chamber gate to the water depth R_Z.

3.1. Responses Corresponding to Area-Ratio R_A of Orifice-Opening

The area-ratio of the orifice cross-section of OWC may have an influence on the performance of an OWC conversion system. Therefore, taken into consideration in this section is a dimensionless ratio of area of the OWC orifice cross-section to the area of water surface in the chamber as was indicated as R_A and listed in Table 1 for which the area of water surface is a constant while the orifice is variable and the ratio R_A is varied from 0.002 to 0.1.

- Velocity of airflow from the OWC

For the examination of the airflow velocity, three periods of the propagating wave corresponding to various wave-heights were applied. Presented in Figure 4 is the average of the first 1/3 maximum (or significant) airflow velocity for various area ratios of orifice-opening of the OWC converter to the area of water surface in the chamber, where Figure 4a–c show the responses for various input waves with periods of 6.0, 8.0 and 10.0 s, respectively. In each figure, three curves are shown representing the responses corresponding to wave heights of 1.0, 2.0 and 3.0 m as indicated in the curve descriptions.

The velocity presented in the figures is the average of the first 1/3 maximum velocity of each time-history analysis. It is observed that corresponding to the increase of the opening ratio R_A, the airflow velocity decreases nonlinearly. The airflow velocity drops very fast during the early stage corresponding to the area-increment of orifice opening. Basically after the opening ratio increases to 0.02, the airflow velocity is reduced to 50 m/s for almost all cases and it becomes smaller and smaller till the last case of $R_A = 0.1$.

A trend that a larger wave-height induces a larger velocity is found and this trend is more significant for an OWC converter with a smaller opening area-ratio. For 3 m wave height during the range where the opening area ratio is smaller than 0.01, the velocity is at a high level of more than 80 m/s and gets higher when the opening area ratio is smaller.

Generally, the OWC converter also has larger velocity when the period of the exerting wave is larger, but the variation becomes less and less significant when the opening area ratio increases. The average of the first 1/3 maximum velocity can reach as high as 143.0 m/s for the case with smallest area ratio as shown in Figure 4b, where the area ratio of the cross-section for the opening is 0.002 when subjected to a wave of 3.0 m wave height and 8.0 s period.

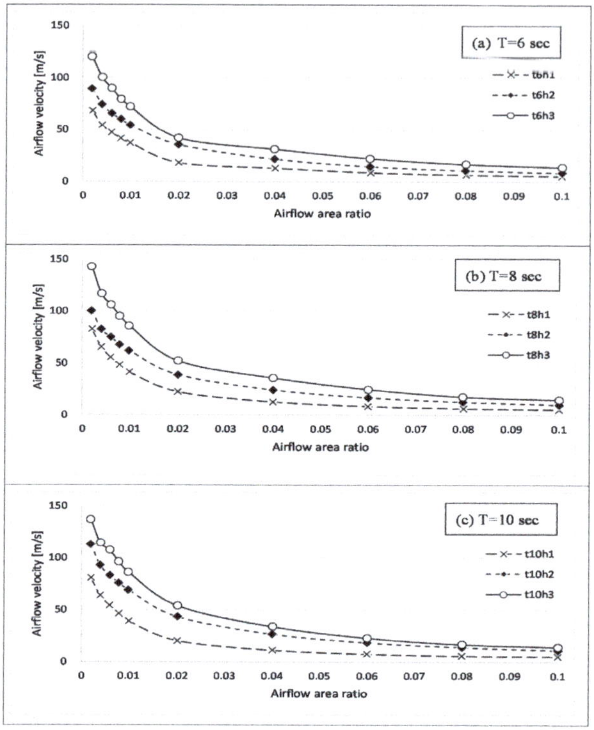

Figure 4. Airflow velocity corresponding to area ratio of orifice cross-section.

- Pneumatic Power from the OWC

For the examination of pneumatic power, three periods of the propagating wave corresponding to various wave-heights were applied. Presented in Figure 5 is the average of 1/3 maximum (or significant) pneumatic power for various OWC converter orifice area ratios, where Figure 5a–c show the responses for various input waves with periods of 6.0, 8.0 and 10.0 s, respectively. Similarly, in each figure, three curves are shown representing the response corresponding to wave heights of 1.0, 2.0 and 3.0 m as indicated in the curve descriptions.

It is observed that corresponding to the increase of ratio of the orifice opening the pneumatic power will increase till it reaches a peak value and then drops rapidly. The peak value occurs at a ratio of R_A = 0.006 for the cross-sectional area of the orifice to the area of water surface in the OWC chamber. Similar to the velocity response, the pneumatic power will decrease at a lower rate corresponding to the area-increment of orifice opening when the opening-ratio is larger than 0.02. Moreover, the pneumatic power is reduced to less than 200 kW for most cases subjected to waves of 3 m height and periods of T = 8 and 10 s, as presented in Figure 5b,c.

A trend whereby a larger wave-height induces a larger power is found and this trend is more significant for OWC converters with smaller opening ratios. The variation of pneumatic power between cases subjected to various wave heights is very significant, especially during the sensitive range of the opening ratio of the cross sectional area, which is between a value of 0.002 and 0.01 when the wave period is larger and the phenomenon is more significant for cases such as for T= 8.0 and T = 10 s as shown in Figure 5b,c.

Generally, the OWC converter also produces larger pneumatic power when the period of the waves is larger, such as for the case of 10.0 s of wave period, but the variation is not very significant between the case of 8.0 s and 10.0 s of wave period while the difference is more significant between 6.0 s and 8.0 s of wave period.

The maximum average pneumatic power is 480.8 kW, which occurs at the case where the OWC is subjected to an incident wave with 3 m wave-height and 10.0 s of wave period as shown in Figure 5c.

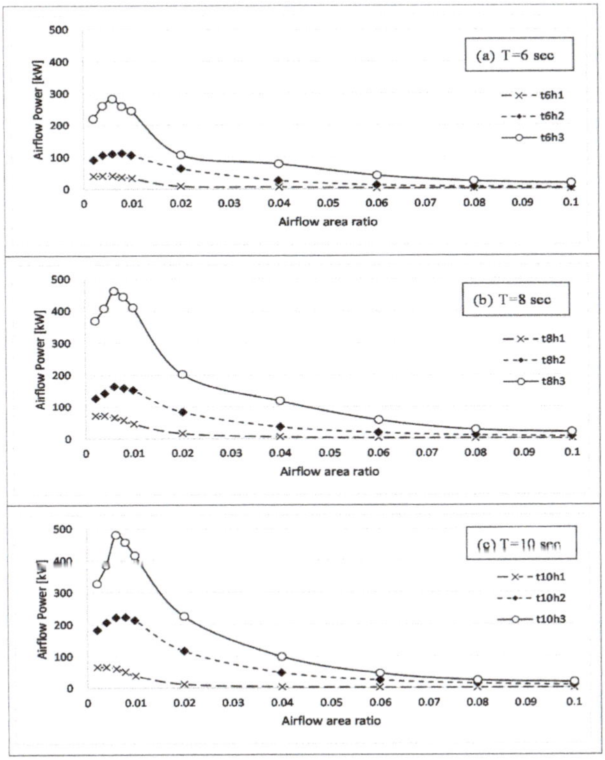

Figure 5. Pneumatic power corresponding to area ratio of orifice cross-section.

- Efficiency of power converted from the OWC

For the examination of efficiency of the power conversion from wave energy to pneumatic power that can drive a turbine generator of an OWC system, an estimation for the ratio between the energy induced by airflow and produced from incident waves is applied as shown in Equations (9)~(15). In this sub-section, similarly three periods of the propagating wave corresponding to three wave heights were applied. Presented in Figure 6 is the conversion efficiency based on the average of the 1/3 maximum airflow velocity for various area ratios of area ratio of the OWC converter, where Figure 6a–c show the responses for various incident waves with periods of 6.0, 8.0 and 10.0 s, respectively. In each figure, three curves are shown, representing the response corresponding to wave heights of 1.0, 2.0 and 3.0 m as indicated in the curve descriptions.

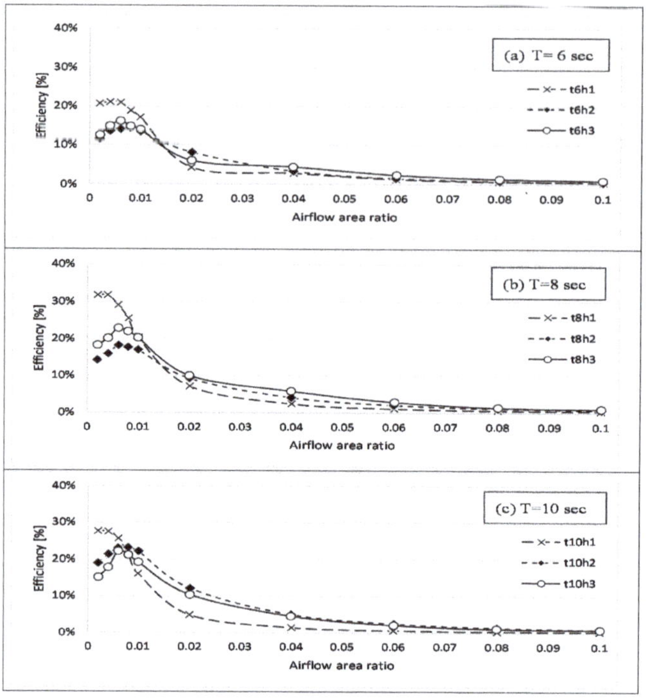

Figure 6. Efficiency corresponding to area ratio of the orifice cross-section.

It is observed that corresponding to the increase of the opening ratio, the conversion efficiency of the OWC decreases. The conversion efficiency stays at a high level first and then drops fast at the area ratio approaches 0.006. Basically, after the opening-ratio increases to 0.02, the airflow velocity is reduced to 10% and the dropping rate becomes less dramatic till the last case of 0.1 opening ratio. However, when the opening ratio is smaller than 0.01, the conversion efficiency is larger than 12% for most cases. When the opening-ratio is larger than 0.04 the conversion efficiency will be less than 5%.

It is also interesting to find that the trend that larger wave-height induces larger velocity or pneumatic power is not found in the conversion efficiency case. In the range of orifice area ratios of less than 0.01, the OWC subjected to a smaller wave height has a larger conversion efficiency for the power conversion. The power conversion efficiency is not positively related to the wave period either.

In some cases the conversion efficiency performance of the OWC system subjected to waves of 8.0 s, as shown in Figure 6b, performs better than the case subjected to 10.0 s of wave period, as shown in Figure 6c. The conversion efficiency can reach as high as 32% for the extreme case with the smallest area ratio, as shown in Figure 6b, where the cross section area ratio for the opening is 0.002 subjected to a wave of 1.0 m wave height and 8.0 s period.

3.2. *Responses Corresponding to the Ratio of Chamber-Length R_B*

During the response analysis of the parameter indicated by the ratio between the length of the air-chamber and the wavelength, represented by R_B, only the chamber-length is varied while the other dimensions of the air-chamber and the water depth (15 m) are set to be constant. Since the periods and wave heights are various during each case of the analysis and the wavelength will vary correspondingly, the chamber length ratio will also change accordingly for each combination of wave period and wave height.

- Velocity of airflow from the OWC

Presented in Figure 7 is the average of first 1/3 maximum (or significant) airflow velocity for various ratios of the chamber length of the OWC converter to the wavelength, where Figure 7 shows the responses for various incident waves with wave heights of 1.0, 2.0 and 3.0 m, respectively as indicated in the curve descriptions, when waves with periods of 6.0, 8.0 and 10.0 s are applied. It is observed that corresponding to the increase of the length ratio of the air chamber, the airflow velocity decreases with a linear variation trend. The airflow velocity also has a positive relationship with the wave height, when the applied wave height is larger the velocity of airflow also has a larger value.

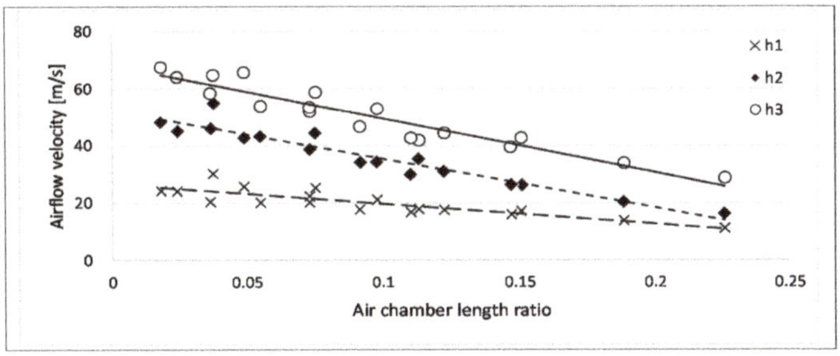

Figure 7. Airflow velocity corresponding to chamber-length ratio of OWC.

A trend that a larger wave height induces s larger velocity is found and this trend is more significant for the OWC converter with a smaller chamber-length ratio. Since the wave length of regular waves is quite large, that a smaller ratio of the chamber-length can produce a larger velocity of the airflow will imply that a small OWC converter may also display an effective performance. However, the influence of the wave period on the velocity of the airflow is not significant for the analysis of the length ratio of the air chamber.

The largest average velocity of the airflow from the OWC chamber in the study for the variation of the chamber-length is 75 m/sec for the case where the chamber-length ratio is 0.02 subjected to a wave of 3 m wave height and 10 s of wave period.

- Pneumatic power from the OWC

Presented in Figure 8 is the average power of airflow at 1/3 maximum velocity for various air chamber length ratios of the OWC converter, where Figure 8a,b show the responses for various incident waves with periods of 8.0 and 10.0 s, respectively. In each figure, similarly three curves are, representing the responses corresponding to wave heights of 1.0, 2.0 and 3.0 m as indicated in the curve descriptions.

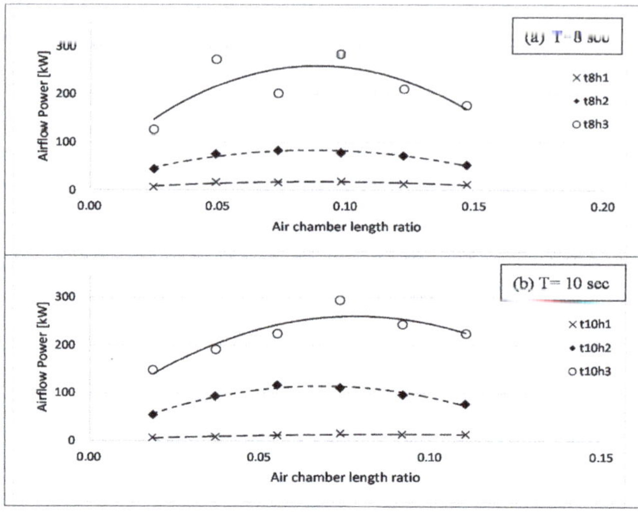

Figure 8. Pneumatic power corresponding to chamber-length ratio of OWC.

It is observed that corresponding to the increase of length ratios of the air chamber the pneumatic power will increase till it reaches a peak value and then it decreases again. The peak value occurs at a length ratio of 0.1 for the OWC converter subjected to waves of 8.0 s of wave period while in the case of 10.0 s of period, the peak occurs at a ratio of 0.075. In these two cases the applied wave height is 3 m and the power values are similar to each other. The largest power obtained in this case is 294 kW that occurs at a length ratio of 0.075 for the OWC converter subjected to a wave of 3 m height and 10.0 s of period.

Therefore, in the analysis of the parameter effect of length ratio to the power produced by the airflow of the OWC converter, the best length ratio will be located in a range of 0.05 to 0.1, where the power can reach a value over 200 kW.

- Efficiency of power conversion by the OWC

Presented in Figure 9 is the efficiency of the pneumatic power conversion by the OWC for various air chamber length ratios, where Figure 9a,b show the responses for incident waves with periods of 8.0 and 10.0 s, respectively. In each figure, similarly three curves are shown, representing the responses corresponding to wave heights of 1.0, 2.0 and 3.0 m as indicated.

It is observed that corresponding to the increase of air chamber length ratio the efficiency of power conversion will increase till it reaches a peak value and then it decreases again. The peak value occurs at a length-ratio of 0.1 for the OWC converter subjected to waves of 8.0 s of wave period while in the case of 10.0 s of period the peak occurs at a ratio of 0.075. For the case subjected to a wave of 10.0 s period, OWC under both 2 m and 3 m wave height conditions has a similar power conversion efficiency performance when the length ratio variation of the OWC converter R_B is less than 0.1. The largest power efficiency obtained in this case is 14.0%, which occurs at a length ratio of 0.1 for the OWC converter subjected to a wave of 3 m height and 8 s of period.

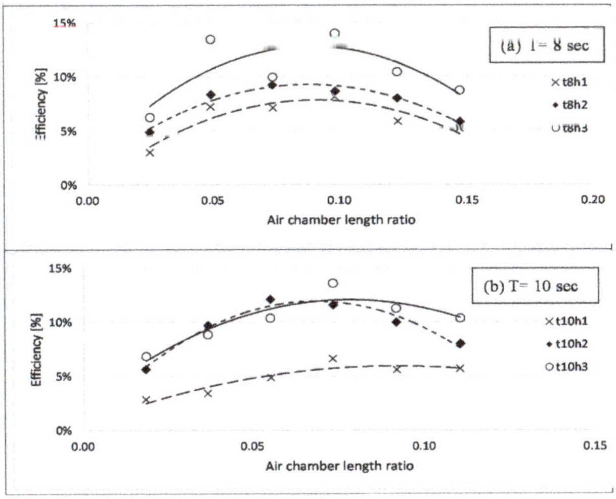

Figure 9. Efficiency corresponding to the chamber length ratio of the OWC.

3.3. *Responses Corresponding to the Opening-Ratio of the OWC Gate R_O*

The opening of the OWC gate on the front wall where the incident wave approaches may influence the conversion efficiency of the system. The opening of the OWC gate is set as a ratio of the gate opening to the water depth and indicated as R_O. The water depth is held at a constant value of 15 m while the opening ratio of the gate is a variable ranging from 10% to 66%, as presented in Table 1.

- Velocity of airflow from the OWC

For the examination of airflow velocity, three periods of the propagating wave corresponding to various wave heights were applied. Presented in Figure 10 is the average of the first 1/3 maximum airflow velocity for various gate opening ratios of the OWC converter to the water depth, where each curve in Figure 10 shows the responses for incident waves with wave heights of 1.0, 2.0 and 3.0 m, respectively as indicated in the descriptions. It is observed that corresponding to the increase of the opening ratio of the air chamber gate, the airflow velocity responds in various ways such as in the cases of wave heights of 1 and 3m, where the airflow velocity seems to not be very influenced by the opening ratio of the OWC gate, but for the case of 2 m wave height the velocity variation is more observable corresponding to the opening ratio of the OWC gate. However, corresponding to the increase of the wave height the power of the airflow velocity increases while the influence from the wave period is not significant.

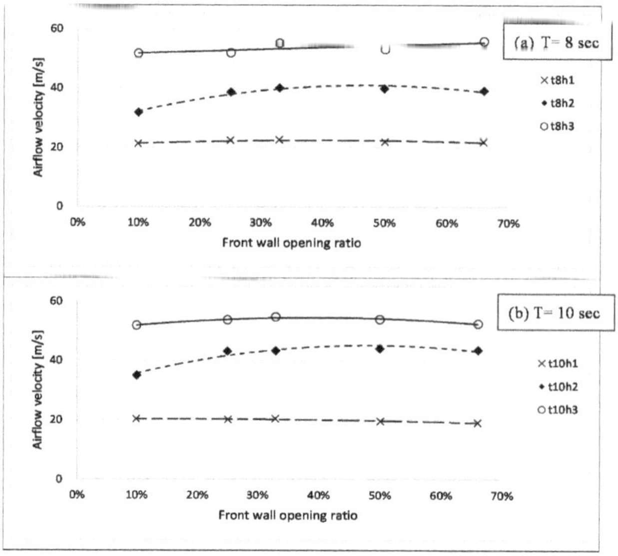

Figure 10. Airflow velocity corresponding to the opening ratio of the OWC gate.

- Power of airflow from the OWC

Presented in Figure 11 is the average power of the airflow at 1/3 maximum velocity for various OWC gate opening ratios to the water depth, where Figure 11a,b show the responses for various incident waves with periods of 8.0 s and 10.0 s, respectively. In each figure, three curves are shown, representing the responses corresponding to wave heights of 1.0, 2.0 and 3.0 m, as indicated.

It is observed that corresponding to the variation of OWC gate opening ratios the pneumatic power performance varies. When the wave height is small, like 1 m, it seems that there is no correspondence between the airflow power and the gate opening ratio, but when the wave height is 3 m a variation of velocity power occurs along with the variation of the opening ratio of the gate, but the variation is not consistent, as shown in Figure 11a where in the case of 8.0 s of wave period the pneumatic power has a linear increment corresponding to the opening ratio of the OWC gate while as shown in Figure 11b, for the case of 10 s of wave period, the power shows only a slight increase at a value of 35% for the opening ratio of the OWC gate and then decreases in a nonlinear way. For the case subjected to waves of 2 m wave height, the variation of velocity power will increase nonlinearly corresponding to the opening ratio of the OWC gate and then stay at a constant level or decrease slightly when the opening ratio of the gate becomes larger.

The largest power obtained from this case is 252 kW that occurs at an opening ratio of 66% for the OWC converter subjected to a wave of 3 m height and 8.0 s of period.

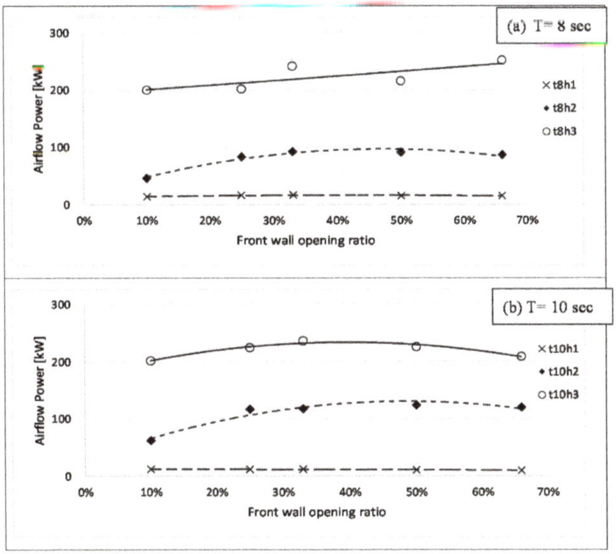

Figure 11. Pneumatic power corresponding to opening ratio of OWC gate.

- Efficiency of the power conversion of the OWC

Presented in Figure 12 is the efficiency of power of airflow converted by the OWC for various OWC gate opening ratios, where Figure 12a,b show the responses for various incident waves with periods of 8.0 s and 10.0 s, respectively. In each figure, three curves are shown, representing the responses corresponding to wave heights of 1.0, 2.0 and 3.0 m as indicated.

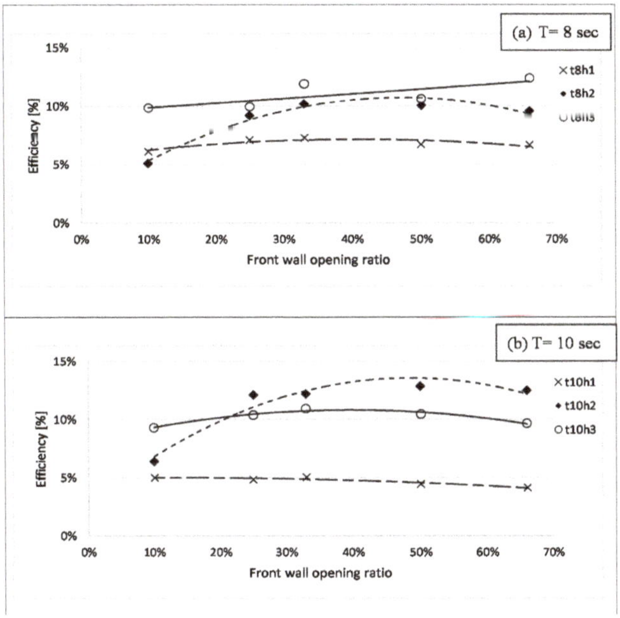

Figure 12. Efficiency corresponding to the opening ratio of the OWC gate.

It seems that there is no obvious trend between the power conversion efficiency and the opening ratio of the OWC gate. When the wave height is small, as in the case where the OWC is subjected to a wave of 1 m wave-height, the conversion efficiency will be more or less than 5% even when the wave has a large period like 10 s. For the cases subjected to waves of 2 m and 3 m wave height, the conversion efficiency will be larger than 10% for most cases for an applied wave of 8.0 s or 10.0 s. In the case of 10.0 s of wave period, the OWC converter subjected to a wave of 2 m wave height has a better conversion efficiency performance as shown in Figure 12b, where the largest efficiency obtained is 13%.

3.4. Responses Corresponding to the Submersion Depth Ratio of the Chamber Gate R_Z

The submersion depth of the OWC gate may have an influence on the performance of an OWC conversion system. Therefore, taken into consideration in this section is a dimensionless ratio of submersion depth of the OWC converter to the water depth, indicated as R_Z and listed in Table 1.

- Velocity of airflow from the OWC

Presented in Figure 13 is the average of the 1/3 maximum (or significant) velocity of airflow for various ratios R_Z of submersion depth of the OWC converter to the water depth, where Figure 13 shows the responses for various incident waves with wave heights of 1.0, 2.0 and 3.0 mm, respectively, as indicated in the description. It is found that corresponding to the increase of the submersion depth of the OWC converter, the airflow velocity generally increases or remains constant. It is clearly observed that the increase of velocity of airflow also has a positive relationship with the increase of wave height. However for the cases subjected to various wave periods the velocity of the airflow seems to not be affected much, as shown in both Figure 13a,b, where the values of the velocity corresponding to same wave height are almost same except that the curve displays some fluctuations. The maximum average velocity of the airflow is 60 m/s, occurring in cases subjected to waves of 3 m height and periods of 8.0 s and 10.0 s.

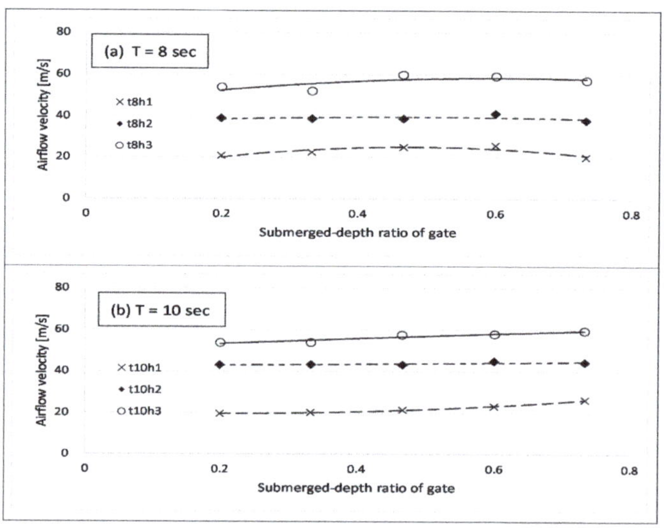

Figure 13. Airflow velocity corresponding to submersion depth ratio R_Z of the OWC gate.

- Power of airflow from the OWC

Presented in Figure 14 is the pneumatic power for various ratios of submersion depth of the OWC gate to the water depth, where Figure 14a,b show the responses for various incident waves with periods of 8.0 s and 10.0 s, respectively. In each figure, three curves are shown, representing the responses corresponding to wave heights of 1.0, 2.0 and 3.0 m, as indicated.

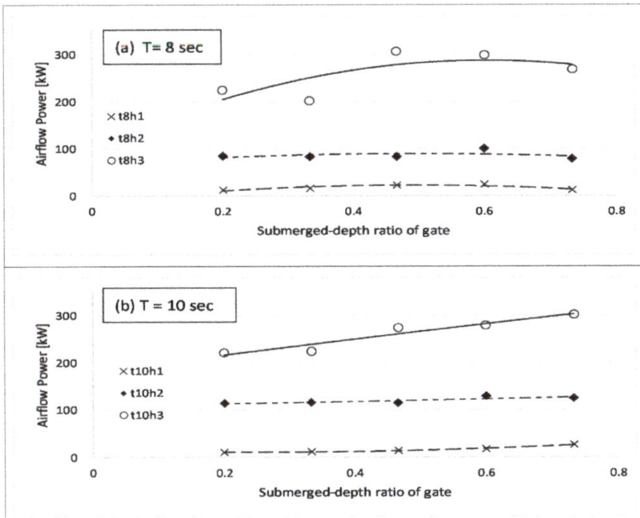

Figure 14. Pneumatic power corresponding to the submersion depth R_Z of the OWC gate.

It is observed that corresponding to the increase of submersion depth ratios of the OWC gate the pneumatic power presents a positive correspondence to the wave height. When the wave height is small like 1 m or 2 m it seems that there is no correspondence between the airflow power and the submersion depth ratio. However when the wave height is 3 m a variation of velocity power occurs along with the variation of the submersion depth ratio of the gate, but the variation is not consistent, as shown in Figure 14b in the case of 10.0 s of wave period where the pneumatic power has a linear increment corresponding to the submersion depth ratio of the OWC gate while as shown in Figure 14a. for the case of 8.0 s of wave period the increment is fluctuating more with the variation of the submersion depth ratio. The maximum average power can reach 300 kW for the OWC subjected to waves of 3 m wave height.

- Efficiency of the OWC power conversion

Presented in Figure 15 is the efficiency of the airflow power conversion by the OWC for various ratios of submersion depth of the OWC gate to the water depth, where Figure 15a,b show the responses for various incident waves with periods of 8.0 s and 10.0 s, respectively. In each figure, three curves are shown, representing the responses corresponding to wave heights of 1.0, 2.0 and 3.0 m, as indicated.

It is found that when the wave period is 8.0 s and the applied wave height is 3 m, the efficiency of the power conversion increases corresponding to the increase of the submersion ratio of the OWC gate. When the wave heights are 2 m or 1 m, the submersion depth seems to not have any influence on the conversion efficiency. However, when the period of the applied wave is 10.0 s, an obvious incremental trend corresponding to the submersion depth ratio is found for all wave heights. The efficiency performance for the 2 m wave height is similar to the performance for a 3 m wave height, where both increase linearly, and when the wave height is 1 m the correspondence between the conversion and the submersion depth seems to be stronger. The largest conversion efficiency is 15%, obtained in the case of an OWC subjected to a wave of 3 m height and 8.0 s of period when the ratio of submersion depth is 0.47.

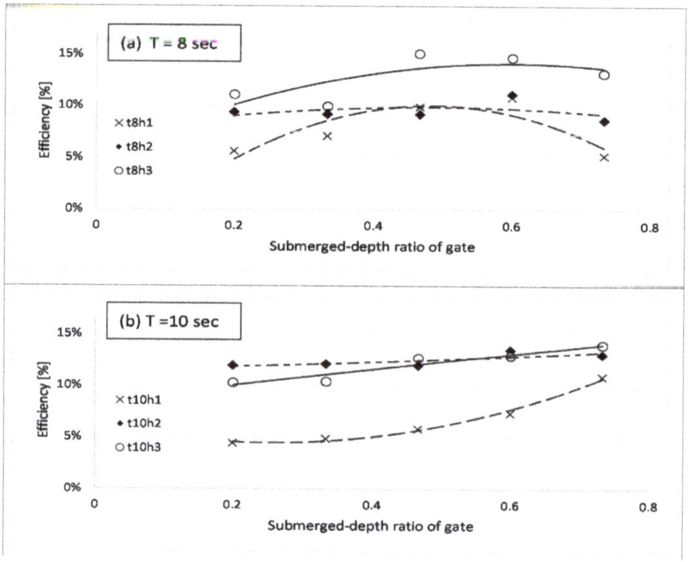

Figure 15. Efficiency corresponding to submersion depth of the OWC gate.

4. Discussion of the Analytical Results

During the parameter analysis based on the dimensional parameters and the corresponding responses, phenomena are observed which are discussed in this section.

4.1. On the Effect of Parameters

During the analysis even though the airflow velocity has a positive relationship corresponding to either the wave height or the wave period, the increment of opening ratio of the orifice will reduce the velocity, but for the response of pneumatic power and the conversion efficiency, there does not seem to be a definite positive relationship corresponding to either the wave height or the wave period. There is a near normal distribution of the pneumatic power in the lower area ratio of orifice opening range between 0.001 and 0.02. During the conversion efficiency analysis when the opening ratio R_A is less than 0.01, the analytical case with a small wave height of 1.0 m has a larger conversion efficiency. This is due to the fact a relatively small wave power is obtained from waves with small wave heights and this leads to a larger conversion efficiency.

For the study on the parameter of the ratio of chamber length R_B, the role of the wave period is not significant. This is also the reason why the responses for a period of 6 s are not shown here. As was shown in the airflow velocity analysis, no matter what the wave period was a linear relationship corresponding to the variation of the ratio of chamber length is presented. As for the pneumatic power and the conversion efficiency a near normal distribution was found corresponding the variation of the ratio of chamber length when the wave height is 3.0 m, but the distribution becomes insignificant when the wave height is 1.0 m. It is also noticed that when the period or height of the wave is small the chamber length ratio parameter does not play an important role in the wave energy conversion. It is because the parameters applied are rather small, about 1/3 compared to the other study [7].

For the study of the opening ratio of the chamber gate, it is clearly shown that when the wave height is small such as 1.0 m, the airflow velocity or pneumatic power or the conversion efficiency are not influenced by the parameter R_O. However, the performances of both the airflow velocity and the pneumatic power are significantly influenced by the wave height while they are not significantly affected by the wave period. The influence from the parameter R_O on the conversion efficiency will happen at a certain wave height such as H = 2.0 m while at other wave heights the influence is not significant.

The influence of the submersion depth of the chamber gate parameter seems to not be significant for the airflow velocity but it shows a positive correspondence for the pneumatic power response when the applied wave height is high, such as 3.0 m. For the power conversion efficiency, when the wave period is large like 10.0 s a positive correspondence is also found for the responses in at all wave heights. When the wave period is 8.0 s, the correspondence is variable for various wave heights and a quadratic nonlinear variation is found for H=1.0 m and 3.0 m while it is almost a constant for a wave height H = 2.0 m.

4.2. A Wave-Power-Based Recommendation for OWC Design

Shown in Figures 16–18 are the responses of air-flow velocity, pneumatic power and conversion efficiency of the OWC wave energy conversion system with specific dimension such as the ones listed in Table 1, referred to as the reference model corresponding to wave power. This is because neither thw wave height nor the wave period may totally represent the nature of a wave and the main objective of the conversion system is the energy that can be obtained. Therefore a design based on wave energy instead of the wave height or wave period might be more suitable for the design of an OWC wave energy conversion system. It is found that the air-flow velocity of the model shows a nonlinear increment trend corresponding to the increase of the wave power as shown in Figure 16, while the relationship between the generated pneumatic power and the wave power is linear as shown in Figure 17. Presented in Figure 18 is a quadratic nonlinear trend for the conversion efficiency corresponding to the wave power, where strong randomness also appears even though the trend of the correspondence to each other is generally positive.

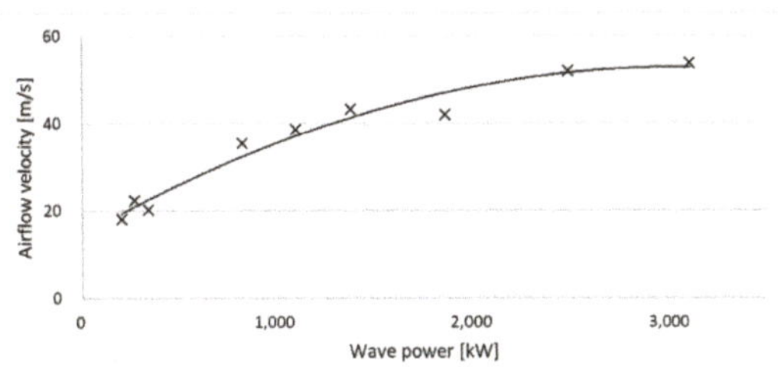

Figure 16. Airflow velocity corresponding to the energy of incident waves (single case).

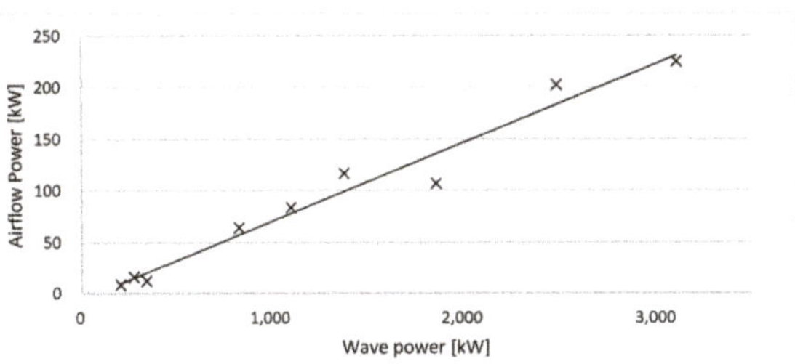

Figure 17. Pneumatic power corresponding to the energy of incident waves (single case).

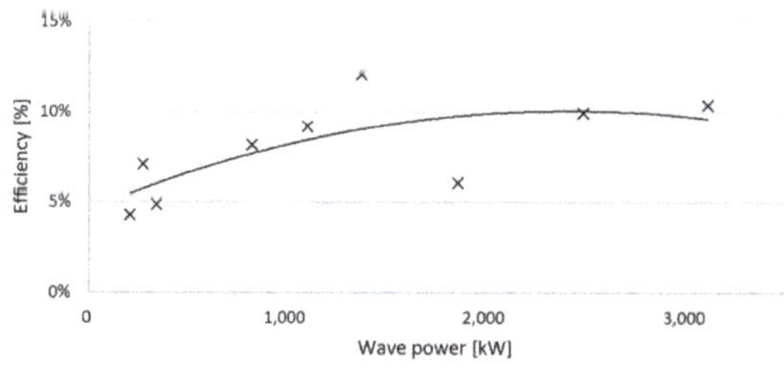

Figure 18. Converting Efficiency corresponding to the energy of incident waves (single case).

This is only for a model case of which every dimension is set and subjected to a combination of waves with three different wave heights and three different periods. It shows that a large amount of transferred pneumatic power does not necessarily represent a good conversion efficiency for the conversion system. Secondly, a strong randomness is still observed in the analysis for the conversion efficiency even though the applied waves are regular and some care was taken with the data such as the first 1/3 maximum responses were averaged as applied in conventional engineering application. This means that a case by case variation exists and analysis will be needed for any design, especially given the variety of available OWC wave energy conversion systems which have evolved tremendously lately.

If a cross examination is to be performed in this study then more than 1500 cases will be analyzed and compared (4500 figures similar to Figures 16 and 18 can be obtained). That will be a tremendous amount of work while the parameters taken into account in this study are only ones based on the local environmental conditions considered to be influential to the design of this specific OWC system. Therefore, what people always want to have, a universal formula for every kind of OWC will be very difficult if not impossible to obtain. Similar types of OWC wave energy conversion systems may have some guidelines for design purposes based on their geometrical dimensions and local wave conditions but some other uncertainties such as the sea bottom conditions, slopes in the way of incident waves and other environmental variables must be also considered during the design process.

5. Conclusions

In this study, a series of investigations of the dimension-related parameters were executed to examine the relationships between the investigated parameters and the wave energy conversion efficiency for an OWC wave energy conversion system. The OWC wave energy converter proposed in this study is a series of sets of independent systems, in which each set of converters is composed of three chambers to capture the wave energy. According to the theoretical and numerical study results some conclusions can be drawn based on the parameters related to the dimensions of each set of OWC conversion systems and discussed as follows:

- For the airflow velocity obtained through the orifice of the chamber, a value of 143 m/s can be found for the average of first 1/3 maximum velocities during the analysis of the orifice opening ratio to the water surface area in the chamber, where the opening ratio is 0.002 and the actual diameter of the orifice will be 26.22 cm, quite small and probably not practical to install an an effective turbine system even though the velocity is pretty high. However, for most cases a velocity of more or less than 50 m/s can be obtained, especially when the wave height applied to the OWC is as large as 3 m. The influence of orifice opening ratio and the applied wave height to the airflow velocity is significant. When the orifice opening ratio is larger than 0.02, increasing the airflow velocity by varying other dimensions in the chamber is very difficult. Therefore, it is recommended that for an optimum design of the OWC power conversion system the range of opening area ratios must be located in a range smaller than 0.02.

- For the pneumatic power study, the highest power that can be generated in this study was 180 kW, also corresponding to an extreme case of a very small orifice while for most cases the pneumatic power will be more or less than 100 kW. For a higher wave height, the pneumatic power will be larger and can reach values as high as 200 kW. The influence of the dimensional parameters on the pneumatic power are generally not significant when the wave height is smaller, but when the applied wave height is larger, such as the case of a 3 m wave height, the increase of the dimensions of the submersion depth of the gate opening has a positive influence on the pneumatic power.
- For the conversion efficiency, unlike the traditional concept that a larger airflow velocity may have better conversion efficiency, in some cases the conversion efficiency is actually better for the case in which the applied wave is smaller, especially for the case where the orifice is small and the velocity is extreme high. The influence of the dimensional parameters on the efficiency performance of the proposed OWC conversion system is also varied. A best efficiency of 32% conversion power from incident waves can be found for the extreme case, where the orifice is very small but for most cases in the study, where the ratio of the orifice to the water surface is set as 0.02, the best efficiency is about 15%. A value of more or less than 10% is obtained for most cases.
- Due to the random nature of waves even though a regular wave was applied in the study, the responses for a specifically designed OWC wave energy conversion system still show irregular uncertainties. It seems that to find a universal rule that can fit for every kind of OWC wave energy conversion system is not possible. Therefore, the dimensions of the OWC chamber are critical during the design stage because they are closely related to the environmental conditions such as the wave height and period of the regular waves. The most critical dimension is the ratio of the cross section of the orifice to the area of the water surface in the chamber. A smaller orifice can produce a larger airflow velocity, however, considering the local environmental conditions, especially the wave height and the cost, an OWC device designed for a smaller wave height but with higher conversion efficiency is possible.

Author Contributions: Conceptualization, H.H.L.; methodology, H.H.L.; software, C.-H.C.; validation, H.H.L.; formal analysis, C.-H.C.; investigation, H.H.L. and C.-H.C.; resources, H.H.L.; data curation, H.H.L.; writing—original draft preparation, H.H.L. and C.-H.C.; writing—review and editing, H.H.L.; visualization, H.H.L. and C.-H.C.; supervision, H.H.L.; project administration, H.H.L.; funding acquisition, H.H.L. All authors have read and agreed to the published version of the manuscript.

Funding: This research was funded in part by the Ministry of Science and Technology (MOST Taiwan under grant "MOST 105-221-E-110-047" and "The APC was funded by MOST Taiwan".

Acknowledgments: This research has been financially supported in part by the Ministry of Science and Technology (MOST Taiwan under grants "MOST 105-221-E-110-047". It is gratefully appreciated.

Conflicts of Interest: The authors declare no conflict of interest.

References

1. Ruddiman, W.F. *Earth's Climate: Past and Future*; W. H. Freeman and Company: New York, NY, USA, 2008; ISBN 9780716784906.
2. Rohli, R.V.; Vega, A.J. *Climatology*, 4th ed.; Jones & Bartlett Learning: Burlington, MA, USA, 2018; ISBN 9781284126563.
3. Boccotti, P. Comparison between a U-OWC and a conventional OWC. *Ocean Eng.* **2007**, *34*, 799–805. [CrossRef]
4. Spanos, P.D.; Strati, F.M.; Malara, G.; Arena, F. An approach for non-linear stochastic analysis of U-shaped OWC wave energy converters. *Probabilistic Eng. Mech.* **2018**, *54*, 44–52. [CrossRef]
5. Sheng, W. Power performance of BBDB OWC wave energy converter. *Renew. Energy* **2019**, *132*, 709–722. [CrossRef]
6. Ning, D.; Zhou, Y.; Zhang, C. Hydrodynamic modeling of a novel dual-chamber OWC wave energy converter. *Appl. Ocean Res.* **2018**, *78*, 180–191. [CrossRef]

7. Bouali, B.; Larbi, S. Contribution to the Geometry Optimization of an Oscillating Water Column Wave Energy Converter. *Energy Procedia* **2013**, *36*, 565–573. [CrossRef]
8. Ansarifard, N.; Fleming, A.; Henderson, A.; Kianejad, S.; Orphin, J. Comparison of inflow and outflow radial air turbines in vented and bidirectional OWC wave energy converters. *Energy* **2019**, *1821*, 159–176. [CrossRef]
9. Liu, Z.; Xu, C.; Qu, N.; Cui, Y.; Kim, K. Overall performance evaluation of a model-scale OWC wave energy converter. *Renew. Energy* **2020**, *149*, 1325–1338. [CrossRef]
10. Elhanafi, A.; Kim, C.J. Experimental and numerical investigation on wave height and power take–off damping effects on the hydrodynamic performance of an offshore–stationary OWC wave energy converter. *Renew. Energy* **2018**, *125*, 518–528. [CrossRef]
11. Simonetti, L.C.; Elsafti, H.; Oumeraci, H. Evaluation of air compressibility effects on the performance of fixed OWC wave energy converters using CFD modelling. *Renew. Energy* **2018**, *119*, 741–753. [CrossRef]
12. Wang, R.; Ning, D. Dynamic analysis of wave action on an OWC wave energy converter under the influence of viscosity. *Renew. Energy* **2020**, *150*, 578–588. [CrossRef]
13. Henriques, J.C.; Portillo, J.C.C.; Gato, L.M.C.; Gomes, R.P.F.; Ferreira, D.N.; Falcao, A. Design of oscillating-water-column wave energy converters with an application to self-powered sensor buoys. *Energy* **2016**, *112*, 852–867. [CrossRef]
14. Babajani, A.; Jafari, M.; Hafezisefat, P.; Mirhosseini, M.; Rosendahl, L. Parametric study of a wave energy converter (Searaser) for Caspian Sea. *Energy Procedia* **2018**, *147*, 334–342. [CrossRef]
15. Elhanafi, A. Prediction of regular wave loads on a fixed offshore oscillating water column-wave energy converter using CFD. *J. Ocean Eng. Sci.* **2016**, *1*, 268–283. [CrossRef]
16. Lopez, I.; Iglesias, G. Efficiency of OWC wave energy converters: A virtual laboratory. *Appl. Ocean Res.* **2014**, *44*, 63–70. [CrossRef]
17. Boccotti, P. Caisson breakwaters embodying an OWC with a small opening. Part I: Theory. *Ocean Eng.* **2007**, *34*, 806–820. [CrossRef]
18. Boccotti, P.; Filianoti, P.; Fiamma, V.; Arena, F. Caisson breakwaters embodying an OWC with a small opening. Part II: A small scale field experiment. *Ocean Eng.* **2007**, *34*, 820–830. [CrossRef]
19. Lee, H.H.; Chiu, Y.-F.; Lin, C.Y.; Chen, C.-H.; Huang, M.-H. Parametric study on a caisson based OWC wave energy converting system. *World J. Eng. Technol.* **2016**, *4*, 213–219. [CrossRef]
20. Goda, Y.; Nakada, H.; Ohneda, H.; Suzuki, M.; Takahashi, S.; Shikamori, M. Results of field experiment of a wave power extracting caisson breakwater. *Proc. Ocean Dev.* **1991**, *7*, 143–148. [CrossRef]
21. Goda, Y.; Shinda, T.; Chiyama, S.; Ohneda, H.; Suzuki, M.; Takahashi, S.; Shikamori, M.; Takaki, Y. Experiment of a wave power extracting caisson breakwater. *Proc. Ocean Dev.* **1989**, *5*, 1–6.
22. Chiu, Y.-F.; Lee, H.H.; Lin, C.-Y.; Chang, F.-T. Study on a caisson based OWC wave energy converting system. *World J. Eng. Technol.* **2014**. [CrossRef]
23. Milne-Thomson, L.M. *Theoretical Hydrodynamics*; The MacMillan Co.: New York, NY, USA, 1960.
24. Sarpkaya, T.; Isaacson, M. *Mechanics of Wave Forces on Offshore Structures*; Van Nostrand Reinhold Company: New York, NY, USA, 1981; ISBN 0-442-25402-2.
25. Schlichting, H. *Boundary-Layer Theory*; McGraw-Hill: New York, NY, USA, 1968.
26. Longuet-Higgins, M.S. Integral properties of periodic gravity waves of finite amplitude. *Proc. R. Soc. Ser. A* **1975**, *342*, 157–174.

energies

MDPI

Article

Optimal Dimensions of a Semisubmersible Floating Platform for a 10 MW Wind Turbine

Giulio Ferri *, Enzo Marino and Claudio Borri

Department of Civil and Environmental Engineering, University of Florence, Via di S. Marta 3, 50139 Firenze, Italy; enzo.marino@unifi.it (E.M.); claudio.borri@unifi.it (C.B.)

* Correspondence: giulio.ferri@unifi.it

Received: 27 April 2020; Accepted: 10 June 2020; Published: 15 June 2020

Abstract: In this paper, an optimal semisubmersible platform is sought considering two key geometry variables: the diameter of the outer cylinders and their radial distance from the platform centre. The goal is to identify a platform configuration able to most efficiently contrast the combined wind-wave action, keeping the platform dimensions as small as possible. The amplitude of the Response Amplitude Operator (RAO) peaks and the integral area of the RAOs in a range of excited frequencies for the selected degrees of freedom are chosen as targets to be minimised. Through an efficient frequency domain simulation approach, we show that upscaling techniques proposed in the literature may lead to overdesigned platforms and that smaller and more performing platforms can be identified. In particular, the optimised platform shows a reduction of about 51% in parked and 54% in power production of the heave RAO peak, and a reduction of about 37% in parked and 50% in power production of the pitch RAO.

Keywords: floating offshore wind turbines; frequency domain model; semisubmersible platform; 10 MW wind turbines; large floating platform; platform optimization

---

1. Introduction

Floating offshore wind turbines (FOWTs) are large and complex structural systems exposed to a variety of simultaneous environmental actions. A reliable prediction of the dynamic behaviour of a FOWT, and the associated loads, requires a proper modelling of the multiple fluid–structure interaction problems involved. Moreover, the design of very large wind turbines (WTs) (10 MW and more), with significantly taller tower and slender rotor blades compared to the average sizes of the currently installed machines, calls for new optimal design techniques of the supporting platforms. Past studies addressing medium-size WT platforms (3–5 MW) focused on different platform stability classes. In [1], Wayman and Sclavounos investigated the stability and the response of a barge-like platform and a Tension Leg Platform (TLP). They provided useful information to assess static stability and developed a simplified frequency domain (FD) model to evaluate the interaction between the wind turbine and the floating system. Robertson and Jonkman [2] compared results from different 5 MW FOWT concepts with a land-based turbine in terms of both ultimate and fatigue loads. They found that for barge-like platforms, the land-to-sea load amplification was unsustainable. Key aspects, such as fatigue loads, ringing phenomena and wind-wave misalignment for different kind of FOWTs have been investigated in [3–5], respectively. Nonlinear wave kinematics and higher-order effects have been studied in [6–9]. Effects that the interaction between the platform oscillations and the turbulent aerodynamic loads may have on the performance of a WT are analysed in [6,10] and [11].

Compared to on-land installations, the design of FOWTs involves a larger number of design variables, which require design optimisation procedures capable of considering the coupled hydro-aero-servo-elastic behaviour of the whole system. Sclavounos et al. [12] optimised different FOWT concepts by using a simplified FD model. Brommundt et al. [13] performed a design optimisation of mooring lines for a 5 MW FOWT under two targeted environmental conditions. Karimi et al. [14] performed an FD multi-objective design optimization of a 5 MW WT supported by a semisubmersible platform and a TLP considering both cost and structural performance as objective functions.

With increasing rotor diameters, above 200 m and 10 MW rated power, rational upscaling techniques [15] based on power ratios have been proposed to directly design platforms for large turbines. Liu et al. [16] evaluated the short-term and long-term dynamic response of an upscaled 13.2 MW wind turbine platform by means of time-domain stochastic simulations. New mooring lines configurations, suitable for such large FOWTs, were investigated in [17], considering wind-wave misalignment. In [18], Hsu et al. incorporated the effects of snap loads in the mooring lines dynamic tension probability distribution. They found that neglecting snap effects could lead to an underestimation of the maximum tension. Additionally, new typologies of platform shapes, such as the tri-spar floater, were studied [19]. Experimental studies of different platforms were also conducted in [20,21] for medium-size wind turbines and in [22] for a 10 MW WT.

Although several critical aspects for very large FOWTs have been addressed to date, the design of optimal platforms supporting 10 MW + WTs is still an open problem. The question we attempt to answer in this work is whether, after a rational upscaling process (see e.g., [15,16]), there is still room for improving the platform dimensions. To answer this question, in the present work, an optimal semisubmersible platform is sought, considering two geometry variables, namely the diameter of the outer cylinders and their radial distance from the platform centre. The goal is to identify a platform configuration able to most efficiently contrast the combined wind-wave action, keeping the platform dimensions as small as possible. Our assumption is that the upscaling techniques may lead to overdesigned platforms, resulting in unnecessary material (and cost) for the supporting system. Since the present paper is the first contribution to a wider project aimed at providing guidance for the optimal design of large semisubmersible platforms involving many design variables, a trade-off between accuracy and computational efficiency is a fundamental requirement. For this reason, we perform simulations in the FD. We consider both radiation/diffraction forces as well as viscous forces from the Morison equation [23]. Mooring lines and wind turbine contributions to the equation of motion are computed as additional mass, damping and stiffness matrices resulting from a linearisation analysis performed with FAST [24] around operational points.

To verify our computational model, first, the OC4 5 MW National Renewable Energy Laboratory (NREL) semisubmersible floating wind turbine [25–27] is studied and the results are compared with the results available in [26]. Then, a parametric study of a semisubmersible platform suitable for a 10 MW wind turbine is performed. The amplitude of the RAOs peaks and the area of the RAOs over the excited band of frequencies for the heave and pitch degrees of freedom (DoFs) are chosen as targets to be minimised. The first gives information directly related to the eigenfrequencies, while the latter provides information on the dynamic behaviour of the platform outside the natural frequencies. The results are discussed and compared with an upscaled 10 MW wind turbine model [28].

2. Frequency Domain Model

In this study, coupled hydro-aero-servo-elastic simulations are performed with FAST [24], an open source code developed at the National Renewable Energy Laboratory (NREL). FAST is capable of performing time domain simulations of the dynamic response of a WT in a stochastic wind and wave environment by coupling aerodynamic, hydrodynamic, control and structural models. In FAST, flexible elements, such as rotor blades and towers, are modelled as Bernoulli–Euler beams; the aerodynamic model is based on the Blade Element Momentum theory [29]; the hydrodynamic model permits considering both first and second order hydrodynamic loads [30], evaluated by external codes (e.g., WAMIT or ANSYS AQWA), as well as Morison forces for slender elements.

FD models neglect or approximate non-linear effects and are restricted to steady load conditions. Several important nonlinear and transient effects on fixed-bottom WTs have been observed in [28–34].

For floating platforms, slow-drift and mean-drift force [7], third-order hydrodynamic contributions [8], and fully nonlinear wave kinematics [9] have been recently investigated, and it was found that nonlinearities significantly affect fatigue damage, as well as mooring line tension at high sea states. Nonlinear effects are important for a safe platform design and must be considered in the actual design process. Uzunoglu and Guedes Soares [35] compared the results obtained by OC5 Project Phase II [27] and demonstrated that the difference in the hydrodynamic modelling has a significant effect on the results. Nevertheless, FD-based approaches, thanks to their efficiency, may offer a guidance in the pre-screening phase of an optimisation-based design. Most of the proposed FD approaches superimpose the turbine contribution in terms of mass, damping and stiffness matrices to the hydrodynamic contribution due to the floating platform. Philippe et al. [36] coupled an in-house radiation/diffraction potential solver with FAST. Wang et al. [37] proposed an alternative FD model which adopts ANSYS AQWA [38] to evaluate hydrodynamics contributions and simplified the formulas to consider the rotor thrust forces and the gyroscopic effects.

The model used in this paper is similar to the one proposed in [36] coupled with ANSYS AQWA as in [37], but incorporates approximated nonlinear effects due to quadratic drag and heave plate damping [14]. The wind turbine and its aerodynamics, as well as the mooring lines, are modelled as mass, damping and stiffness matrices. They are evaluated trough a linearisation analysis in FAST_v7 around the operating point. The dynamics of the system are linearised, neglecting phase shifting. The periodic response amplitude at each wave frequency is calculated and then superimposed.

2.1. Hydrodynamic Model

For the sake of completeness, in this section, we briefly summarise the standard theories [39,40] we used to model the fluid–structure interaction problem.

The platform hydrodynamic model requires a potential flow solver to calculate the frequency-dependent added mass, radiation damping and diffraction forces; the first and the second do not depend on the wave heading direction and amplitude, while the third does. Assuming an inviscid and incompressible fluid, a velocity potential $\Phi(x, y, z, t)$ exists. Under the hypothesis of small displacements and rotations, the body boundary condition and the free surface conditions can be linearised, isolating the time dependency of the potential, as follows:

$$-\omega^2\varphi + g\frac{\partial\varphi}{\partial z} = 0, \tag{1}$$

where ω is the circular frequency, g is the gravity acceleration, z is the vertical coordinates and φ is the time-independent part of the velocity potential $\Phi(x, y, z, t)$. Considering a steady-state condition, the velocity potential can be rewritten, isolating the time dependence and splitting into three contributions:

$$\Phi(x, y, z, t) = \varphi(x, y, z)e^{-i\omega t} = (\varphi_D + \varphi_R)e^{-i\omega t}, \tag{2}$$

where φ_D is the diffraction potential, which collects the incident wave and scattering potential, and φ_R is the radiation potential. The dynamic pressure p at the wetted body surface is then calculated by means of the linearised Bernoulli equation:

$$p(x, y, z, t) = -\rho\frac{\partial\Phi(x, y, z, t)}{\partial t}, \tag{3}$$

where ρ is the water density. Integrating the dynamic pressure on the wetted surface, the added mass (Equation (4)), radiation damping (Equation (5)), and diffraction forces (Equation (6)) are evaluated, respectively, as follows:

$$A_{jk}(\omega) = \mathrm{Re}\left\{\iint_{S_B} p_R \cdot n dS\right\} = -\rho\mathrm{Re}\left\{\iint_{S_B} \varphi_{jR} n_k dS\right\}, \tag{4}$$

$$B_{jk}(\omega) = \mathrm{Im}\left\{\iint_{S_B} p_R n dS\right\} = \rho\omega \mathrm{Im}\left\{\iint_{S_B} \varphi_{jR} n_k dS\right\}, \quad (5)$$

$$X_j(\omega, \beta) = \iint_{S_B} p_D n dS = -\rho\omega \iint_{S_B} \varphi_D n_j dS. \quad (6)$$

These three contributions are dominant on bluff bodies, such as barge-like elements or cylinders where $\lambda/D < 5$, λ being the wavelength and D, the cylinder diameter. When dealing with slender elements, such as cross braces of a semisubmersible platform, inertia and viscous drag forces due to flow separation become dominant. On such elements, the wave-induced force per unit-length can be expressed by Morison equation as

$$\mathbf{dF}^{\mathbf{Morison}} = \frac{\pi}{4}\rho C_M D^2 \dot{\mathbf{u}} - \frac{\pi}{4}\rho C_A D^2 \ddot{\mathbf{q}} + \frac{1}{2}\rho C_D (\mathbf{u} - \dot{\mathbf{q}})|\mathbf{u} - \dot{\mathbf{q}}|, \quad (7)$$

where $\mathbf{u}$ is the wave velocity component normal to the member axis, $\dot{\mathbf{q}}$, $\ddot{\mathbf{q}}$ are the cylinder velocity and acceleration, C_M is the inertial coefficient, C_A is the added-mass mass coefficient and C_D is the viscous drag coefficient. The last two terms of the right-hand side of Equation (7) are expressed as a function of the six DoFs of the platform regarded as a rigid body. The viscous drag term being quadratic requires a linearization. The Borgman linearisation [41] is herein adopted, namely, the nonlinear drag term is approximated with an equivalent linear term written as

$$\mathbf{dF}_{\mathbf{Drag}} = \frac{1}{2}\rho C_D (\mathbf{u} - \dot{\mathbf{q}})|\mathbf{u} - \dot{\mathbf{q}}| \cong \frac{1}{2}\rho C_D \sqrt{\frac{8}{\pi}}\sigma_u (\mathbf{u} - \dot{\mathbf{q}}), \quad (8)$$

where σ_u is the standard deviation of the relative velocity. As a first approximation, in this study, σ_u is replaced by the standard deviation of the exciting wave. Denoting with $\boldsymbol{\eta}$ the vector of the six rigid platform DoFs $\boldsymbol{\eta} = \begin{bmatrix} \eta_1 & \eta_2 & \eta_3 & \eta_4 & \eta_5 & \eta_6 \end{bmatrix}$—referring to surge, sway, heave, roll, pitch and yaw, respectively—the linearised Morison force can be rewritten as a function of $\boldsymbol{\eta}$ as follows:

$$\begin{aligned} \mathbf{dF}^{\mathbf{Morison}} &= \tfrac{\pi}{4}\rho C_M D^2 \dot{\mathbf{u}} - \tfrac{\pi}{4}\rho C_A D^2 \ddot{\mathbf{q}} + \tfrac{1}{2}\rho C_D \sqrt{\tfrac{8}{\pi}}\sigma_u (\mathbf{u} - \dot{\mathbf{q}}) = \\ &= \tfrac{\pi}{4}\rho C_M D^2 \dot{\mathbf{u}} - \tfrac{\pi}{4}\rho C_A D^2 \mathbf{T}\ddot{\boldsymbol{\eta}} + \tfrac{1}{2}\rho C_D \sqrt{\tfrac{8}{\pi}}\sigma_u \mathbf{u} - \tfrac{1}{2}\rho C_D \sqrt{\tfrac{8}{\pi}}\sigma_u \mathbf{T}\dot{\boldsymbol{\eta}}, \end{aligned} \quad (9)$$

where $\mathbf{T}$ is a transport matrix, which permits expressing the velocity of the members as a function of the six platform DoFs.

The hydrodynamic loads on the heave plates are described with a modified Morison equation [25], able to take into account inertial, viscous and dynamic effects:

$$dF_z^{Morison} = \rho C_{Az} V_R D^2 (\dot{w} - \ddot{q}_z) + \frac{1}{2}\rho C_{Dz} \frac{\pi}{4} D^2 (w - \dot{q}_3)|w - \dot{q}_3| + \frac{\pi}{4} D_h^2 p_b - \frac{\pi}{4}(D_h^2 - D_c^2) p_t, \quad (10)$$

where $C_{Az} = 1$ is the added mass coefficient in the heave direction; C_{Dz} is the drag coefficient in the heave direction, equal to 4.8 according to [25]; w is the vertical component of the wave particle velocity; $\dot{q}_3$ is the heave velocity of the heave plates; D_h is the diameter of the heave plates; D_c is the diameter of the upper column; and p_b and p_t are respectively the dynamic pressures, projected in the direction of the normal vectors, at the bottom and at the top of the heave plates. V_R is the reference volume, i.e., the volume of fluid that is displaced by the oscillations of the heave plates, calculated as

$$V_R = \frac{A_{33}(0)}{3\rho}, \quad (11)$$

where $A_{33}(0)$ is the zero-frequency added mass coefficient in the heave direction. The viscous drag term is linearised according to [42] as follows:

$$\mathbf{dF}_{\mathbf{Drag,z}}^{\mathbf{Heaveplates}} \cong \frac{1}{3}\rho C_{Dz} D^2 \omega a (\mathbf{w} - \dot{\mathbf{q}}_3), \quad (12)$$

where ω is the wave frequency and a is the heave plate oscillation amplitude. By expressing the velocity and the acceleration of the heave plates as a function of the six platform DoFs, the contributions of the heave plates to the mass and damping matrices are calculated.

2.2. Equation of Motion

The equation of motion of the FOWT is

$$\begin{gathered}(\mathbf{A}(\omega)+\mathbf{M}_{\mathbf{Platform}}+\mathbf{M}_{\mathbf{Turbine}}+\mathbf{A}_{\mathbf{Morison}}(\omega,\beta))\ddot{\eta}(t)+(\mathbf{B}(\omega)+\mathbf{B}_{\mathbf{Turbine}}+\\ \mathbf{B}_{\mathbf{Morison}}(\omega,\beta,a))\dot{\eta}(t)+\left(\mathbf{C}_{\mathbf{Hydrostatic}}+\mathbf{C}_{\mathbf{Mooring}}+\mathbf{C}_{\mathbf{Turbine}}\right)\eta(t)=\\ \left(\hat{\mathbf{F}}_{\mathbf{D}}(\omega,\beta)+\hat{\mathbf{F}}_{\mathbf{Morison}}(\omega,\beta)\right)e^{i\omega t},\end{gathered} \tag{13}$$

$$\eta(t)=\hat{\eta}(\omega,\beta)e^{i\omega t}, \tag{14}$$

where $\hat{\eta}(\omega,\beta)$ is the complex amplitude displacement vector of the platform. $\mathbf{A}(\omega)$ is the added mass matrix, $\mathbf{B}(\omega)$ is the radiation damping matrix, $\mathbf{C}_{\mathbf{Hydrostatic}}$ is the hydrostatic stiffness matrix, $\mathbf{C}_{\mathbf{mooring}}$ is the linearised mooring lines matrix, $\mathbf{F}_{\mathbf{D}}$ is the diffraction force acting on bluff cylinders and $\mathbf{F}_{\mathbf{Morison}}$ is the Morison force acting on slender cylinders. Considering Equation (14), the equation of motion can be rewritten as

$$\left(-\omega^2\mathbf{M}_{\mathbf{TOT}}(\omega,\beta)+i\cdot\omega\cdot\mathbf{B}_{\mathbf{TOT}}(\omega,\beta,a)+\mathbf{C}_{\mathbf{TOT}}\right)\hat{\eta}(\omega,\beta)=\hat{\mathbf{F}}_{\mathbf{D}}(\omega,\beta)+\hat{\mathbf{F}}_{\mathbf{Morison}}(\omega,\beta). \tag{15}$$

By solving the above linear system for $\hat{\eta}(\omega,\beta)$, the Response Amplitude Operator (RAO), i.e., the linear transfer function of the dynamic system, is evaluated as follows:

$$\mathbf{RAO}(\omega,\beta)=\left(-\omega^2\mathbf{M}_{\mathbf{TOT}}(\omega,\beta)+i\cdot\omega\cdot\mathbf{B}_{\mathbf{TOT}}(\omega,\beta,a)+\mathbf{C}_{\mathbf{TOT}}\right)^{-1}\left(\hat{\mathbf{F}}_{\mathbf{D}}(\omega,\beta)+\hat{\mathbf{F}}_{\mathbf{Morison}}(\omega,\beta)\right), \tag{16}$$

2.3. Verification with the 5 MW DeepCwind Semisubmersible Platform

To verify our FD model, in this section, we compare our results with those available in the literature referring to the 5MW OC4 DeepCwind [25] semisubmersible platform (see Table 1 and Figure 1).

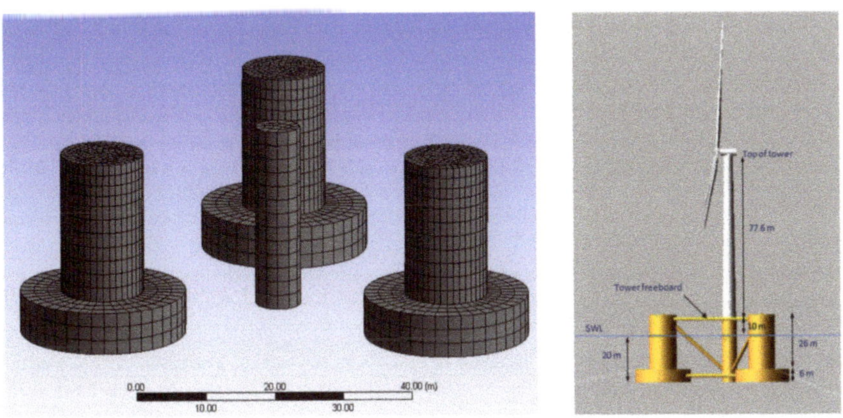

(a) (b)

Figure 1. ANSYS AQWA hydrodynamic model of the 5 MW wind turbine (**a**); 5 MW NREL semisubmersible platform [25] (**b**).

Table 1. 5 MW National Renewable Energy Laboratory (NREL) DeepCwind semisubmersible platform specifications, SWL indicates Sea Water Level and CM indicates Centre of Mass.

5MW NREL DeepCwind Platform	
Depth of platform base below SWL	20 m
Elevation of main column above SWL	10 m
Elevation of offset columns above SWL	12 m
Length of upper columns	26 m
Length of heave plates	6 m
Depth to top base columns below SWL	14 m
Diameter of main columns	6.5 m
Diameter of offset columns	12 m
Diameter of heave plates	24 m
Diameter of pontoon and cross-braces	1.6 m
Platform CM location below SWL	13.46 m
Water depth	200 m

A mesh-independence study was carried out to evaluate mesh sensitivity. The wetted surface is discretized with 2 m × 2 m panels (2173 elements). Morison forces acting on slender members and heave plates, added mass and viscous drag matrices are computed with an in-house algorithm developed in MATLAB based on the following procedure: each member is divided in 1 m-wide strips. A banded white noise spectrum between 0.05 and 0.25 Hz is applied to the system in the surge direction, i.e., 0°. RAOs, and the power spectral densities (PSDs) in surge, heave and pitch are calculated and compared with results presented in [26] (see Figures 2 and 3). Figure 2 shows a very good agreement of our FD model with the majority of the results reported in the wide code-to-code comparison campaign available in [26]. Given the large number of results available in [26], to facilitate the comparison, in Figure 2, Figure 14 (top row) of [26] has been placed in the background with the colours slightly faded, whereas our results are plotted in the foreground with a more vivid colour (blue solid line). Underneath the figure, we report the original legend that associates each background line to a specific model/institution. The goal of this graphical presentation is not to make clearly distinguishable each single background line, which can be much better interpreted directly in [26], but rather to provide an at-a-glance overall assessment of our formulation. It is pointed out that the reference results in [26] are obtained from time domain simulations based on different coupled FOWT dynamic solvers employing different theories and models, ranging from Morison and potential flow only to second-order formulations. For a detailed description of each formulation, we refer to [26]. Eigenfrequencies of the system are well captured in the PSDs shown in Figure 3, which are also in good agreement with the reference results of [26] (not shown here). The surge eigenfrequency is 0.00856 Hz, the pitch eigenfrequency is 0.0408 Hz, and the heave eigenfrequency is 0.056 Hz. The second peak of the PSD in surge is the pitch eigenfrequency, and vice versa for the PSD in pitch. This is due to the coupling between these two DoFs. The PSD results are influenced in terms of amplitude by the shape of the white noise spectrum as the PSD is calculated as

$$S_\eta(\omega) = S_{wave}(\omega)RAO(\omega)^2, \tag{17}$$

where $S_{wave}(\omega)$ is the input wave spectrum. Thus, considering Figures 2b and 3b, we note that as the heave RAO approaches 1 when the frequency tends to 0, the heave spectrum shape is completely determined by the shape of the wave spectrum, namely by how the white noise spectrum goes to 0, as outside the excited frequency range (below 0.05 Hz), its value must be close to 0. This is not true for surge and pitch spectra as they are coupled. RAOs below 0.05 Hz do not approach 1 or a constant value because of such coupling effects.

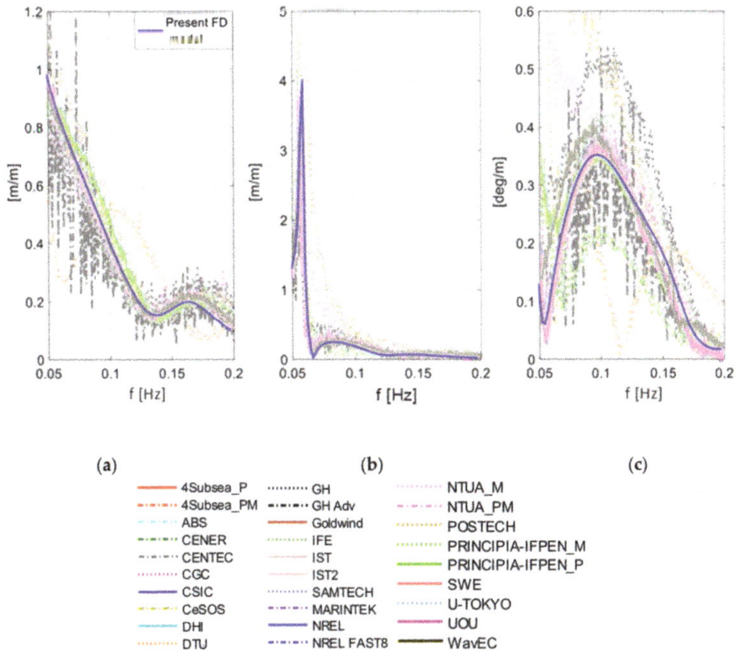

Figure 2. Comparisons for the NREL 5 MW semisubmersible Response Amplitude Operators (RAOs): surge (**a**), pitch (**b**) and heave (**c**). The background figures with the Offshore Code Comparison Collaboration Continuation Within IEA Wind Task 30: Phase II results are from [23].

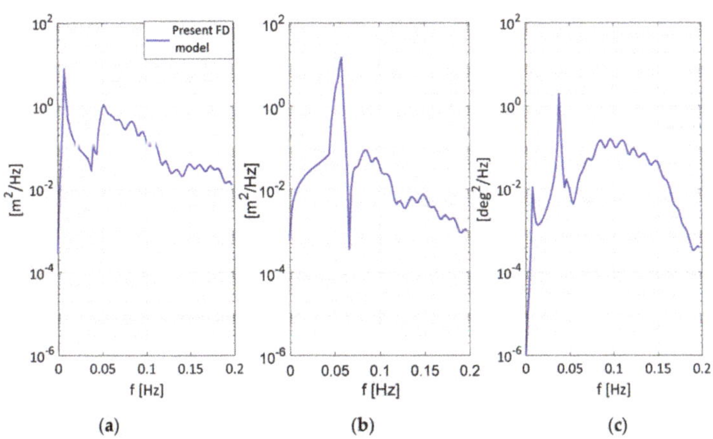

Figure 3. Comparisons for the NREL 5 MW semisubmersible RAOs: surge (**a**), pitch (**b**) and heave (**c**).

3. 10 MW DTU Upscaled Semisubmersible Platform

The 5 MW NREL semisubmersible platform is upscaled to support the DTU 10 MW wind turbine [28]. Following [15], the scaling factor is determined by the power rating between the two turbines, namely

$$\text{sf} = \sqrt{\frac{10\text{ MW}}{5\text{ MW}}} = \sqrt{2}, \tag{18}$$

The main geometry dimensions of the new platform obtained by upscaling each length by sf are reported in Table 2. The diameter of the central column was scaled with a different scaling factor to match the turbine tower base diameter.

Table 2. 10 MW DTU upscaled semisubmersible specifications.

10 MW DTU Upscaled Platform	
Depth of platform base below SWL (draft)	28.30 m
Elevation of main column above SWL	14.14 m
Elevation of offset columns above SWL	16.97 m
Length of upper columns	36.77 m
Length of heave plates	8.48 m
Depth to top base columns below SWL	19.80 m
Diameter of main columns	8.30 m
Diameter of offset columns	16.97 m
Diameter of heave plates	33.94 m
Diameter of pontoon and cross-braces	1.6 m
Platform CM location below SWL	21.27 m
Water depth	200 m

The ballast is recalculated in order to maintain the upscaled draft of 28.30 m. Moreover, due to the higher loads, ballast is supposed to be made of concrete to increase the draft of the platform centre of mass (COM).

3.1. Mooring Lines System

The mooring lines are kept as in the 10 MW DTU platform [43], specifically chosen to restrain a semisubmersible platform for a 10 MW turbine. The anchor points are kept in the same position as in the 5 MW NREL semisubmersible platform, but the fairlead points are adjusted in order to keep the connection point at the top of each heave plate. Considering the change of the fairlead position, the unstretched cable length is modified to avoid a reduction of the mooring system stiffness in the surge direction. The mooring lines' linearised stiffness is highly dependent on the catenary shape of the cables—a small modification in the hanging points would lead to high variation of the overall mooring lines stiffness. Therefore, to maintain almost constant their stiffness contribution, the unstretched length of each cable is reduced in order to let the 10 MW cable catenary shape at the equilibrium point match the 5 MW one. This is done by implementing a quasi-static cable model [44]. Once the non-linear system of equations of the catenary is solved with a Newton–Raphson iterative procedure, the simplified mooring system stiffness k is calculated as the first order series expansion of the horizontal tension at the fairlead in the neighbourhood of the equilibrium [39]. Namely,

$$k = \mu\left[-2\left(1+\frac{2H_F}{\mu z_F}\right)^{-\frac{1}{2}} + \cosh^{-1}\left(1+\frac{\mu z_F}{H_F}\right)\right]^{-1}\Bigg|_{X_{F,EQ}} \quad (19)$$

where μ is the cable weight in water per unit length, H_F is the horizontal tension at the fairlead and (x_F, z_F) are the horizontal and vertical coordinates of the fairlead at the equilibrium in the cable refence frame. The simplified restoring geometric stiffness of the mooring stiffness in the surge direction for oscillations around the equilibrium point is then given by

$$K_{11} = \sum\nolimits_{i=1}^{n} k_i \cos^2(\phi_i), \quad (20)$$

3.2. Results

The parked state is analysed first. Linearisation is performed around the initial position to evaluate the system mass, damping and stiffness contributions stemming from platform, turbine and mooring lines. The results are inserted into the equation of motion for the platform (Equation (13)), which is solved for a 0° heading broad banded white noise sea of PSD 1 m^2/Hz.

For power production case, a constant wind speed of 11 m/s is applied at the hub height. Such a wind speed corresponds to the case of maximum thrust for the 10 MW DTU wind turbine. A time-domain simulation in the power production condition is performed to identify the steady operating point around which the linearisation is performed. The "Multi-blade Coordinate Transformation Utility for 3-Bladed Wind Turbines" [45] is adopted to express rotor-rotating quantities in the fixed refence system of the platform. The results are again adopted to solve the equation of motion for the same input sea.

The results in terms of surge, pitch and heave RAOs are presented and compared in Figure 4, where the blue and red lines refer to the parked and power production states, respectively. The three eigenfrequency peaks are clearly visible, surge at 0.00488 Hz, pitch at 0.0428 Hz, and heave at 0.04883 Hz. In Figure 4b, it is possible to see the strong damping effect in the pitch motion of the wind turbine in power production. On the contrary, in surge (Figure 4a) and heave (Figure 4c), the hydrodynamic damping is dominant, and no significant difference is seen between parked and operating states. Another noticeable dynamic effect of the rotor can be seen by comparing the stiffness matrix terms of the system in parked and power production states. A stiffening of the system in yaw is the result of the rotational effect of the rotor. This is expected since the rotor rotation around its axis results in an increase in damping and stiffness around the orthogonal axis of yaw. The stiffness **C** and damping $\mathbf{B_t}$ coefficients in yaw (6,6) and pitch (5,5) for the parked and operating states are reported in Table 3.

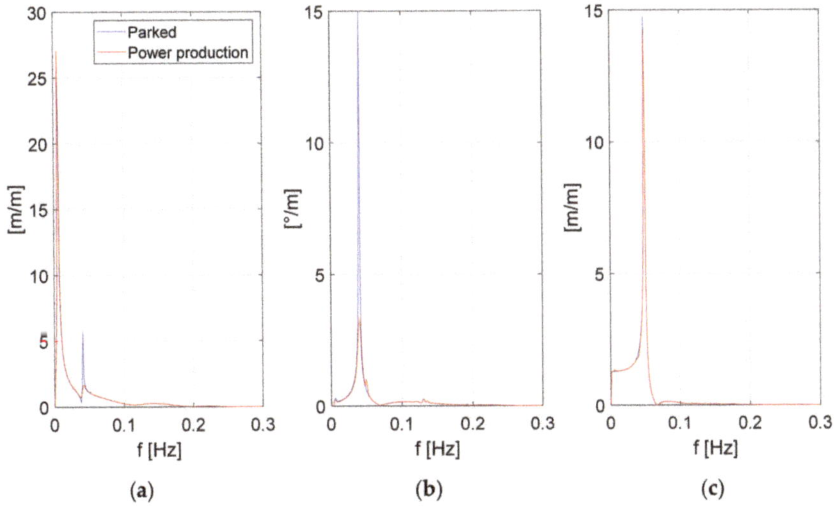

Figure 4. RAOs of the 10 MW DTU upscaled platform: (**a**) surge, (**b**) pitch and (**c**) heave degrees of freedom (DoFs).

Table 3. Stiffness and turbine damping contribution in pitch and yaw motions.

Load Case	C (5,5) [Nm/rad]	C (6,6) [Nm/rad]	B_t (5,5) [kgm^2/s]	B_t (6,6) [kgm^2/s]
Parked	$5.62{\cdot}10^9$	$2.04{\cdot}10^8$	0	0
Power production	$5.637{\cdot}10^9$	$2.65{\cdot}10^8$	$1.31{\cdot}10^9$	$1.31{\cdot}10^8$

4. Optimal Column Diameter and Platform Radius for the 10 MW Wind Turbine

In this section, we investigate the effects of two key design variables, namely the outer column diameter d and platform radius r, on the dynamic response of a semisubmersible platform supporting a 10 MW wind turbine. By the platform radius is meant the distance from the central column to the outer columns. The ratio between the heave plates and the column diameter is set to 2, keeping the same ratio of the platform for the 5 MW WT. The heave plate height is fixed to 6 m. The draft is kept constant at 28 m for all the combinations analysed by modifying the concrete ballast.

Eight different column diameters and platform radii are chosen in order to create a grid of 32 combinations. The grid spacing for d and r are 1 m and 5 m, respectively. The results for all 32 combinations are then interpolated using spline surfaces.

To assess the dynamic response of the system, heave and pitch response have been considered. The surge RAO is not considered as a control quantity because the surge eigenfrequency is usually very low (0.004 to 0.008 Hz). Both heave and pitch are indicators of turbine performance. From the heave and pitch response, the difference between the platform draft windward and the local free surface wave elevation can be evaluated according to [12]. This is a fundamental quantity that must be considered to prevent bottom-slamming and, therefore, to reduce fatigue loads on mooring lines and turbine tower. Pitch response is also related to another indicator of wind turbine performance, nacelle acceleration. Nacelle acceleration has been widely used to estimate wind turbine performance in the frequency domain from platform performance ([9,11]). It can be calculated as

$$\mathrm{RAO}_{\ddot{\delta}nac}(\omega) = -\omega^2(\mathrm{RAO}_1(\omega) + \mathrm{RAO}_5(\omega)h_{nac}), \tag{21}$$

where ω is the circular frequency, $\mathrm{RAO}_1(\omega)$ is the surge RAO, $\mathrm{RAO}_5(\omega)$ is the pitch RAO and h_{nac} is the nacelle height with respect to the sea water level.

Two scalar quantities, one local and one global, are considered for the two DoFs of pitch and heave. The first quantity is the peak of the RAO, and the second is the integral of the RAO in the range of frequencies from 0 to 0.2 Hz excited by a constant 1 m^2/Hz white noise.

First, for each combination pair (d, r), the mean pitch rotation angle and the mean surge displacement are checked as to whether they are suitable for a power production state (see Figure 5). As it is possible to observe from Figure 5a, a rather high average operating pitch angle of 5° is reached for the semisubmersible platform with column diameters of 13 m to 14 m and a radius equal to 30 m. This is mainly due to the lightness of the platform and to its reduced hydrostatic stiffness in roll and pitch DoFs. Such angles may not be compatible with the power production state; therefore, they are discarded in the following analyses (note that 5° is chosen as the maximum allowable value of the average operating pitch angle [1]). The surge operating displacement (Figure 5b) is almost constant for all the platforms considered. This confirms that by varying the unstretched cable length it is possible to keep constant the translational stiffness of the mooring lines.

In Figures 6 and 7, heave and pitch RAO peaks for the whole set of platforms are presented. The heave peak does not vary significantly between parked (Figure 6) and power production (Figure 7). Focusing on the parked condition, a region of minimum values for the heave peak of the RAO (see Figure 6a) can be identified between 16 m and 17 m for the outer columns' diameter, and for almost all the platform radii analysed. An optimal candidate appears to be 16/17 m for the column diameter and 35 m for the platform radius. Parked state pitch minimum RAO peaks (Figure 6b) compatible with power production can be identified in four regions: between 13 m and 14 m for the column diameter and a platform radius of 35 m, between 16 m and 17 m for the column diameter and a platform radius of 35 m, between 18 m and 19 m for the column diameter and 40–45 m for the platform radius, and between 19 m and 20 m for the column diameter and 33–40 m for the platform radius. The last two minimum points are reached with heavier and larger solutions than the upscaled platform (marked as a red dot in the figures), so they are discarded. Platforms with 16–17 m column diameters and 35 m platform radii appear to be the best choice, which minimises both heave and pitch peaks; moreover, they offer a relatively lower mean pitch angle (around 3°) than the 13–14 m column diameter and 35 m radius solutions (around 4°) (see Figure 5a). Heave peaks (Figure 7a) remain almost unchanged with respect to the parked state condition, the minimum region appears to be again around 16–17 m for the column diameter, and no significant influence of the platform radius can be noticed. Pitch peak, instead (Figure 7b), is strongly influenced by the damping effect stemming from the rotor blades, which is capable of drastically reducing the corresponding pitch peak in the parked state. We emphasise the importance of examining not only the power production but also the parked or low rotational speed conditions. During the shut-down or parked state with severe sea state conditions, the rotor drastically reduces its rotation speed, and thus its damping effects on the platform are significantly reduced.

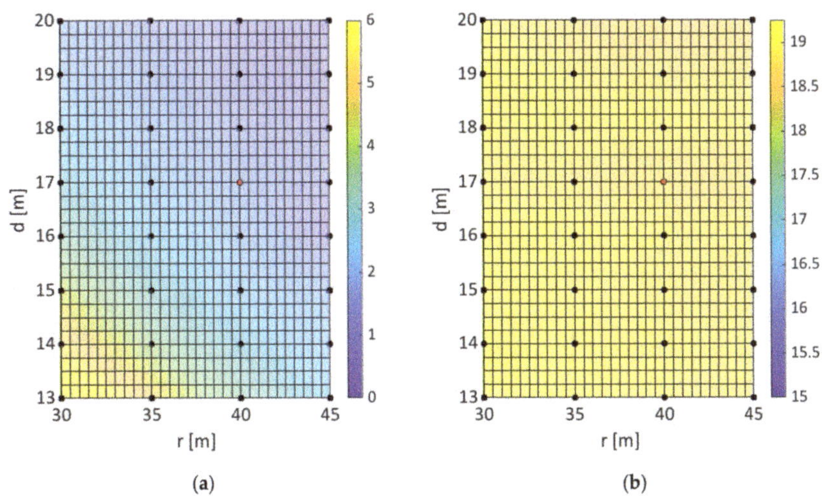

Figure 5. Power production operating point, pitch (deg) (**a**) and surge (m) (**b**). Red dots refer to the upscaled configuration.

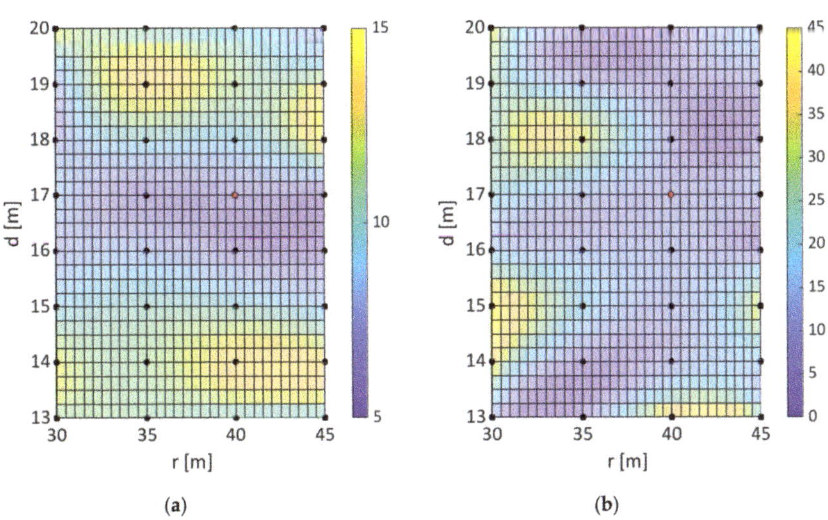

Figure 6. Parked condition; RAO peaks of heave (m/m) (**a**) and pitch (deg/m) (**b**). Red dots refer to the upscaled configuration.

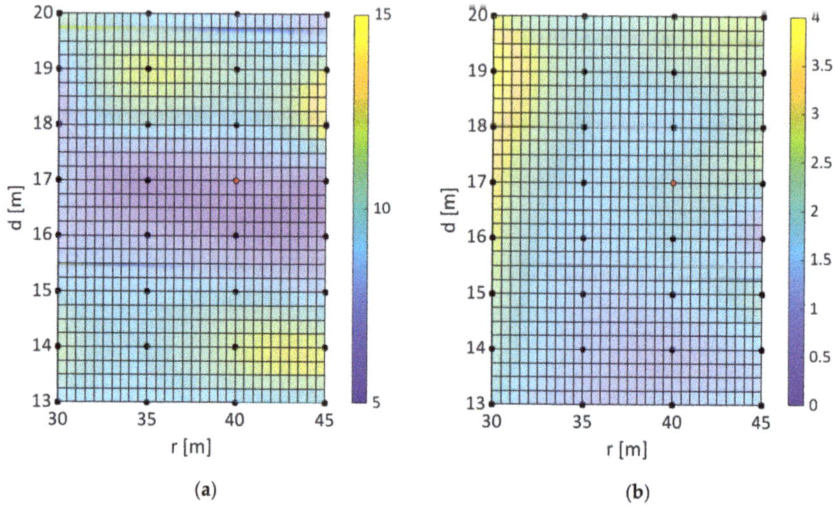

Figure 7. Power production condition; 11 m/s wind speed, RAO peaks of heave (m/m) (**a**) and pitch (deg/m) (**b**). Red dots refer to the upscaled configuration.

In Figures 8 and 9, RAO integrals of heave and pitch DOFs are presented. Figures 8a and 9a reveal a lower minimum region between 16 m and 17 m for the outer column diameter. Local maxima in the bottom and upper part of the domain are caused by an increasing of the eigenfrequency peak and by heave–pitch coupling. Rotor damping contribution does not have an appreciable effect as heave plate damping dominates. The pitch RAO integral is again influenced positively by the rotating blades so that the parked RAO integral is almost always higher than the power production one (compare Figures 8b and 9b and note that the two colour bars have different bounds). A smaller platform radius appears to produce a negative effect, making the pitch RAO integrals increase both in parked and operating conditions. This is related to both the increasing of the RAO pitch peaks and to the change of the shape of the RAO immediately outside the pitch eigenfrequency. This can be observed in Figure 9a where the RAO of the pitch DoF for combinations (d, r) (16, 30) and (16, 35) are compared. A small increase in the heave RAO integral in power production for a 30 m radius platform is due to a stronger coupling effect between heave and pitch degrees of freedom (Figure 10b).

Considering all the platforms analysed, the optimal solution that presents a low heave response in the parked state and a low pitch peak in both parked and power production states (although not the lowest values) lies between 16 m and 17 m for the column diameter and 35 m for the radius of the platform. These two configurations are compared in Figure 11. We observe that in the parked state for the combination (16 m, 35 m), the peak in pitch is reduced by about 25% with respect to the configuration (17 m, 35 m) (see Figure 11a), whereas the heave peak is increased by 18% (Figure 11b). In power production, the reduction of the pitch peak (see Figure 11c) is about 10%. The smaller reduction compared to the parked case is associated with the reduced sensitivity to the platform geometry caused by damping effects stemming from the rotor, which is now relevant. On the contrary, the increase in the heave peak (see Figure 11d) remains almost constant at 18%.

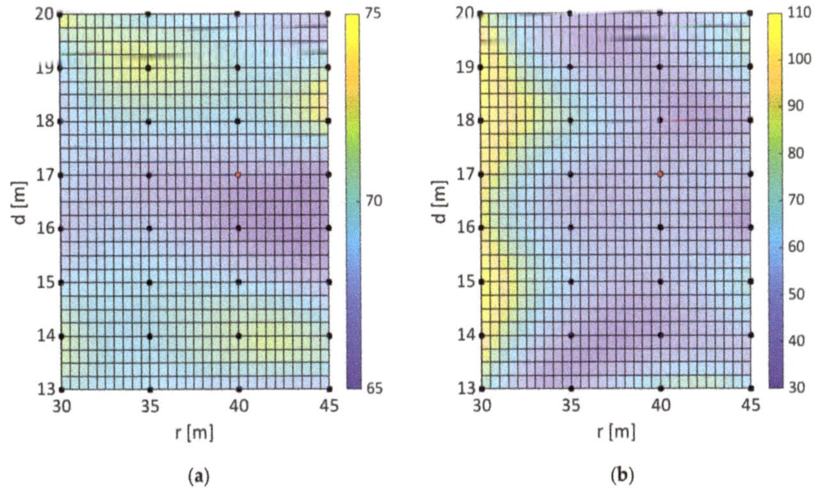

Figure 8. Parked condition; RAO integral of heave (**a**) and pitch (**b**). Red dots refer to the upscaled configuration.

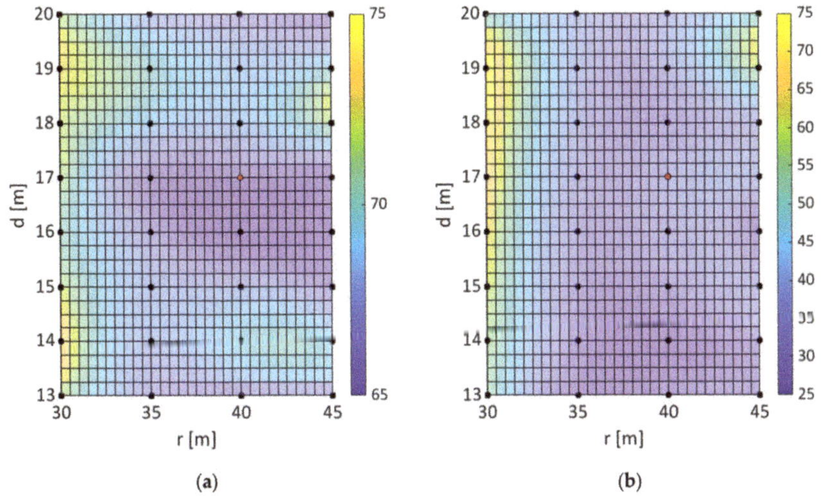

Figure 9. Power production condition; 11 m/s wind speed, RAO integral of heave (**a**) and pitch (**b**). Red dots refer to the upscaled configuration.

Considering that the platform response in pitch would affect more the fatigue behaviour of both turbine tower and mooring lines than the response in heave, the configuration with a 16 m column diameter and 35 m platform radius is preferred to the other option (17 m, 35 m).

In Figures 12 and 13, the RAOs in the surge, pitch and heave of the configuration chosen as optimal (16 m, 35 m) are compared with the RAOs of the upscaled platform (17 m, 40 m). In the figures, also shown are the ranges of typical exciting frequencies of the sea states 1–6 and 7–9. Only very severe sea states can excite the pitch and heave of both the upscaled and optimised platforms, but for the optimised one, a higher damping is achieved, both in heave and pitch in the neighbourhood of the natural frequency.

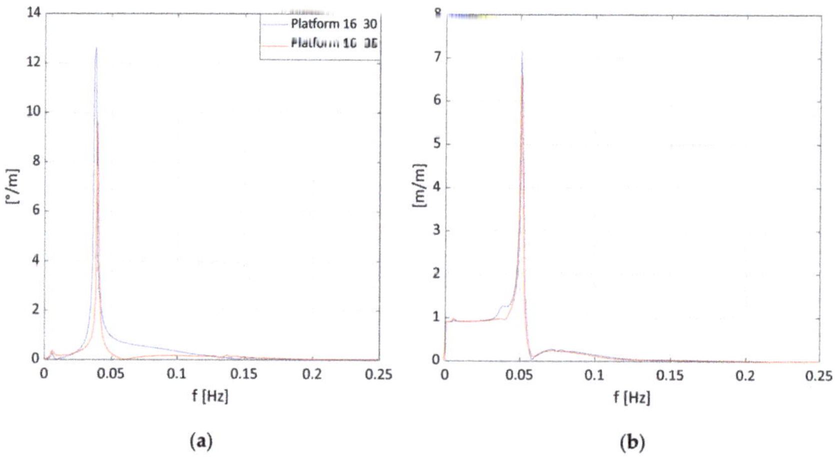

Figure 10. Pitch parked (**a**) and heave power production (**b**) RAOs of platforms with 16 m column diameter, 35 m radius (red line) and 30 m radius (blue line).

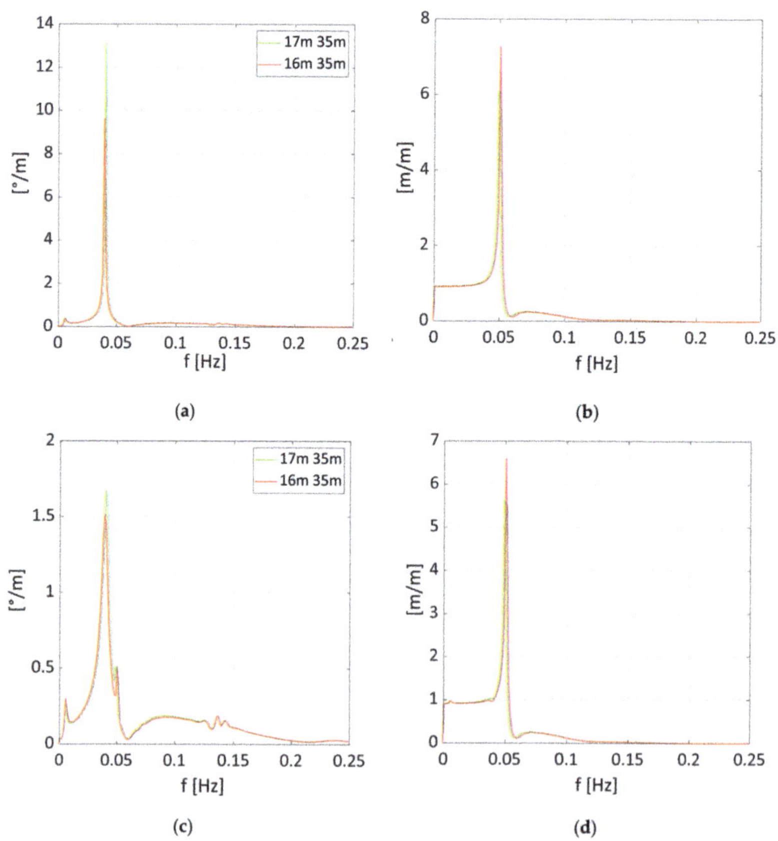

Figure 11. Pitch parked (**a**), heave parked (**b**), pitch power production (**c**) and heave power production (**d**) RAOs with 16 m (red line) and 17 m (green line) column diameters and a 35 m radius.

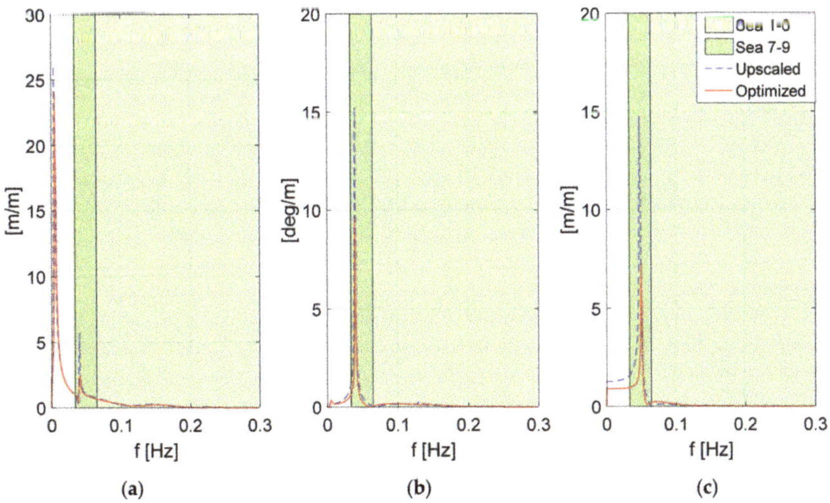

Figure 12. RAO comparison in the parked state between the upscaled and optimised platforms for surge (**a**), pitch (**b**) and heave (**c**) DoFs.

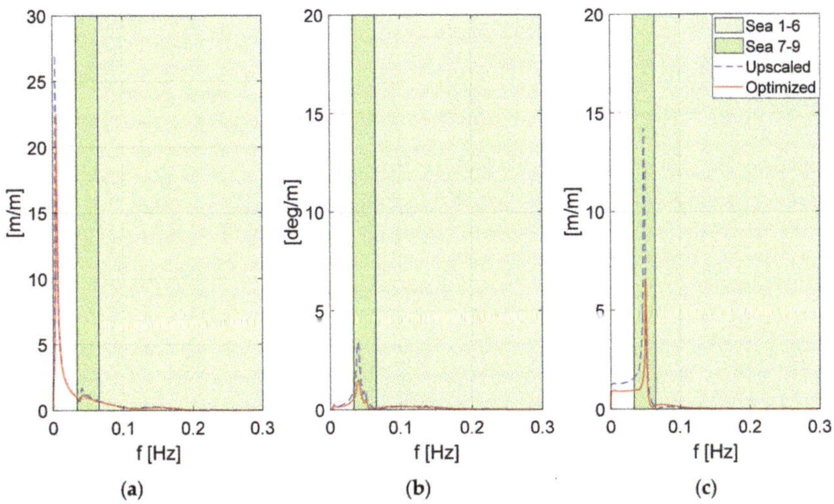

Figure 13. RAO comparison in power production state between upscaled and optimised platforms for surge (**a**), pitch (**b**) and heave (**c**) DoFs.

In order to assure a better performance of the wind turbine, the nacelle acceleration RAO has been evaluated for the optimised platform and compared to the one obtained for the upscaled one.

The results are presented in Figure 14 for the case of power production. In the figure, also shown are the ranges of the typical exciting frequencies of the sea states 1–4 and 5–9. As it is possible to observe, in a wide band of realistic seas (Sea 5 to 9) compatible with power production wind speeds, better turbine performance is achieved with the optimised platform. On the contrary, the upscaled solution provides better turbine performance for mild sea states associated with low wind speeds where the optimisation process is less relevant compared with sea states from 5 to 9.

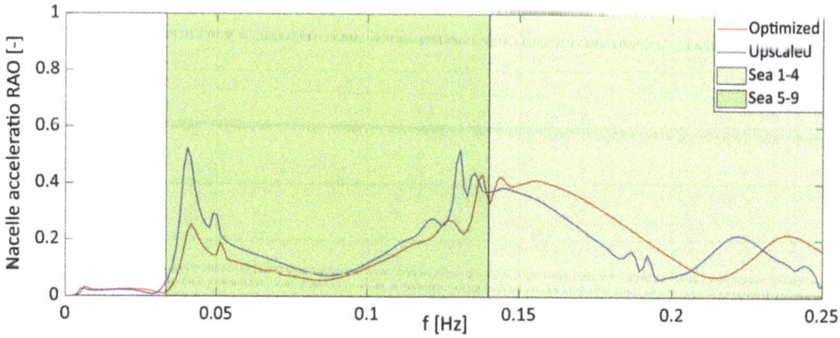

Figure 14. Nacelle acceleration RAO comparison in power production state between upscaled and optimised platforms.

In Table 4, the numerical values of the RAO peaks of the upscaled and the optimised platform are compared. As it is possible to observe, a low reduction is achieved in surge because the dynamic behaviour is mainly dominated by mooring lines, which are not optimised in this study.

Table 4. Comparisons of RAO peak values between the upscaled and optimised platforms.

	Surge Peak		Heave Peak		Pitch Peak	
	Parked	**Power Production**	**Parked**	**Power Production**	**Parked**	**Power Production**
Upscaled	26.12	27.04	14.74	14.24	15.31	3.39
Optimised	24.08	22.51	7.27	6.59	9.62	1.51
Reduction [%]	7.78	16.75	50.69	53.69	37.12	55.51

Considering the main differences between the upscaled and the optimised platforms, heave plate thickness appears not to produce a significant benefit in the heave and pitch responses (for the upscaled platform, the heave plate thickness is equal to 8.48 m, while for the optimised ones, it is 6 m).

5. Conclusions

In the present work, a frequency domain based-investigation aimed at finding the optimal configuration of a semisubmersible platform for a 10 MW WT is presented. Firstly, a frequency domain model is formulated and verified through comparisons with benchmark results available in the literature. Then, the frequency domain model is applied to an upscaled platform designed to support the DTU 10 MW wind turbine. Both parked and power production states are simulated. Two design variables are considered, namely, the diameter of the outer columns and the platform radius. The quality of the dynamic response is assessed through selected RAO peaks and RAO integral areas over a prescribed frequency range. The results show that an optimal configuration is achieved for a diameter of the outer columns d = 16 m and a platform radius r = 35 m. Such a configuration is lighter and smaller than the upscaled one (d = 17 m and r = 40 m) and presents significantly better performances in surge, pitch and heave. Moreover, it is found that for a certain combination of diameter and radius, a coupling of heave and pitch motion arises; likewise, for spar-like platforms. The results also highlight how in the operating state, aerodynamic damping can reduce the pitch excitation. On the other hand, in the parked state, high sensitivity to pitch excitation is observed. The absence of rotor damping makes the peak in pitch about five times higher than that in power production.

We remark that the present study is purely linear and cannot capture important nonlinear effects. However, due to its efficiency, the approach proposed in this work can be of valuable support in the pre-screening phase of the optimisation-based design of very large wind turbine platforms. Moreover, the present study is limited to two design variables. A forthcoming study will extend the optimisation process to more design variables, such as platform draft and mooring line geometry.

Author Contributions: Conceptualisation, G.F. and E.M.; methodology, G.F. and E.M.; software, G.F.; validation, G.F.; investigation, G.F. and E.M.; resources, E.M.; data curation, G.F.; writing—original draft preparation, G.F.; writing—review and editing, E.M.; visualisation, G.F.; supervision, E.M. and C.B.; funding acquisition, E.M. and C.B. All authors have read and agreed to the published version of the manuscript.

Funding: This research was funded by the European Union H2020 Programme under the MaRINET2 Project, grant agreement No 731084.

Conflicts of Interest: The authors declare no conflict of interest.

References

1. Wayman, E.N.; Sclavounos, P.D. Coupled Dynamic Modeling of Floating Wind Turbine Systems. In Proceedings of the Offshore Technology Conference, Houston, TX, USA, 1–4 May 2006.
2. Robertson, A.N.; Jonkman, J.M. Loads analysis of several offshore floating wind turbine concepts. *Proc. Int. Offshore Polar Eng. Conf.* **2011**, 443–450.
3. Kvittem, M.I.; Moan, T. Time domain analysis procedures for fatigue assessment of a semi-submersible wind turbine. *Mar. Struct.* **2015**, *40*, 38–59. [CrossRef]
4. Bachynski, E.E.; Moan, T. Ringing loads on tension leg platform wind turbines. *Ocean Eng.* **2014**, *84*, 237–248. [CrossRef]
5. Bachynski, E.E.; Kvittem, M.I.; Luan, C.; Moan, T. Wind-wave misalignment effects on floating wind turbines: Motions and tower load effects. *J. Offshore Mech. Arct. Eng.* **2014**, *136*, 1–12. [CrossRef]
6. Li, L.; Liu, Y.; Yuan, Z.; Gao, Y. Dynamic and structural performances of offshore floating wind turbines in turbulent wind flow. *Ocean Eng.* **2019**, *179*, 92–103. [CrossRef]
7. Chuang, Z.; Liu, S.; Lu, Y. Influence of second order wave excitation loads on coupled response of an offshore floating wind turbine. *Int. J. Nav. Archit. Ocean Eng.* **2020**, 1–9. [CrossRef]
8. Orszaghova, J.; Taylor, P.H.; Wolgamot, H.; Madsen, F.J.; Pegalajar-Jurado, A.; Bredmose, H. Second and third order sub-harmonic wave responses of a floating wind turbine. In Proceedings of the 35th International Workshop on Water Waves and Floating Bodies, Seoul, Korea, 26–29 April 2020.
9. Xu, K.; Shao, Y.; Gao, Z.; Moan, T. A study on fully nonlinear wave load effects on floating wind turbine. *J. Fluids Struct.* **2019**, *88*, 216–240. [CrossRef]
10. Fang, Y.; Duan, L.; Han, Z.; Zhao, Y.; Yang, H. Numerical analysis of aerodynamic performance of a floating offshore wind turbine under pitch motion. *Energy* **2020**, *192*, 116621. [CrossRef]
11. Jeon, M.; Lee, S.; Lee, S. Unsteady aerodynamics of offshore fl oating wind turbines in platform pitching motion using vortex lattice method. *Renew. Energy* **2014**, *65*, 207–212. [CrossRef]
12. Sclavounos, P.; Tracy, C.; Lee, S. Floating offshore wind turbines: Responses in a seastate Pareto optimal designs and economic assessment. *Proc. Int. Conf. Offshore Mech. Arct. Eng. OMAE* **2008**, *6*, 31–41.
13. Brommundt, M.; Krause, L.; Merz, K.; Muskulus, M. Mooring system optimization for floating wind turbines using frequency domain analysis. *Energy Procedia* **2012**, *24*, 289–296. [CrossRef]
14. Karimi, M.; Hall, M.; Buckham, B.; Crawford, C. A multi-objective design optimization approach for floating offshore wind turbine support structures. *J. Ocean Eng. Mar. Energy* **2017**, *3*, 69–87. [CrossRef]
15. Leimester, M.; Bachynski, E.E.; Muskulus, M.; Thomas, P. Rational Upscaling of a Semi-submersible Floating Platform Supporting a Wind Turbine. *Energy Proc.* **2016**. [CrossRef]
16. Liu, J.; Thomas, E.; Goyal, A.; Manuel, L. Design loads for a large wind turbine supported by a semi-submersible fl oating platform. *Renew. Energy* **2019**, *138*, 923–936. [CrossRef]
17. Liu, J.; Manuel, L. Alternative mooring systems for a very large offshore wind turbine supported by a semisubmersible floating platform. *J. Sol. Energy Eng. Trans. ASME* **2018**, *140*, 1–9. [CrossRef]
18. Hsu, W.T.; Thiagarajan, K.P.; Manuel, L. Extreme mooring tensions due to snap loads on a floating offshore wind turbine system. *Mar. Struct.* **2017**, *55*, 182–199. [CrossRef]

19. Azcona, J.; Vittori, F.; Schmidt, U.; Svanije, F.; Kapogiannis, G.; Karvelas, X. Design Solutions for 10 MW Floating Offshore Wind Turbines Document information. *INNWIND* **2017**, 308974.
20. Tomasicchio, G.R.; D'Alessandro, F.; Avossa, A.M.; Riefolo, L.; Musci, E.; Ricciardelli, F.; Vicinanza, D. Experimental modelling of the dynamic behaviour of a spar buoy wind turbine. *Renew. Energy* **2018**, *127*, 412–432. [CrossRef]
21. Wan, L.; Greco, M.; Lugni, C.; Gao, Z.; Moan, T. A combined wind and wave energy-converter concept in survival mode: Numerical and experimental study in regular waves with a focus on water entry and exit. *Appl. Ocean Res.* **2017**, *63*, 200–216. [CrossRef]
22. Pegalajar-Jurado, A.; Hansen, A.M.; Laugesen, R.; Mikkelsen, R.F.; Borg, M.; Kim, T. Experimental and numerical study of a 10 MW TLP wind turbine in waves and wind. *J. Phys. Conf. Ser.* **2016**, *753*, 092007. [CrossRef]
23. Morison, J.R.; Johnson, J.W.; Schaaf, S.A. The Force Exerted by Surface Waves on Piles. *J. Pet. Technol.* **1950**, *2*, 149–154. [CrossRef]
24. Jonkman, J.; Buhl, M.L. *FAST User's Guide*; NREL/EL-500-38230 (Previously NREL/EL-500-29798); National Renewable Energy Laboratory: Golden, CO, USA, 2005.
25. Robertson, A.; Jonkman, J. *Definition of the Semisubmersible Floating System for Phase II of OC4*; Technical Report NREL/TP-5000-60601; National Renewable Energy Laboratory: Golden, CO, USA, 2014.
26. Robertson, A.; Jonkman, J.; Vorpahl, F.; Popko, W.; Qvist, J.; Frøyd, L. Offshore code comparison collaboration continuation within IEA wind task 30: Phase II results regarding a floating semisubmersible wind system. *Proc. Int. Conf. Offshore Mech. Arct. Eng. OMAE* **2014**, *9B*. [CrossRef]
27. Robertson, A.N.; Wendt, F.; Jonkman, J.M.; Popko, W.; Dagher, H.; Gueydon, S. OC5 Project Phase II: Validation of Global Loads of the DeepCwind Floating Semisubmersible Wind Turbine. *Energy Proc.* **2017**, *137*, 38–57. [CrossRef]
28. Bak, C.; Zahle, F.; Bitsche, R.; Kim, T.; Yde, A.; Henriksen, L.C.; Andersen, P.B.; Natarajan, A.; Hansen, M.H. Design and performance of a 10 MW wind turbine. *J. Wind. Energy* **2013**.
29. Jonkman, J.M.; Hayman, G.J.; Jonkman, B.J.; Damiani, R.R.; Murray, R.E. *AeroDyn User's Guide and Theory Manual*; NREL: Golden, CO, USA, 2015; p. 46.
30. Jonkman, J.M.; Robertson, A.N.; Hayman, G.J. *HydroDyn User's Guide and Theory Manual*; Technical Report NREL/TP; National Renewable Energy Laboratory: Golden, CO, USA.
31. Marino, E.; Lugni, C.; Borri, C. The role of the nonlinear wave kinematics on the global responses of an OWT in parked and operating conditions. *J. Wind Eng. Ind. Aerodyn.* **2013**, *123*, 363–376. [CrossRef]
32. Paulsen, B.T.; Bredmose, H.; Bingham, H.B.; Jacobsen, N.G. Forcing of a bottom-mounted circular cylinder by steep regular water waves at finite depth. *J. Fluid Mech.* **2014**, *755*, 1–34. [CrossRef]
33. Marino, E.; Giusti, A.; Manuel, L. Offshore wind turbine fatigue loads: The influence of alternative wave modeling for different turbulent and mean winds. *Renew. Energy* **2017**, *102*, 157–169. [CrossRef]
34. Mockutė, A.; Marino, E.; Lugni, C.; Borri, C. Comparison of nonlinear wave-loading models on rigid cylinders in regular waves. *Energies* **2019**, *12*, 4022. [CrossRef]
35. Uzunoglu, E.; Soares, C.G. On the model uncertainty of wave induced platform motions and mooring loads of a semisubmersible based wind turbine. *Ocean Eng.* **2018**, *148*, 277–285. [CrossRef]
36. Philippe, M.; Babarit, A.; Ferrant, P. Comparison of time and frequency domain simulations of an offshore floating wind turbine. *Proc. Int. Conf. Offshore Mech. Arct. Eng. OMAE* **2011**, *5*, 589–598.
37. Wang, K.; Ji, C.; Xue, H.; Tang, W. Frequency domain approach for the coupled analysis of floating wind turbine system. *Ships Offshore Struct.* **2017**, *12*, 767–774. [CrossRef]
38. *AQWA User Manual*; ANSYS, Inc.: Canonsburg, PA, USA, 2012.
39. Faltinsen, O.M. *Sea Loads on Ships and Offshore Structures*; Cambridge University Press: Cambridge, UK, 1993.
40. Journèe, J.M.; Massie, W. *Offshore Hydromechanics*; TU Delft: Delft, The Netherlands.
41. Savenije, F.J.; Peeringa, J.M. *Aero-Elastic Simulation of Offshore Wind Turbines in the Frequency Domain*; Technical Report; Energy Research Center of the Netherlands: Petten, The Netherlands, 2009.
42. Tao, L.; Dray, D. Hydrodynamic performance of solid and porous heave plates. *Ocean Eng.* **2008**, *35*, 1006–1014. [CrossRef]

43. Müller, K.; Lemmer, F.; Yu, W.; Norbek, J.A.; Berthelsen, P.A. LIFES50+ Qualification of Innovative Floating Substructures for 10 MW Wind Turbines and Water Depths Greater than 50 m. In *D4.2 Public Definition of the Two LIFES50+ 10 MW Floater Concepts Disclaimer*; University of Stuttgart: Stuttgart, Germany, 2017.
44. Jonkman, J.M. Dynamics of offshore floating wind turbines-model development and verification. *Wind Energy* **2009**, *12*, 459–492. [CrossRef]
45. Bir, G. *NWTC Design Codes (MBC)*; NWTC: Green Bay, WI, USA, 2008; Volume 3.

energies

MDPI

Article

Fatigue Life Assessment for Power Cables in Floating Offshore Wind Turbines

Mohsen Sobhaniasl [1], Francesco Petrini [1,*], Madjid Karimirad [2] and Franco Bontempi [1]

[1] Department of Structural an Geotechnical Engineering, Sapienza University of Rome, 00184 Rome, Italy; mohsen.sobhaniasl@uniroma1.it (M.S.); franco.bontempi@uniroma1.it (F.B.)

[2] Civil Engineering, School of Natural and Built Environment, Queen's University Belfast (QUB), Belfast BT7 1NN, UK; madjid.karimirad@qub.ac.uk

* Correspondence: francesco.petrini@uniroma1.it

Received: 6 May 2020; Accepted: 12 June 2020; Published: 15 June 2020

Abstract: In this paper, a procedure is proposed to determine the fatigue life of the electrical cable connected to a 5 MW floating offshore wind turbine, supported by a spar-buoy at a water depth of 320 m, by using a numerical approach that takes into account site-specific wave and wind characteristics. The effect of the intensity and the simultaneous actions of waves and wind are investigated and the outcomes for specific cable configurations are shown. Finally, the fatigue life of the cable is evaluated. All analyses have been carried out using the Ansys AQWA computational code, which is a commercial code for the numerical investigation of the dynamic response of floating and fixed marine structures under the combined action of wind, waves and current. Furthermore, this paper applies the FAST NREL numerical code for comparison with the ANSYS AQWA results.

Keywords: wind energy; floating offshore wind turbine; dynamic analysis; fatigue life assessment; flexible power cables

1. Introduction

Power cables are wideley used in power transmission lines and the electrification of floating oil and gas production infrastructures, where they have to withstand considerable cyclic loads induced by the combination of floating body dynamics with wind, waves and current effects [1,2].

One of the main design issues for power cables in marine applications is the fatigue strength; fatigue assessment studies have demonstrated that accurate analysis of the complex dynamic behavior induced by the offshore environment must be carried out for fatigue assessment purposes. Dai et al. (2020) [3] presented experimental and numerical studies on dynamic stress and curvature in steel tube umbilicals, and the results show that curvature, governs the fatigue problem, even when it is small. Yang et al. (2018) [4] presented a parametric study of the dynamic motions and mechanical characteristics of power cables for a wave energy conversion (WEC) system. The results show that the large curvature responses of the cable typically occur at high wave heights and near the wave period of resonance. Yang et al. (2017) [5], regarding the same WEC system, demonstrated that minor fouling can increase the fatigue life of the power cable.

A relatively new application area for marine power cable elements is floating offshore wind turbines (FOWTs) [6] and farms. In fact, wind energy has become one of the most important renewable energy sources in recent decades. Its development requires an increasing electric transmission capacity and better ways to preserve system reliability. Wind-generated power continues to grow rapidly throughout the world, and offshore wind farms are expected to account for a large portion of the total wind energy output and may even contribute significantly to the total electricity production in some countries [7]. The cost-effective operation of marine energy conversion systems needs the reliable design of marine power cables, which are currently the components for which design procedures mainly need further development, due to their limited fatigue strength. FOWTs in particular, together with their mooring systems and connected power cables, must withstand various environmental actions, such as correlated variability in the wind, current, and wave loads, which cause variable and uninterrupted dynamic motion and stress in the power cables that connect the single FOWTs with each other and/or with the land. The accumulation of variable stress results in cumulative damage to the fatigue [8], which, for this kind of system, is a weak point in the design panorama.This is something that, together with other design or operating issues, currently prevents the large-scale installation and diffusion of floating offshore wind farms as reliable systems for energy generation and as an evolution of the currently widely diffused fixed-foundation offshore wind farms. These weaknesses are particularly relevant for the electrical cables which connect the single turbines to one another in the farm [9], named "inter-array" or "umbilical" cables.

Therefore, the definition of advanced and reliable tools for the investigation of fatigue damage is necessary, and this topic is highly recognized by the research community [9] and the industry [10,11]. Design approaches in this area intend to ensure the required fatigue life for the structure or key parts of it [12], and they are devised toghether with appropriate inspection programs for monitoring fatigue damage initiations or progress.

This paper proposes a procedure for the fatigue life evaluation of a power cable which is attached to a single FOWT in a lazy wave geometrical configuration by considering (i) the correlated intensities of wind, ocean current and wave actions [13]; (ii) the different operating conditions of the turbine; (iii) the complete non-linear dynamics of the turbine-cable coupled system [14]; and (iv) the difference between fatigue actions in different locations along the cable. Several modeling issues are faced in order to increase confidence in the results obtained by the multiphysics modeling of the system. The wind field sampled by the rotor in operating conditions, the fluid-structure interaction between the wind and the FOWT due to the platform's large displacements, the joint bending and axial induced strain in the cable and the non-linearity involved in the problem (due to both large displacements and the appearance of plastic strains) are included in the analyses by making reliable assumptions and by calibrating the procedures employing sensitivity analyses. The objective of this paper is to provide a reliable procedure for assessing the fatigue failure of the power cables of FOWTs. The proposed procedure is applied to the umbilical cable attached to the spar-buoy supported 5 MW NREL wind turbine [15] in a water depth of 320 m, considering site-specific wind, waves and current for a short-term sea state of the China Sea. The case study is modeled both in the FAST NREL code [15] and in the ANSYS AQWA Commercial code [16]. The obtained results offer supporting data for the a-priori identification of weaknesses and critical components for fatigue strength and for the evaluation of the desired level of reliability before deployment.

2. Analysis Method for Fatigue Life Evaluation

As is well known, the simplified relationship between the stress of amplitude Si and the associated number of cycles Ni that leads to the fatigue failures of mechanical components under constant-amplitude oscillating loads can be represented for different intensities by the S-N (σ-N) fatigue curve [17]. The S-N curve is a characteristic of the materials, of the shape of the component or of the design configuration of the structure. Under the indications provided in the Det Norske Veritas (DNV) design standards [18,19], the fatigue damage for different levels of stress oscillation magnitude can be calculated for the pertinent components of FOWTs by the Palmgren–Miner assumption, using Equation (1) [17]:

$$Dm = \Sigma \frac{ni}{Ni_{cycles}} \tag{1}$$

where Dm is the accumulated fatigue damage due to all considered stress oscillation magnitudes, ni is the number of the stress oscillations with magnitude Si and Ni_{cycles} is the number of cycles of stress oscillation with magnitude Si which lead to fatigue failure.

The electrical cables that connect the FOWTs in floating offshore wind farms are of the umbilical typology, providing a service-support from the main station to the single FOWTs. Such umbilical cables are of the optical powered submarine type and are provided to supply electrical power or to send operating control inputs and data to underwater elements in offshore oil and gas infrastructures [20]. Umbilical cables are implemented in floating offshore wind farms and connected to each FOWT for the same purpose; therefore, their integrity and service ability are crucial for the correct coordination and operation of the FOWTs within the farm. The stress conditions in such non-homogeneous cross-sections of the umbilical cables is quite complex, with non-linear local stress due to contact and friction between the components of the tube, and the model and aleatory uncertainty due to this complexity is critical for the reliability of any fatigue analysis. For the sake of simplicity, and in line with state of the art literature on the topic [9], the cable will be modeled in this paper with a homogeneous-material equivalent to the cross-section.

As an alternative to the σ-N plane, fatigue curves may also be represented by the total strain (ε) versus the number of cycles (N), as detailed below. Equation (1) is useful for both the stress versus number of cycles curve and the strain versus number of cycles curve (ε-N). The evaluation of the total strain time history in a critical fatigue-prone location of the cable, and corresponding to a specific sea state, is conducted by a non-linear numerical analysis of the coupled FOWT-electrical cable system, and ε accounts for both elastic (ε_e) and plastic strain (ε_p). This may then be expressed as follows [21]:

$$\varepsilon(t) = \varepsilon_e(t) + \varepsilon_p(t) = \frac{\sigma_{el}(t)}{E} + \varepsilon_p(t) \tag{2}$$

where E is the Young's elastic modulus assigned to the equivalent homogenous material modeling the cable, and σ_{el} is the elastic tension stress of the electrical cable, evaluated step-by-step as the maximum cross-section tensile stiffness in the homogeneous material cross-section by considering both the axial and bending contributions:

$$\sigma_{el}(t) = \frac{N(t)}{A} + \chi(t) \cdot E \cdot \frac{D}{2} \tag{3}$$

in which, by referring to the circular cross-section of the cable, N is the axial force, A is the area, χ is the bending curvature and D is the diameter.

The plastic strain contribution ε_p enters the game only when the total stress, as evaluated by Equation (3), is larger than the yielding stress σ_y of the cable, which can be evaluated from the strain–stress constitutive law assumed for the equivalent homogeneous material used for modeling the cable (see Figure 1), as detailed in Equation (4)

$$\varepsilon_p(t) = \frac{\sigma_y}{E} + \frac{\sigma_y - \sigma_{el}(t)}{E'} \tag{4}$$

where E' is the material plastic modulus.

The non-linear stress–strain behavior and relaxation characteristics mean that the impact of plastic strains plays a non-negligible role in the electrical cable's fatigue analysis. Inside the umbilical cable, the conductors, which are commonly made of pure electrolytic tough pitch (ETP) copper, are vulnerable to fatigue accumulation and mechanical failure, as highlighted by Karlsen and co-authors (2009) [11], who state that "the material has excellent conductivity but poor mechanical properties, which include stress-relaxation (creeping) and non-linear stress-strain behaviour, and they suggests using ε-N curves (referring to strain) instead of traditional σ-N curves".

The Monograph 177 [22] from the National Institute of Standards and Technology (NIST) summarizes the results of a total of 126 plastic strain-controlled tests and 150 stress-controlled tests that were performed on a large set of coppers made by annealed material and cold-worked material. The stress-controlled test data may be converted to elastic strain values according to Equation (2) with $\varepsilon_p = 0$.

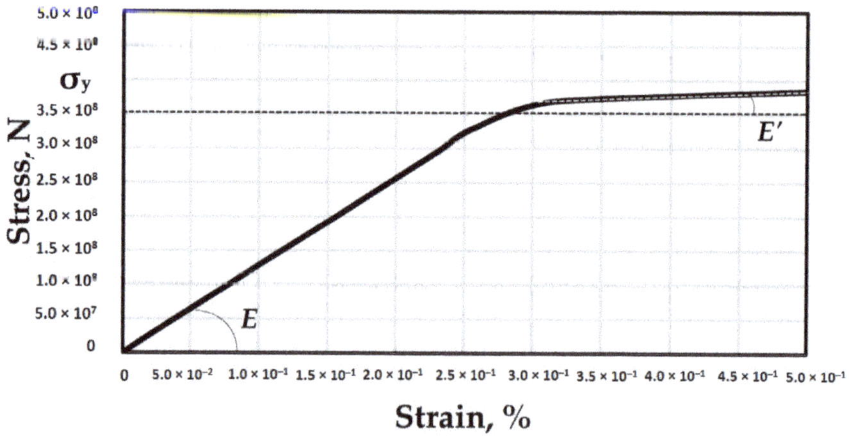

Figure 1. Stress–strain constitutive relationship of the cable.

The ε-N fatigue curve considering the effect of both elastic and plastic strain can be expressed by the Coffin–Manson relationship [21]:

$$\varepsilon\left(N_{cycles}\right) = \varepsilon_e\left(N_{cycles}\right) + \varepsilon_p\left(N_{cycles}\right) == C_1 \cdot N_{cycles}^{-\beta_1} + C_2 \cdot N_{cycles}^{-\beta_2} \tag{5}$$

where $\varepsilon\left(N_{cycles}\right)$ are the strain amplitudes leading to failure at N_{cycles} number of cycles, and C_i and β_i (with $i = 1$ or 2) are appropriate material constants, provided by Thies et al. (2011) [23], for a typical FOWT umbilical cable as $C_1 = 0.7692$, $\beta_1 = 0.5879$, $C_2 = 0.0219$ and $\beta_2 = 0.1745$. This curve is shown in Figure 2, and it is the same that is used in the present paper for the fatigue analyses.

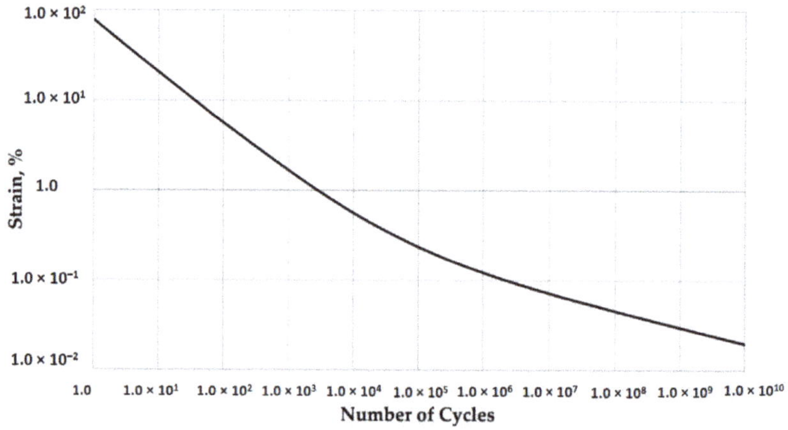

Figure 2. Strain cycle (ε-N) fatigue curve for electrolytic tough pitch (ETP) copper. Adapted from Reference [11].

According to the above-mentioned calculations, the total strain time history in a critical cross-section of the cable has to be evaluated and then processed by cycle counting, leading to a set of amplitudes—number of cycle histograms which are compared with the fatigue curve of the cable in order to assess the fatigue damage Dm associated with each single time history. One of the most popular cycle counting methods is the rainflow method [24].

In the context of FOWTs, the fatigue damage was calculated for each considered sea state. Then, from the climate statistics of the site where the floating offshore wind farm was located, the joint probability distribution of the sea states in the reference period of the analysis *Tr* (e.g., 1 year) was estimated, and then the occurrences in *Tr* of the damage associated to each sea state could be evaluated.

3. Case Study

The NREL 5-MW wind turbine on the OC3-Hywind spar-buoy FOWT was chosen as a case study for the evaluation of the fatigue life of an inter-array turbine (umbilical) cable. This case study was chosen for its simplicity in design, suitability for modeling and the existence of numerical results in the literature [25,26]. As already stated, the case study was modeled in both FAST and ANSYS AQWA software. The structural, hydrodynamic and mooring line properties of the model are reported in Appendix A.

Figure 3 illustrates the model of the 5 MW NREL wind turbine on the OC3-Hywind spar-buoy in FAST (right) and the model with the same properties in ANSYS AQWA (left). It is worth noting that the electrical cable was modeled in ANSYS AQWA but it was not modeled in FAST; the motivations for this are given in Section 4. The tables in the appendix show the structural and hydrodynamic properties and the mooring properties of the model. The lazy wave configuration of the electrical cable is taken from Rentschler et al. [9], in which a hydrostatic optimization of the layout of the cable is carried out with a genetic algorithm, with the goal of minimizing the cable fatigue, and by chosing the position of the buoyancy elements as design variables, this results in a lower waveform position as the optimal solution, with an accumulative length of the buoyancy parts from 18% to 23% of the total cable length.

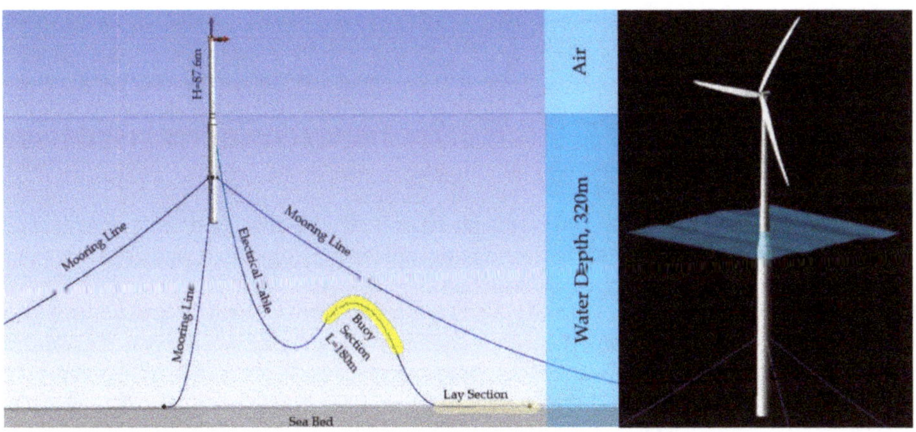

Figure 3. Left: model in ANSYS AQWA, Right: model in NREL FAST [26].

For the simulations in this paper, a typical cross-section of the electrical cable was utilized, with a single layer of galvanized steel armor wires and extruded cross-linked polyethylene (XLPE) isolators. The mechanical characteristics of the cable were assumed from the literature [6,27]. The values for the diameter, weight and strength are shown in Figure 4, with reference to Figure 1; the linear and non-linear elastic modules E and E' and the yielding stress σ_y were equal to 128 GPa, 6.4 GPa and 350 MPa, respectively.

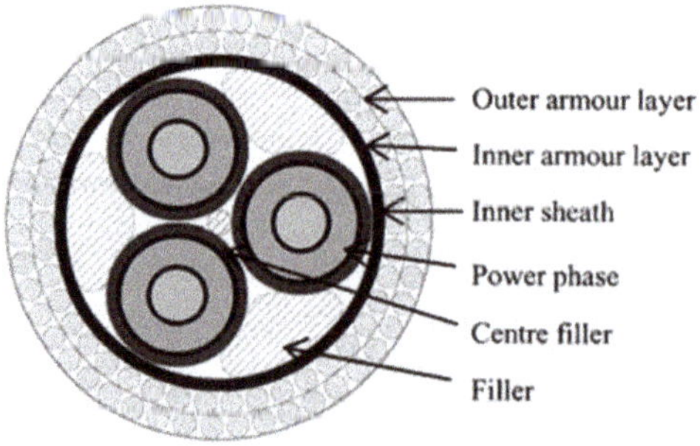

Cable Specification	
Nominal Diameter	0.2 m
Nominal weight in seawater	39 Kg
Bending Stiffness	10 kNm²
Axial stiffness	700 MN
Minimum Breaking Load	100 kN

Figure 4. Cable specifications [27].

4. Modeling the Spar-Buoy Supported Platform and the Cable

Depending on the wind speed and the environmental conditions, the configuration of the rotor changes: it is in the operational, rotating blade condition when the wind speed is between the cut-in and the cut-out values, while the wind turbine is parked for wind speeds higher than the cut-out value, with the blades parallel to the wind [28]. Time history analyses were conducted for 4000 s for different environmental conditions, and the first 400 s were removed in order to eliminate the transitory effects of the simulation and wait for to the model to reach the stationary operation state.

The version of the FAST Code (version 8) that was used did not allow the researchers to add extra non-structural elements (like power cables, etc.), but there were some possibilities to add extra elements in ANSYS AQWA. This was the motivation for conducting the electrical cable and the fatigue analyses in ANSYS AQWA. On the other hand, one of the issues we faced with the version of ANSYS AQWA (version 18.2) that was used was the difficulty of modeling the rotating blades in the operating conditions; consequently, the rotor-nacelle assembly was modeled there as a concentrated mass at the hub height. To mitigate this weakness, the total rotor aerodynamic load time history (force in the x-direction), named the "*RtAeroFxh*" output [15], as extracted from a preliminary FAST simulation (where the electrical cable was not modeled) in a certain sea state case, was successively applied to the hub in the model in ANSYS AQWA (where the electrical cable was then modeled and meshed by 100 finite elements). Since the hydrodynamic loads were evaluated in ANSYS AQWA based on the implemented wave theory and by making use of Morrison's equation, in order to avoid the double-counting of the hydrodynamic effects during the simulation in ANSYS AQWA, the *RtAeroFxh* aerodynamic force was evaluated by not taking into account the three following effects: (i) presence of waves; (ii) action of sea current and; (iii) inertial effect of water volumes moving with the platform. This was because all these effects were taken into account in the subsequent ANSYS AQWA simulation.

One of the procedures that can be implemented to avoid double-counting is running the preliminary FAST analysis by fully restraining the bottom Degrees Of Freedom (DOFs) of the platform and maintaining the over-water dynamics (named the "fixed DOFs at spar" option). This procedure would completely neglect the double-counting of all the three listed effects, but, on the other hand, this option would also neglect some fluid-structure interaction (FSI) effects that have a non-negligible impact on the *RtAeroFxh* (e.g., the wind's incoming velocities and angles of attack are not combined with the correct translational velocities and rotational displacements experimented by the rotor during the large pitch, thrust and heave displacements of the floating platform when relative wind-rotor velocities and rotational angles are obtained for aerodynamic load calculation purposes). The second strategy for resolving the double-counting problem is to run the preliminary FAST analysis by assuming a "still water" (no sea current and no waves) and by allowing the floating platform DOFs. By adopting this strategy, the FSI effects are mostly taken into account (the relative velocities between the wind and the rotor and tower will take into account the large displacements of the floating platform induced by the wind itself); this is also important if the impact on the relative translational velocities of the hydrodynamic forces (less important than that of wind, which is due to both the translation and rotation of the tower) is not considered and, at the end of the process, there will also be a (partial) double-counting of the inertial effect of the water volumes moving with the platform. In this case, a compromise should be put in place to run the analysis by the joint use of the two computational codes. In order to show the differences found in the *RtAeroFxh* by running the preliminary analysis with three different strategies, i.e., the above-mentioned "fixed DOFs at spar" and "still water" options, and the analysis carried out by considering all the loads in FAST ("wave and current" mode, something that would cause the above-mentioned double-counting effect if the extracted *RtAeroFxh* is successively applied to ANSYS AQWA), the time histories of the *RtAeroFxh* values obtained for the three cases are compared in Figure 5 for one of the sea states considered in the following fatigue analysis: sea state 15, as identified in Table 2 by grey highlighting (wind velocity at the hub height equal to 21.7 m/s, current velocity equal to 0.5 m/s, wave height equal to 4.5 m with period of 10 s).

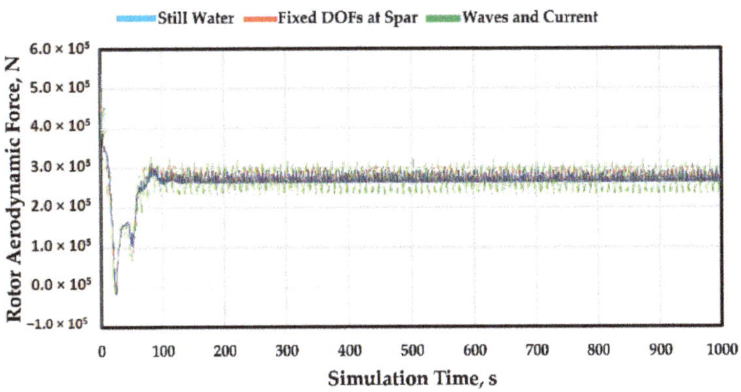

(a)

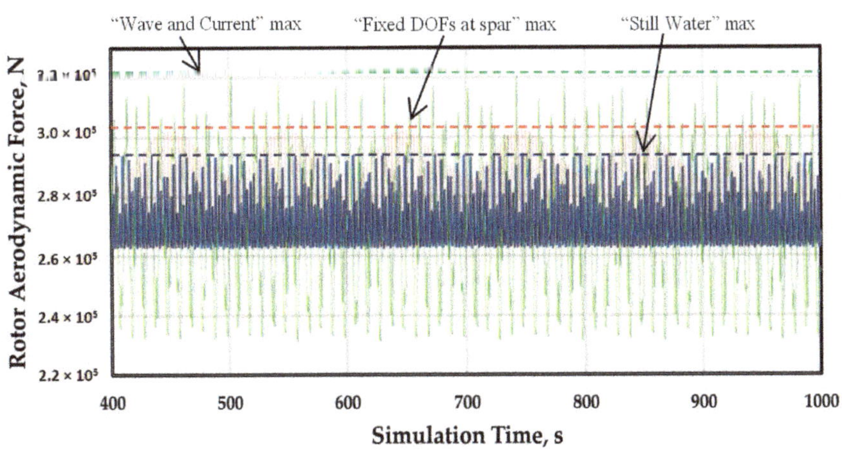

(b)

Figure 5. *Cont.*

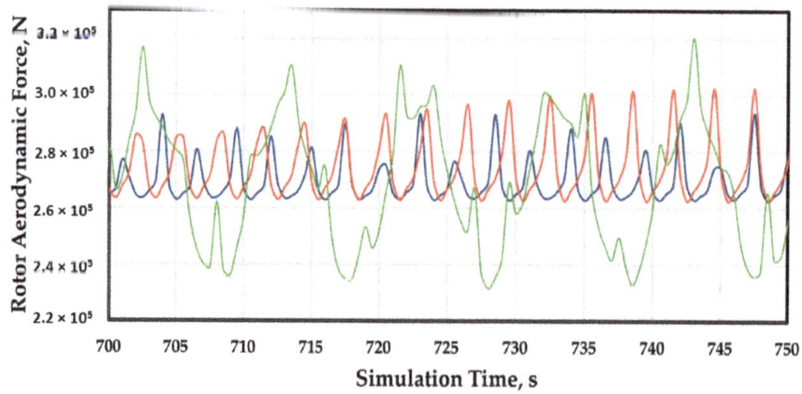

(c)

Figure 5. Rotor aerodynamic force *RtAeroFx* as extracted from FAST for sea state 15 and the three strategies that would be adopted for neglecting the double-counting of the hydrodynamic effects. 0–100 sec scale (**a**); 400–1000 sec scale (**b**); 700–750 sec time interval (**c**).

From Figure 5b, it is clear that the "fixed DOFs at spar" option provides conservative results with respect to the "still water" option, while it is also obvious that the inclusion of the waves and current would produce an additional contribution (the one that, if included, would be double counted in ANSYS AQWA) that is relevant for the examined load case. In Figure 5c, it is shown that the frequency content slightly changed between the three options (with different frequency contents for the "wave and current" case compared to the other two). Based on this analysis, the final choice was made to opt for the "fixed DOFs at spar" strategy in the subsequent fatigue calculations as this is on the conservative side with respect to the "still water" model.

In order to further discuss the adopted strategy in coupling the two codes, it is worthwhile to say that a step-by-step "two-way" coupling, that is, the continuous data exchange between the two codes in order to take into account the interactions of the sub-domains (e.g., aerodynamics–hydrodynamics) which are evaluated in the two different numerical codes, would be able to solve the above-mentioned problems regarding the double-counting or the neglect of some effects. When, as in this paper, some "one-way" coupling is put in place, the two codes elaborate separately the sub-domains of the problem, and for just one of them the elaboration takes into account the data coming from the other domain/code (in this paper, for example, we lost the effect of the ANSYS AQWA results on the FAST evaluations). Therefore, the use of the one-way coupling implies the acceptance of a compromise regarding the effects we lost and the accuracy of the obtained results. The correct strategy to pursue the one-way coupling is to adapt/calibrate the separate sub-domain models in a way that minimizes the aspects we lost but, more importantly, keeps the design/analysis on the conservative side. In our case, we adapted the FAST model, used for the aerodynamic sub-domain evaluation, with the main goal of being on the conservative side but with the secondary goal of not being too conservative (and far from accuracy) by trying to eliminate the double-counting of the displacements induced by the hydrodynamic forces.

Model Verification

Due to the above-mentioned joint use of the two codes, this procedure needs to be validated, especially regarding the evaluation of the hydrodynamic loads which are obtained by the two codes: having the same response of the model under hydrodynamic loads is a primary condition in order to assume that the obtained results are reliable and that the above-mentioned procedure of extracting the aerodynamic loads from FAST leads to a consistent dynamics of the system. With this purpose, a set of validation analyses (comparison of the platform dynamics obtained by FAST and ANSYS AQWA and comparison with the results in the literature for the same problem under different load conditions) are carried out, based on the published results of Phase IV of the NREL 5 MW FOWT [26]. Fully non-linear analyses in ANSYS AQWA and FAST are performed with all wind turbine DOFs (surge, sway, heave, pitch, roll, yaw). The environmental conditions for the validation simulation are shown in Table 1. The hydrodynamic, structural and mooring line properties were considered the same as in [26], while the electrical cable was not modeled in this validation phase.

Table 1. Environmental conditions used for model verification purposes [26].

Enabled DOFs	Platform, Tower
Wind Condition	None
Wave Condition	Regular Airy: $H = 6$ m and $T = 10$ s
Analysis Type	Time-series solution

Figure 6 shows the time histories of the platform surge, heave and pitch displacement considering the load case which is shown in Table 1 as obtained in the present study by ANSYS AQWA and FAST, compared with the results published in [26] as obtained with FAST ("OC3" in the figure) for validation purposes. The simulation shows there is a good agreement between the results of FAST and ANSYS AQWA.

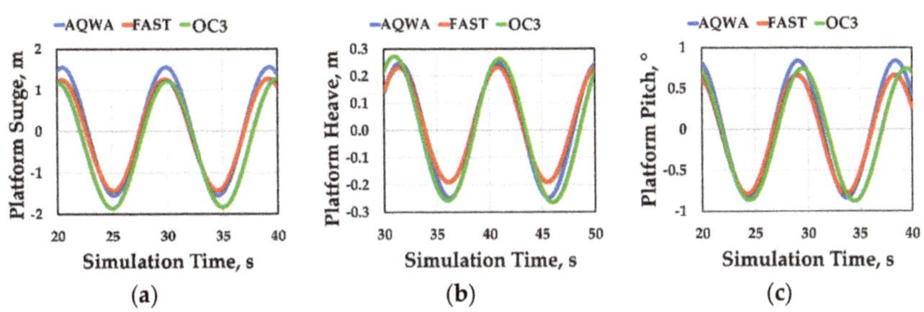

Figure 6. Comparison of the time histories of the platform main DOFs under the load case shown in Table 1, as obtained by FAST and ANSYS AQWA models. Surge (**a**), heave (**b**), pitch (**c**).

In the fatigue analysis presented in the next section, the motion responses of the spar-buoy platform were calculated in ANSYS AQWA under the combination of wind, current and wave loads. Concerning the wind, in each analysis, in addition to the time history of the total force acting on the rotor *RtAeroFxh* as extracted from FAST, the mean wind (Equation (6) below) is applied to the tower above the sea. The Airy theory for regular waves [26] is applied for the waves' dynamics computation, while Morrison's equations [26] are applied for the hydrodynamic forces evaluation. To model the environmental condition, regular sea states are assumed.

5. Fatigue Analysis

The above-mentioned general procedure for fatigue life evaluation has been applied to the case study by using the validated numerical models described above. The complete flowchart of the fatigue analysis procedure adopted in this study is shown in Figure 7. The goal was to determine the fatigue life in the cross-section of the cable that is located at a critical position, along with the cable development for fatigue.

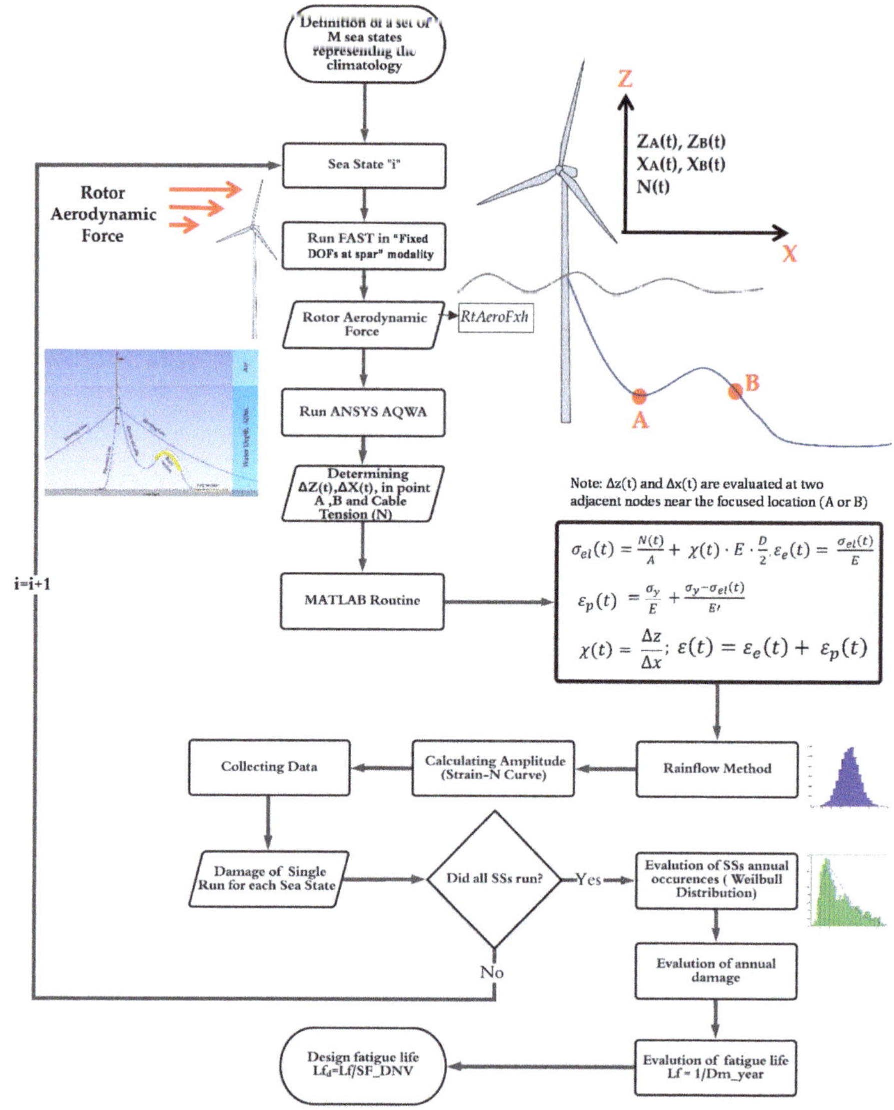

Figure 7. Fatigue analysis of dynamic electrical cable.

As already stated, for each considered loading condition (sea state "SS" in what follows), the aerodynamic forces' time histories are extracted from FAST (in "fixed DOFs at spar" conditions and without modeling the electrical cable) and applied to the top of the tower in the ANSYS AQWA model. Then, the structural response is evaluated in ANSYS AQWA by a fully non-linear time-domain analysis, considering also the model provided by the electrical cable. The outputs that are relevant for fatigue damage assessment in the electrical cable with the method described in the previous section (axial force and bending curvature in the cross-section of the cable at the location which is relevant for fatigue) are then extracted to feed an in-house developed MATLAB routine which implements the evaluation of the total strain time history as described in Equations (2)–(4). The bending curvature of the cable is obtained step-by-step as a spatial first derivative of the bending deflection, starting from the time histories of the displacements of the two nodes adjacent to the focused location. When the time history of the total strain in the given location for a certain SS is obtained, a MATLAB routine for the rainflow cycle counting method [24] is applied and the fatigue damage for the considered SS is evaluated at the focused location by the comparison of the cycle-counting histogram with the ε-N fatigue curve shown in Figure 2. The procedure is repeated for all the relevant SSs (for a total of 20 SSs reported in Table 2). Finally, the damages obtained are multiplied by their pertinent annual occurrences, evaluated by a fitted Weibull distribution indicating the climate of the FOWT location. The total annual fatigue damage *Dm_year* is then evaluated as the sum of the annual fatigue damages induced by the single SSs, and the fatigue life of the cable *Lf* is evaluated as the inverse of the annual fatigue damage, while the design fatigue life *Lf_d* is obtained by considering a safety factor SF_DNV equal to 10 as indicated in DNV standards [29].

Table 2. Short-term sea atates (in grey, we show the sea state used for comparison in Figure 5) [30].

SS	U_{10} (m/s)	*U(H)* (m/s)	*Hs* (m)	*T* (s)	*Cv* (m/s)	*P* (%)
1	5.6	7.279722555	0.675	4	0.168	2.24096
2	6	7.799702738	0.675	5	0.180	8.68372
3	7	9.099653194	1.050	4	0.210	1.96084
4	7.80	10.13961356	1.050	6	0.234	14.006
5	8.5	11.04957888	1.550	4	0.255	1.4006
6	9	11.69955411	1.550	5	0.270	10.36444
7	9.40	12.21953429	1.550	7	0.282	20.16864
8	10.8	14.03946493	2.175	5	0.324	5.32228
9	11.2	14.55944511	2.175	7	0.336	15.4066
10	12	15.59940548	2.875	6	0.360	8.96384
11	13.2	17.15934602	3.625	6	0.396	3.08132
12	14.5	18.84928162	4.000	6	0.432	0.56024
13	15.0	19.49925684	4.500	7	0.450	3.64156
14	16.1	20.92920235	5.000	7	0.483	0.84036
15	16.7	21.70917262	4.500	10	0.501	0.84036
16	17.2	22.35914785	4.500	11	0.516	0.28012
17	17.4	22.61913794	5.500	10	0.522	0.56024
18	18	23.39910821	5.500	11	0.540	0.56024
19	19.1	24.82905372	6.750	10	0.573	0.84020
20	20	25.99900913	3.625	12	0.600	0.280
		Sum of Probability				100

Figure 8 shows the electrical cable configuration. As already stated, the electrical cable configuration provided in [9] was used in the simulations. The cable hung from a fixed point on the spar-buoy platform at the sea surface, and the electrical cable was modeled by 100 elements. The fatigue of the cable was calculated at different locations along the cable to individuate the most critical cross-section for fatigue. The fatigue damage was highly influenced by the position of the cross-section along the cable. For demonstration purposes, in the following, the fatigue life was evaluated for locations A and B along the cable, as shown in Figure 8.

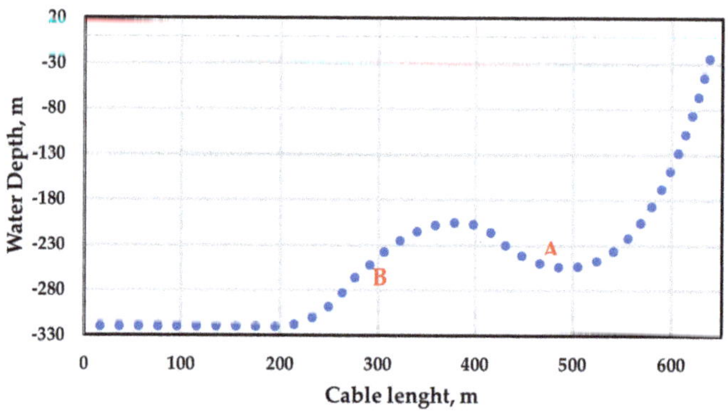

Figure 8. Electrical cable layout.

5.1. Environmental Conditions and Considered SSs

The SS conditions for fatigue life evaluations are selected from the South China Sea S4 area, the angle of incidence of wind and waves is fixed to the direction coinciding with to the X-axis, that is orthogonal to the cable longitudinal extension, since this is recognized to be the most critical configuration for fatigue in umbilical cables. The chosen sea states are listed in Table 2 in accordance with the wave scatter diagram of the South China Sea S4 [30] in a one year period. The wind speed profile, representing the variation of the mean wind with the height above the still water level $U(z)$, is evaluated by the power-law model shown in Equation (6), where U_{10} is 10 min average wind speed at a 10 m height above the sea still water level, and α is the power-law profile [29]:

$$U(z) = U_{10} \cdot \left(\frac{z}{10}\right)^{\alpha} \tag{6}$$

Additionally, in Table 2, Hs is the wave height and T is the wave period, Cv is the current velocity, and P is the annual probability of each sea state as evaluated by a Weibull probability distribution function, which is adopted in fitting with the number of occurrences of the sea states. Equation (7) describes the Weibull probability density function, while Figure 9 shows the Weibull probability distribution function.

$$PDF_{weibull} = \frac{\beta}{V}\left(\frac{U_{10}}{\gamma}\right)^{\beta} e^{-\left(\frac{U_{10}}{\gamma}\right)^{\beta}} \tag{7}$$

where, in this paper, $\beta = 2.49$ and $\gamma = 10.4$ m/s are the values assumed for the shape parameter and scale parameter [31] in order to fit the available SS data.

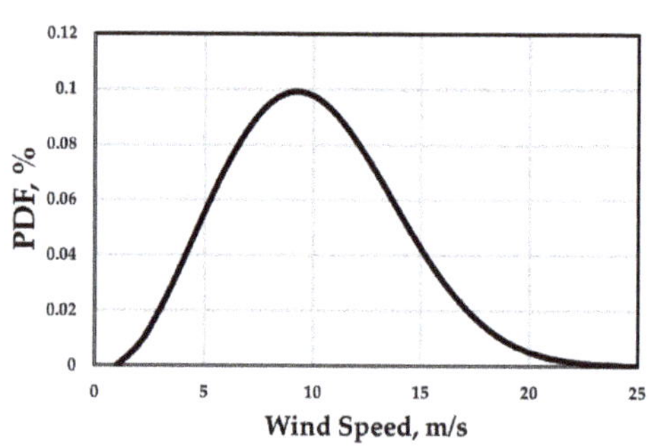

Figure 9. Weibull distribution adopted for U_{10}.

Wind turbines are typically designed for 20–25 years [32]. Under average wind conditions, an onshore wind turbine can produce electricity for 4000–7000 h a year, corresponding to 70%–80% of the total hours in the year [33]. On the above basis, an average number of 15 windy hours each day for a total of 365 days during the year (5475 h per year) is considered in evaluating the occurrence of the SSs.

5.2. Fatigue Life Estimation

As already stated, in order to avoid the transitory effects of the simulation in the rainflow method calculations, the first 400 s are removed from the ANSYS AWQWA analysis output before feeding them into the cycle-counting MATLAB routine. As an example, Figure 10 shows the time histories of the cable's axial force as obtained at location A from the SS 10 simulation in Table 2. The results obtained for the annual fatigue damage at location A for the 20 SSs are shown in Figure 11, while Figure 12 shows the percentage of the contribution of each SS to the annual fatigue damage. As stated above, the annual fatigue damage due to a particular SS is evaluated by multiplying the fatigue damage obtained by the non-linear time history analysis conducted in ANSYS AQWA by the number of annual occurrences of that SS.

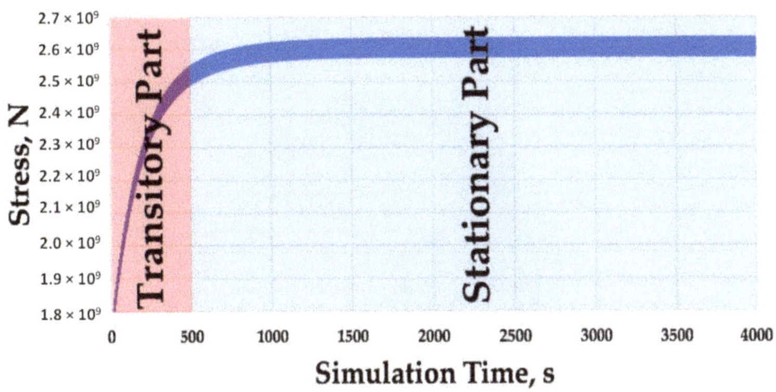

(a)

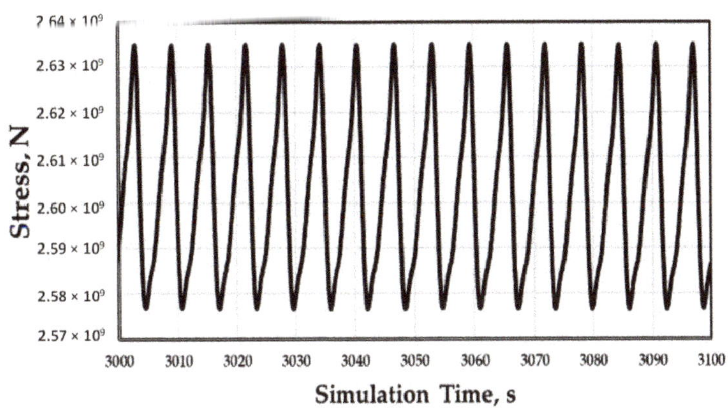

(b)

Figure 10. Example of axial stress time history in the cable at location A and for SS (sea state) 10. Whole (**a**); focus (**b**).

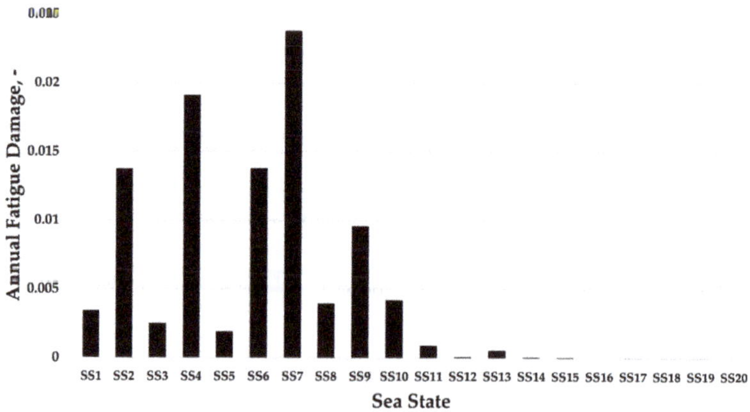

Figure 11. Annual fatigue damage induced at location A for all considered SSs.

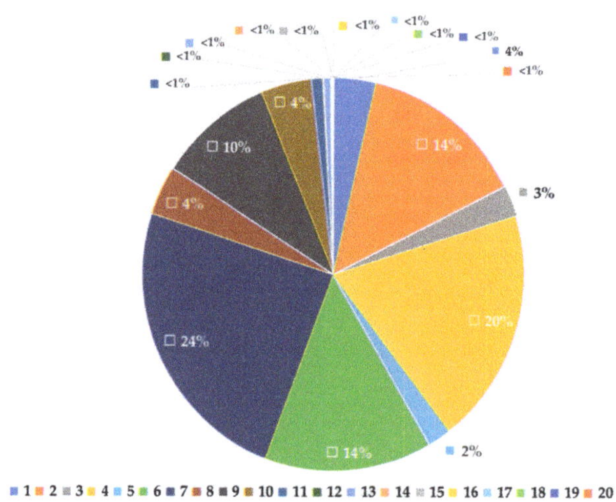

Figure 12. Relative contribution at location A to the annual fatigue damage of all considered SSs.

Finally, Table 3 represents the evaluation of the annual fatigue damage *Dm_year* due to all 20 of the SSs considered with their occurrences, obtained as the sum of the annual fatigue damages obtained for all SSs. From these results, it is evident that location A is critical for fatigue and that it suffers important damage which quantifies the fatigue life *Lf* of the umbilical cable to about 10 years, if scaled by the safety factor SF_DNV which is reduced to 1 year. This result confirms the notion that fatigue in electrical cables is a weak point for this kind of structure.

Table 3. Annual fatigue damage and fatigue life estimation.

Sea State	Annual Fatigue Damage. Location A	Annual Fatigue Damage. Location B
1	3.442737×10^{-3}	2.68358×10^{-4}
2	1.3747238×10^{-2}	9.76374×10^{-4}
3	2.526371×10^{-3}	2.0901×10^{-4}
4	1.9108678×10^{-2}	1.10222×10^{-3}
5	1.923422×10^{-3}	1.13001×10^{-4}
6	1.37916×10^{-2}	8.4595×10^{-4}
7	2.380188×10^{-2}	1.44312×10^{-3}
8	3.962855×10^{-3}	2.18672×10^{-4}
9	9.60747×10^{-3}	3.96218×10^{-4}
10	4.204048×10^{-3}	1.88023×10^{-4}
11	9.2328×10^{-4}	3.25774×10^{-5}
12	1.00209×10^{-4}	3.50426×10^{-6}
13	5.72823×10^{-4}	1.28245×10^{-5}
14	8.0497×10^{-5}	1.65752×10^{-6}
15	6.27443×10^{-5}	1.28641×10^{-6}
16	1.97191×10^{-5}	3.1258×10^{-7}
17	3.50939×10^{-5}	5.99213×10^{-7}
18	2.66189×10^{-5}	5.45779×10^{-7}
19	2.54863×10^{-5}	9.23608×10^{-7}
20	5.53736×10^{-6}	4.84702×10^{-7}
Sum of the annual damage for all SSs	9.796831×10^{-2}	5.81566×10^{-3}
Fatigue life of the cable Lf (years)	10.2074	171.9495
Fatigue design life of the cable Lf_d (with the application of the SF_DNV safety factor (years)	1.02074	17.19495

From Table 3, it is also evident that the evaluated fatigue damage is strongly dependent on the location along the cable: location B experiences much less damage (more than 10 times lower) than location A.This difference in the fatigue damage is due to the fact that the experimented oscillations and bending deformations of the cable decrease from A to B due to the presence of the buoy section of the cable development, the cable being stabilized by the buoyancy (see Figure 3).

Fatigue life assessment is highly dependent on the cable diameter and the location of the calculation point for the fatigue analysis. As already mentioned above, two different points are used in the fatigue analysis in this research in order to calculate the strain amplitude of the points (Figure 8). The suggested safety factor (SF) of 10, based on recommendations [29], is applied in the present study to assess the fatigue life.

6. Conclusions

This paper has assessed the fatigue life expected for an umbilical cable that is subjected to the loading regime of a typical marine environment, and under the non-linear dynamic conditions as they can be expected in a spar-buoy floating offshore wind turbine.

The main novelty of the paper lies in bringing together the mostly available knowledge to provide a comprehensive procedure/tool for the fatigue analysis of the vulnerable electrical cables of FOWTs. Such a tool cannot avoid complex structural dynamics analyses by putting in place simplistic assumptions about loads, their occurrences and numerical models. In fact, in this paper, the fatigue life of the cable's copper conductor was estimated for the FOWT site by setting out an ad-hoc complete procedure that: (i) computes the total strain time histories in the cable cross-section by the avail of fully non-linear time histories and numerical analyses of the whole turbine system including the electrical cable; (ii) evaluates the fatigue damage induced by a number of different sea states and by covering the wide range of climate and operating conditions of the turbine; (iii) defines annual occurrences of the different considered sea states for annual damage evaluation purposes based on the sea state statistics.

Numerical analyses have been conducted by the joint use of two well-established numerical codes for offshore floating structures, which is important for allowing multi-physic and multi-scale analyses of such a complex structural system: critical points arising from this coupling have been pointed out (e.g., correct consideration of the FSI and large displacements) and the associated analytical steps/choices for ensuring that the design remains on the conservative side under the current integration level of the two codes have been discussed, which is also a point of novelty for the paper.

Regarding the fatigue design of FOWT electrical cables, our general engineering findings are:

- floating offshore wind turbine power cables must withstand a dynamic loading regime when they are attached to the support platform (e.g., spar-buoy) and are susceptible to fatigue failures. The fatigue life of the examined case is 10 years if not scaled with the safety factor prescribed by the DNV standards. When a value of 10 is applied as a safety factor, the design fatigue life reduces to 1 year only;
- the correct evaluation of all the operating (parked versus rotating blades) and climate conditions (different sea states with their occurrences) is fundamental to obtain reliable estimations of the annual fatigue damage. This is evidenced by the large difference obtained by the damage evaluated by different conditions;
- due to the concurrent contribution of both bending and axial strains, and both elastic and plastic strains to the total strains used for damage evaluation, the fatigue damage evaluated for the cable is strongly dependent on the location along the cable development where the analysis is focused, which means that evaluating the fatigue damage in different locations along the cable is crucial for the analysis.

All the above general considerations suggest that the development of reliable advanced tools for fatigue damage calculation, like the one presented in this paper, which can include non-linearities of the structural behaviours, statistics of the climatology at the site, complex dynamic behaviours of rotating blades and aero-hydrodynamics under large displacements, is an important topic for the reliable evaluation of the fatigue of electrical cables, something that is a design weakness in floating offshore wind turbine systems.

Further developments for the research are under development and include:

- consideration of the effects of irregular sea states;
- refinement of the wind action and the fluid-structure interaction model by a two way-coupling between the two codes used for the analysis, with the inclusion of aerodynamic damping;
- refinement of the electrical cable modeling (currently modeled as an equivalent homogeneous element);
- exploration of more complex dynamic configuration of the electrical cable (modeling of more than one turbine, different cable layouts).

Author Contributions: Conceptualization, M.K., F.P.; methodology, M.K., F.P. and F.B.; software, M.S. and F.P.; formal analysis, M.S.; writing—original draft preparation, M.S.; writing—review and editing, M.K. and F.P.; supervision, F.B. All authors have read and agreed to the published version of the manuscript.

Funding: This research was funded by Sapienza University of Rome, "Fondi di avvio alla Ricerca", grant number AR118164360855CD.

Conflicts of Interest: The authors declare no conflict of interest.

Appendix A

Table A1. Structural properties of the model [25].

Description	Unit
Gravitational acceleration (m/s^2)	9.80665
Hub mass (kg)	56780
Hub inertia about rotor axis [3 blades] or teeter axis [2 blades] (kg m^2)	115926
Generator inertia about HSS (kg m^2)	534.116
Nacelle mass (kg)	240000
Nacelle inertia about yaw axis (kg m^2)	60789.2×10^6
Yaw bearing mass (kg)	0
Platform mass (kg)	7.46633×10^6
Platform inertia for roll tilt rotation about the platform CM (kg m^2)	4.22923×10^9
Platform inertia for pitch tilt rotation about the platform CM (kg m^2)	4.22923×10^9
Platform inertia for yaw rotation about the platform CM (kg m^2)	1.6423×10^8

Table A2. Hydrodynamic properties of the model [25].

Description	Unit
Water density (kg/m^3)	1025
Water depth (meters)	320
Displaced volume of water when the platform is in its undisplaced position (m^3)	8029.21
Incident wave kinematics model	Regular
Analysis time for incident wave calculations (s)	3630
Time step for incident wave calculations	0.25
Significant wave height of incident waves (meters)	6
Peak-spectral period of incident waves	10
Range of wave directions(degrees)	90
Wave Type	Stokes 2nd-order wave theory
Low frequency cutoff used in the summation frequencies (rad/s)	0.1
High frequency cutoff used in the summation frequencies (rad/s)	1.9132
Current profile model	No Current
Analysis time for wave (s)	2000
Time step for wave (s)	0.0125
Additional linear damping in surge N/(m/s)	100,000
Additional linear damping in sway N/(m/s)	100,000
Additional linear damping in heave N/(m/s)	130,000
Additional linear damping in yaw Nm(rad/s)	13,000,000
Hydrostatic restoring in heave (N/m)	332,941
Hydrostatic restoring in roll (Nm/rad)	−4,999,180,000
Hydrostatic restoring in pitch (Nm/rad)	−4,999,180,000

Table A3. Mooring line properties of the model [25].

Description	Unit
The mass per unit length of the line (kg/m)	77.7066
The line stiffness, product of elasticity modulus and cross-sectional area (N)	384.243×10^6
Diameter (m)	0.09

References

1. de Alegria, I.M.; Martín, J.L.; Kortabarria, I.; Andreu, J.; Ereño, P.I. Transmission alternatives for offshore electrical power. *Renew. Sustain. Energy Rev.* **2009**, *13*, 1027–1038. [CrossRef]
2. Green, J.; Bowen, A.; Fingersh, L.J.; Wan, Y.H. Electrical collection and transmission systems for offshore wind power. *Offshore Technol. Conf. Proc.* **2007**, *4*, 2215–2221.
3. Dai, T.; Sævik, S.; Ye, N. Experimental and numerical studies on dynamic stress and curvature in steel tube umbilicals. *Mar. Struct.* **2020**, *72*, 102724. [CrossRef]

4. Yang, S.H.; Ringsberg, J.W.; Johnson, E. Parametric study of the dynamic motions and mechanical characteristics of power cables for wave energy converters. *J. Mar. Sci. Technol.* **2018**, *23*, 10–29. [CrossRef]
5. Yang, S.H.; Ringsberg, J.W.; Johnson, E.; Hu, Z. Biofouling on mooring lines and power cables used in wave energy converter systems—Analysis of fatigue life and energy performance. *Appl. Ocean Res.* **2017**, *65*, 166–177. [CrossRef]
6. Nasution, F.P.; Sævik, S.; Gjøsteen, J.K.Ø. Fatigue analysis of copper conductor for offshore wind turbines by experimental and FE method. *Energy Proc.* **2012**, *24*, 271–280. [CrossRef]
7. Watson, S.; Moro, A.; Reis, V.; Baniotopoulos, C.; Barth, S.; Bartoli, G.; Bauer, F.; Boelman, E.; Bosse, D.; Cherubini, A.; et al. Future emerging technologies in the wind power sector: A European perspective. *Renew. Sustain. Energy Rev.* **2019**, *113*, 109270. [CrossRef]
8. Lotsberg, I. *Fatigue Design of Marine Structures*; Cambridge University Press: Cambridge, UK, 2016.
9. Rentschler, M.U.T.; Adam, F.; Chainho, P. Design optimization of dynamic inter-array cable systems for floating offshore wind turbines. *Renew. Sustain. Energy Rev.* **2019**, *111*, 622–635. [CrossRef]
10. Deltares. Joint Industry Project Cables Lifetime Monitoring. 2018. Available online: https://www.deltares.nl/app/uploads/2018/04/PB_Joint-Industry-Project-Cables-Life-Time-Monitoring_v2.pdf (accessed on 10 June 2020).
11. Karlsen, S.; Slora, R.; Heide, K.; Lund, S.; Eggertsen, F.; Osborg, P.A. Dynamic deep water power cables. In Proceedings of the 9th International Conference and Exhibition for Oil and Gas Resources Development of the Russian Arctic and CIS Continental Shelf, RAO/CIS Offshore, St Petersburg, Russian, 15 September 2009.
12. Homb, H.R. Fatigue Analysis of Mooring Lines on the Floating Wind Turbine Hywind Demo. Master's Thesis, Norges Teknisk-Naturvitenskaplige University, Trondheim, Norway, 2013.
13. Manenti, S.; Petrini, F. Dynamic analysis of an offshore wind turbine: Wind-waves nonlinear interaction. In Proceedings of the 12th Biennial ASCE Aerospace Division International Conference (Earth & Space 2010), Honolulu, HI, USA, 14–17 March 2010.
14. Karimirad, M. Modeling aspects of a floating wind turbine for coupled wave-wind-induced dynamic analyses. *Renew. Energy* **2013**, *53*, 299–305. [CrossRef]
15. Jonkman, J.M.; Buhl, M.L., Jr. *FAST User's Guide*; National Renewable Energy Laboratory: Golden, CO, USA, 2005.
16. Canonsburg, T.D. *AQWA User Manual*; ANSYS, Inc.: Cannon Sburg, PA, USA, 2012.
17. DNV GL. *DNVGL-RP-C203: Fatigue Design of Offshore Steel Structures*; DNV GL: Oslo, Norway, 2016.
18. DNV GL. *DNV-OS-J101–Design of Offshore Wind Turbine Structures*; DNV GL: Oslo, Norway, 2014.
19. Veritas, D.N. *DNV-OS-J103: Design of Floating Wind Turbine Structures, Offshore Standard*; DNV GL: Oslo, Norway, 2013.
20. Cárdenas, N.O.; Machado, I.F.; Gonçalves, E. Cyclic loading and marine environment effects on the properties of HDPE umbilical cables. *J. Mater. Sci.* **2007**, *42*, 6935–6941. [CrossRef]
21. Dieter, G.E.; Bacon, D. *Mechanical Metallurgy*; McGraw-Hill: New York, NY, USA, 1986; Volume 3.
22. Drexler, E.S.; Simon, N.J.; Reed, R.P. *Properties of Copper and Copper Alloys at Cryogenic Temperatures. No. NIST-MN-177. NIST*; National Institute of Standards and Technology (MSEL): Buolder, CO, USA, 1992.
23. Thies, P.R.; Johanning, L.; Smith, G.H. Assessing mechanical loading regimes and fatigue life of marine power cables in marine energy applications. *J. Risk Reliab.* **2012**, *226*, 18–32. [CrossRef]
24. Rychlik, I. A new definition of the rainflow cycle counting method. *Int. J. Fatigue* **1987**, *9*, 119–121. [CrossRef]
25. Jonkman, J. *Definition of the Floating System for Phase IV of OC3. No. NREL/TP-500-47535*; National Renewable Energy Laboratory (NREL): Golden, CO, USA, 2010.
26. Jonkman, J.; Musial, W. *Offshore Code Comparison Collaboration (OC3) for IEA Task 23 Offshore Wind Technology and Deployment (NREL/TP-5000-48191)*; National Renewable Energy Laboratory (NREL): Golden, CO, USA, 2010.
27. Loos, B. Operability Limits Based on Vessel Motions for Submarine Power Cable Installation. Master's Thesis, Royal Boskalis Westminster NV, Papendrecht, The Netherlands, 2017.
28. Karimirad, M.; Moan, T. Stochastic dynamic response analysis of a tension leg spar-type offshore wind turbine. *J. Wind Energy (Wiley)* **2013**, *16*, 953–973. [CrossRef]
29. DNV GL. *DNV-RP-F401, Electrical Power Cables in Subsea Applications*; DNV GL: Oslo, Norway, 2012.
30. Qiao, D.; Yan, J.; Ou, J. Fatigue analysis of deepwater hybrid mooring line under corrosion effect. *Pol. Marit.* **2014**, *21*, 68–76. [CrossRef]
31. Chang, R.; Zhu, R.; Badger, M.; Hasager, C.B.; Xing, X.; Jiang, Y. Offshore wind resources assessment from multiple satellite data and WRF modeling over South China Sea. *Remote Sens.* **2015**, *7*, 467–487. [CrossRef]

32. Bouty, C.; Schafhirt, S.; Ziegler, L.; Muskulus, M. Lifetime extension for large offshore wind farms: Is it enough to reassess fatigue for selected design positions? *Energy Procedia* **2017**, *137*, 523–530. [CrossRef]
33. Jensen, P.C.; Jacobsen, S.H. *Wind Turbines in Denmark*; Danish Energy Agency: København, Denmark, 2009.

MDPI
St. Alban-Anlage 66
4052 Basel
Switzerland
Tel. +41 61 683 77 34
Fax +41 61 302 89 18
www.mdpi.com

Energies Editorial Office
E-mail: energies@mdpi.com
www.mdpi.com/journal/energies

www.ingramcontent.com/pod-product-compliance
Lightning Source LLC
Chambersburg PA
CBHW061819210926
51850CB00006B/748
9783039369126